Imperial College London
2407645344

AF598251

INTERNATIONAL UNION OF CRYSTALLOGRAPHY
MONOGRAPHS ON CRYSTALLOGRAPHY

INTERNATIONAL UNION OF CRYSTALLOGRAPHY
BOOK SERIES

This volume forms part of a series of books sponsored by the International Union of Crystallography (IUCr) and published by Oxford University Press. There are three IUCr series: IUCr Monographs on Crystallography; which are in-depth expositions of specialized topics in crystallography; and IUCr Texts on Crystallography, which are more general works intended to make crystallographic insights available to a wider audience than the community of crystallographers themselves; and IUCr Crystallographic Symposia, which are essentially the edited proceedings of workshops or similiar meetings supported by the IUCr.

IUCr Monographs on Crystallography

1 *Accurate molecular structures: Their determination and importance*
A. Domenicano and I. Hargittai, *editors*

2 *P. P. Ewald and his dynamical theory of X-ray diffraction*
D. W. J. Cruickshank, H. J. Juretschke, and N. Kato, *editors*

3 *Electron diffraction techniques, Volume 1*
J. M. Cowley, *editor*

4 *Electron diffraction techniques, Volume 2*
J.M. Cowley, *editor*

5 *The Rietveld method*
R.A. Young, *editor*

IUCr Texts on Crystallography

1 *The solid state: From superconductors to superalloys*
A. Guinier and R. Jullien, *translated* by W.J. Duffin

2 *Fundamentals of crystallography*
C. Giacovazzo, *editor*

IUCr Crystallographic Symposia

1 *Patterson and Pattersons: Fifty years of the Patterson function*
J.P. Glusker, B.K. Patterson, and M. Rossi, *editors*

2 *Molecular structure: Chemical reactivity and biological activity*
J.J. Stezowski, J. Huang, and M. Shao, *editors*

3 *Crystallographic computing 4: Techniques and new technologies*
N.W. Isaacs and M.R. Taylor, *editors*

4 *Organic crystal chemistry*
J. Garbarczyk and D.W. Jones, *editors*

5 *Crystallographic computing 5: From chemistry to biology*
D.Moras. A.D. Podjarny, and J.C. Thierry, *editors*

Electron Diffraction Techniques

VOLUME 2

Edited by

John M.Cowley

Department of Physics and Astronomy
Arizona State University

INTERNATIONAL UNION OF CRYSTALLOGRAPHY
OXFORD UNIVERSITY PRESS

This book has been printed digitally and produced in a standard specification in order to ensure its continuing availability

OXFORD
UNIVERSITY PRESS

Great Clarendon Street, Oxford OX2 6DP

Oxford University Press is a department of the University of Oxford.
It furthers the University's objective of excellence in research, scholarship,
and education by publishing worldwide in

Oxford New York

Auckland Cape Town Dar es Salaam Hong Kong Karachi
Kuala Lumpur Madrid Melbourne Mexico City Nairobi
New Delhi Shanghai Taipei Toronto

With offices in

Argentina Austria Brazil Chile Czech Republic France Greece
Guatemala Hungary Italy Japan South Korea Poland Portugal
Singapore Switzerland Thailand Turkey Ukraine Vietnam

Oxford is a registered trade mark of Oxford University Press
in the UK and in certain other countries

Published in the United States
by Oxford University Press Inc., New York

Reprinted 2007

ISBN 978-0-19-855733-3

Printed and bound by CPI Antony Rowe, Eastbourne

PREFACE

The proposal for a book on the techniques of electron diffraction arose from the discussions of the Commission on Electron Diffraction of the International Union of Crystallography, held at the time of the IUCr Congress in Perth, Western Australia in 1987. Correspondence with Commission members, Consultants of the Commission, representatives of the IUCr and various prominent members of the electron diffraction community confirmed that the need for such a book existed. There have been some excellent books on particular aspects of electron diffraction, including low-energy electron diffraction (LEED) and gas diffraction (GED) and the various books on non-biological electron microscopy have, necessarily, included sections on electron diffraction; but no comprehensive volume on electron diffraction as such has been produced for many years. The subject has expanded enormously in the last ten or twenty years and many aspects of the theory and practice have been developed to the stage that the earlier comprehensive volumes appeared to be quite inadequate as representations of the current status of the subject.

The initial plan was for a multi-author volume of about 500 pages, prepared in camera-ready form by the authors in order to speed publication, with a publication date set for 1989. The agreement between the International Union of Crystallography and Oxford University Press for the publication of a series of books related to crystallography provided a convenient and appropriate publishing medium.The question of who would be editor was resolved by default. As Chairman of the CED it was my responsibility to find a suitable volunteer but no one was nominated by the commission members except myself. It was proposed that the book should be concerned mostly with the diffraction of high energy (20keV or more) electrons by solids, but chapters on LEED and GED should be included for purposes of comparison and correlation, emphasizing the unity of the electron diffraction field which is often more apparent in principle than in practice.

The intended audience for the book comprises all those embarking on research which involves the use of electron diffraction methods, including graduate students and more experienced researchers who wish to add electron diffraction to their array of research tools. The background in mathematical techniques which is assumed is limited to that of a bachelor's degree graduate in the physical sciences, although some familiarity with the elementary theory of X-ray crystallography may be helpful. The introductory Chapter 1 may well be omitted by those readers having some familiarity with the subject matter gained, for example, in an introductory course, or experience, in electron microscopy.

The production of the book has been delayed mostly because the selected chapter authors, being outstanding international authorities in their subject areas, are also very busy people and in some cases unexpected complications or changes of their careers have distracted them from their original purpose and it has been necessary to co-opt associate authors to help get the chapters completed. Several proposed sections, including the separate chapter on LEED, had to be abandoned as a result of difficulties experienced by the chosen authors. To some extent the resulting deficiencies have been compensated by inclusion of short sections in the introductory survey, Chapter 1, or in other chapters.

The major change that has been made to the original plan for the book is to divide the material into two volumes. It became apparent, before long, that the original assignment of about 40 pages per chapter was not realistic. Most authors found it impossible to give an adequate account of their subject matter within that limited space. Although editorial advice led to some reductions, the decision was made that there should be no enforcement of reductions in length that would eliminate valuable material. The consequence was a total of about 900 pages in the standardized format and hence the decision to produce two volumes instead of one.

A reasonable basis for dividing the material presented itself as a consequence of the relative delays in submission of some of the chapters. Volume 1 contains the introductory chapters and the sections on the aspects of electron diffraction which are less dependent on considerations of imaging in electron microscopes. Volume 2 deals with those aspects when there is stronger correlation of the diffraction phenomena with the electron microscope imaging. However the distinction is not always clear-cut and we must apologize for any lack of convenience for the reader which results from the separation.

Volume 2 now contains the full texts of the Chapter 1 on Diffraction Contrast and High Resolution Electron Microscopy of Structures and Structural Defects and of Chapter 5 on Identification of Unknowns. The gaps left in both the scientific content and the chapter-numbering when these chapters were extracted from the original sequence were filled by very short chapters in Volume 1, intended only as bridging material to compensate for the delay of the complete chapters.

Our sincere thanks go to those chapter authors who provided excellent, defect-free text, conforming to all of the instructions and specifications on format and style: also to those authors who were not equipped to follow these specifications but helped with the careful proof-reading and updating of their chapters. We are particularly grateful for the word-processing efforts of Chula Eslamieh, Mary Ryan, and Carolyn Frederick who laboured cheerfully through many revisions and final versions of most of the chapters.

Tempe, Arizona
July 1992

J.M.C.

Contents

Contents for Volume 1

Contributors

S. Amelinckx Department of Physics, University of Antwerp (RUCA), Groenenborgerlaan 171, B-2020 Antwerpen, Belgium

M.J. Carr Sandia National Laboratories, Albuquerque, New Mexico, 87185, USA

J.K. Gjønnes Department of Physics, University of Oslo, N-0316 Oslo 3, Norway

C.E. Lyman Department of Materials Science and Engineering, Lehigh University, Bethlehem, PA 18015-3195, USA

D. Van Dyck Department of Physics, University of Antwerp (RUCA), Groenenborgerlaan 171, B-2020 Antwerpen, Belgium

K. Yagi Physics Department, Tokyo Institute of Technology, Oh-Okayama, Meguro-ku, Tokyo 152, Japan

1
Diffraction Contrast and High Resolution Electron Microscopy of Structures and Structural Defects

S. Amelinckx and D.Van Dyck

Diffraction Contrast

1.1 Introduction

In volume I of this treatise the kinematical and dynamical theories of electron diffraction by perfect crystals were discussed [1]. In the present chapter we will discuss the application of these theories to the imaging of defects in crystals. We shall first recall the basic equations of both theories and in some cases provide an alternative derivation of these equations, which might shed some additional light on their physical contents. We shall formulate them furthermore in a form which is well adapted to the study of defects.

Although the kinematical theory is in itself not sufficient as a basis for the realistic computation of images, it is nevertheless of some importance in providing insight and making possible analytical treatments which lead to informative, albeit qualitative conclusions.

We shall also provide intuitive insight whenever possible since this is of great help in the qualitative interpretation of images as a preliminary to more detailed quantitative calculations using computer programs. This is especially valuable since in such computations, in view of defect identification, one still uses the "trial and error" method which implies that one must start with a reasonable model, which can subsequently be refined.

1.2 The Kinematical Theory of Diffraction by Perfect Crystals

The kinematical theory can be formulated in a variety of ways. We shall summarize the results for a perfect crystal using essentially three different approaches: a geometrical one, a graphical method and the first Born approximation.

1.2.1 The geometrical approach

In this approach the amplitudes of the waves scattered by the different scattering units are summed taking the phase differences among scatterers, due to their geometrical configurations, properly into account. The scattering units may be single atoms, atom clusters or unit cells. Each scattering unit is assumed to see the same incident beam, which is not depleted on scattering.

Let the scattering units be placed at positions $\mathbf{r}_j$ and let their electrostatic potential responsible for the scattering be represented by $V_j(\mathbf{r})$. The total potential is then

$$V(\mathbf{r}) = \sum_j V(\mathbf{r} - \mathbf{r}_j) \tag{1.1}$$

The phase difference with respect to the origin, on diffraction in the direction $\mathbf{h}$, due to the volume element $d\mathbf{r}$ at $\mathbf{r}$ is $2\pi\mathbf{h}.\mathbf{r}$ and the amplitude $A(\mathbf{h})$ of the scattered beam is then obtained by summing or integrating over all scattering units

$$A(h) = \int \sum_j V_j(r - r_j) \exp(+2\pi i \mathbf{h}.\mathbf{r})\, d\mathbf{r} \tag{1.2}$$

or

$$A(\mathbf{h}) = \sum_j f_j(\mathbf{h}) \exp(+2\pi i \mathbf{h}.\mathbf{r}_j) \tag{1.3}$$

with

$$f_j(\mathbf{h}) = \int V_j(\mathbf{r}) \exp(+2\pi i \mathbf{h}.\mathbf{r})\, d\mathbf{r} \tag{1.4}$$

where $f_j(\mathbf{h})$ is the scattering amplitude of the unit j. In the literature one often finds a formulation which amounts to changing $\mathbf{h} \rightarrow -\mathbf{h}$; the final results are independent of this choice. The relations (1.3) and (1.4) show that $A(\mathbf{h})$ is the Fourier transform of $V(\mathbf{r})$ and $f_j(\mathbf{h})$ that of $V_j(\mathbf{r})$.

In a periodic crystal it is possible to group the atomic scatterers in units which are identical but related by a lattice translation $\mathbf{r}_L = \sum_i l_i \mathbf{a}_i$ ($\mathbf{a}_i$ = base vectors of the lattice; l_i = integers). These units are <u>unit cells</u>. The scattering amplitude due to one unit cell (i.e. the <u>structure amplitude</u> or <u>structure factor</u>) is then

$$F(h) = \sum_{K=1}^{N} f_K(\mathbf{h}) \exp(+2\pi i \mathbf{h}.\rho_K) \tag{1.5}$$

where the vectors ρ_K describe the N atomic positions within the unit cell as referred to its origin. It is clear that $\mathbf{r}_j = \mathbf{r}_L + \rho_K$ and from (1.3):

$$A(\mathbf{h}) = F(\mathbf{h}) \sum_L \exp(+2\pi i \mathbf{h}.\mathbf{r}_L) \tag{1.6}$$

The summation over the lattice L yields delta functions at the nodes **g** of the reciprocal lattice, which is based on the vectors $\mathbf{b}_j$, defined as $\mathbf{b}_i.\mathbf{b}_j = \delta_{ij}$ Kronecker delta) (i = 1,2,3). The reciprocal lattice vectors are defined by $\mathbf{g} = \sum_j h_j \mathbf{b}_j$ where the h_j are integers (Miller indices). One obtains finally

$$A(\mathbf{h}) = N\, F(\mathbf{h})\, \delta(\mathbf{g}-\mathbf{h}) \tag{1.7}$$

where N is the number of unit cells; the Dirac δ expresses the Bragg or Ewald condition, leading to the Ewald construction (see [1]. The diffracted amplitude can thus be described by the reciprocal lattice nodes weighted by the structure factors F(**h**) = F(**g**). If the diffraction condition is not exactly satisfied **h** = **g** + **s**, where **s**, the deviation parameter, was defined in [1]. One then obtains from (1.6):

$$\begin{aligned} A(\mathbf{h}) &= F(\mathbf{h}) \sum_L \exp[+2\pi i(\mathbf{g}+\mathbf{s}).\mathbf{r}_L] \\ &= F(\mathbf{g}) \sum_L \exp(+2\pi i \mathbf{s}.\mathbf{r}_L) \end{aligned} \tag{1.8}$$

since $\mathbf{g}.\mathbf{r}_L$ = integer and $F(\mathbf{h}) \approx F(\mathbf{g})$. We introduce the shorthand $F_h \propto F_g$.

This summation over the lattice was performed in [1]. If the crystal contains a large number of unit cells in the three directions of the basic lattice, the result was shown to be well approximated, apart from a phase factor, by

$$A(\mathbf{h}) = F_g(\Omega / V_r)\ \delta(s_1)\ \delta(s_2)\ \delta(s_3) \tag{1.9}$$

where Ω is the volume of the crystal and V_r the volume of the unit cell. s_1, s_2 and s_3 are the components of the vector **s** If the crystal is a thin foil, as used in electron diffraction experiments, it contains large numbers N_1 and N_3 of unit cells in the **x** and **y** direction but only a finite number $N_3 (N_3 << N_1, N_2)$ in the z direction normal to the foil surface. The expression (1.9) then becomes

$$A(\mathbf{h}) = F_g(\Omega / V_r)\ \delta(s_1)\ \delta(s_2)\ \frac{\sin \pi N_3 s_3}{\pi N_3 s_3} \tag{1.10}$$

The last factor follows from the summation, along the z-direction, of the contribution to the amplitude of a finite number N_3 of unit cells, i.e. (with $s_3 \equiv s$)

$$A(\mathbf{h}) \propto F_h \sum_{n=0}^{N_3-1} \exp(+2\pi i s \mathbf{z}_n) \tag{1.11a}$$

with $z_n = na_3$, or in a continuum approximation: $z_n = n\Delta z$, and $N_3 a_3 = z_0$. In the limit $\Delta z \rightarrow 0$ this becomes:

$$A(\mathbf{h}) \propto F_h \int_0^{z_0} \exp(+2\pi i s z) dz \tag{1.11b}$$

where the integration is now performed along a column parallel to the z-axis; z_0 is the thickness of the foil, i.e.

$$A(\mathbf{h}) \propto F_h [\sin(\pi s z_0) / (\pi s z_0)] \tag{1.12}$$

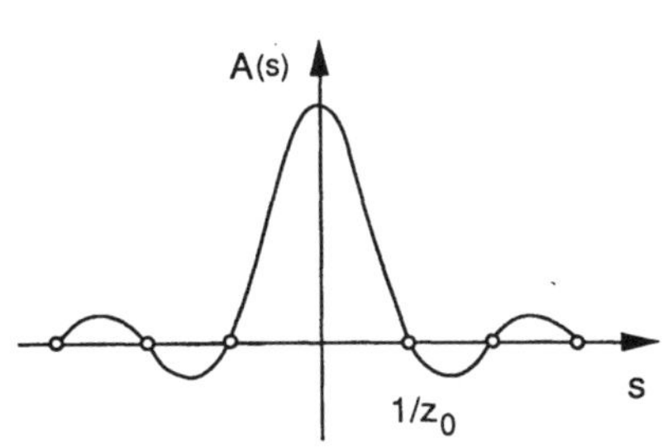

Figure 1.1. Diffracted amplitude as a function of the deviation from Bragg's condition.

The vector **s**, which reduces to a point in reciprocal space if the crystal is infinite in the three directions, is now found to become a segment or rod of finite length, oriented along $\mathbf{b}_3$, i.e. along the normal to the foil plane, and having a weight profile $\sin \pi N_3 s_3 / \pi N_3 s_3$. The corresponding intensity profile is called the "rocking curve" according to the kinematical theory (Figure 1.1). An amplitude can be associated with each intersection point of the Ewald sphere with this segment, the amplitude being given by the value of this profile at the intersection point. It is customary to describe the diffraction geometry by saying that the reciprocal lattice nodes have become "relrods" with a length profile given by the above mentioned function but infinitely sharp in the directions x and y. The non-vanishing component is called s_g or simply s in what follows.

In a thin foil the vector **s** is thus oriented along the normal to the foil plane. By convention **s** is counted positive in the sense of the propagating electrons, i.e. if the reciprocal lattice node G is inside Ewald's sphere (see [1]).

It is worthwhile pointing out that the function $(\sin \pi s z_0 / \pi s z_0)$ can be considered as the envelope of a wave packet, such as is often used in wave mechanics to represent particles and which is formed by the interference of waves with slightly different wavevectors, i.e. with a spread in the linear impulse $\hbar k$ given by $\Delta k \approx s_0$, when limiting the peak width to the first zero $(s=s_0)$. The peak width expressed in reciprocal space can thus be interpreted as the uncertainty in linear impulse of the scattered electrons, i.e. after interaction with the foil. The uncertainty on the position of the electron at the time of its interaction with the foil is the foil thickness z_0. The well known uncertainty relation of Heisenberg would then state that $\hbar \Delta k . \Delta z \approx \hbar$ or $s_0 z_0 = 1$, which is the result of the kinematical theory [2].

1.2.2. A graphical approach: the amplitude-phase diagram

A plane wave represented by $A \exp i(kx - \omega t + \psi) \equiv Ae^{i\psi} \exp i(kx - \omega t)$ is characterized by a complex amplitude $A \exp i\psi$, a wavevector **k** and an angular frequency ω.

The interference between two waves of this type, assuming the wavevector **k** and the angular frequency ω to be the same for both, produces a resultant wave with the same ω and **k** but with different A and ψ. The sum thus reduces to the summing of complex amplitudes $A \exp i\psi$, the propagation factor being common to all waves to be summed.

The complex amplitude can be represented in the complex plane by a vector with modulus A and argument ψ. It is easy to show that the sum of two vectors representing such waves is again a vector representing the resultant wave. Waves with the same **k** and ω can thus be summed graphically by adding vectors in the complex plane.

The amplitude scattered by a column of crystal along the z-axis, in the kinematical approximation, is given by the sum (1.11a) or (1.11b). This sum can be considered as consisting of terms $F_h.\Delta z.\exp(2\pi isz)$, i.e. the amplitude is $A = F_h\Delta z$; and the phase $y=2\pi isz$, corresponding with slices Δz of the column. Theamplitude-phase diagram then consists of vectors, all of the same length $F_h\Delta z$ and enclosing angles of $2\pi s\Delta z$. In the limit of $\Delta z \to 0$ the locus of the endpoints is a circle with radius:

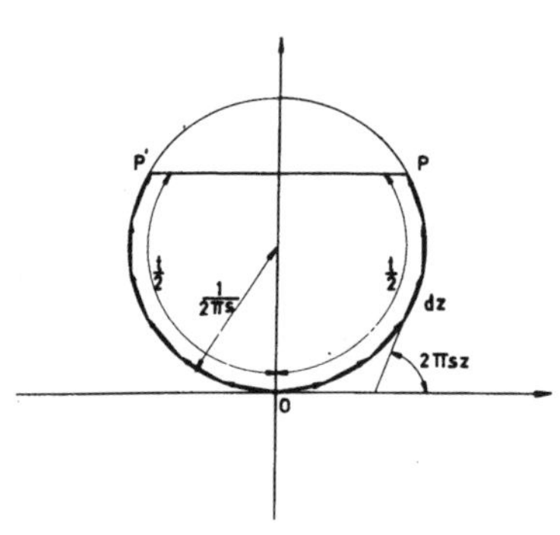

Figure 1.2. Amplitude-phase diagram for a perfect crystal foil.

$$R = \lim_{\Delta z \to 0} (\Delta z / 2\pi s\Delta z) = (2\pi s)^{-1}$$

(Figure 1.2). The length of the circular arc is equal to the column length, i.e. to the foil thickness z_0. The amplitude A(**h**) is obtained by joining P to P′. The diagram, of which the origin is assumed to represent the center of the foil, shows that the diffracted amplitude will be zero if the circular arc is a number of complete circles, i.e. for k/s; there will be maxima if $z_0 = 1/s\,(k+1/2)$ (k = integer); the maximum amplitude being equal to a diameter of the circle, i.e $A_{max} = F_h / \pi s$.

1.3 Column Approximation

This approximation was already mentioned in Ref. 1; it becomes particularly important when discussing diffraction contrast images.

The formulae (1.6) and (1.11) can be understood intuitively as being due to the summation of the contributions due to the scattering elements dz in "columns" parallel to the beam direction. The factor exp $(2\pi isz)$ takes into account the "phase" of the element dz due to its depth z behind the entrance face; the factor F_h describes the strength and the phase of the reflection due to the diffracting material in the element dz at z. The size of the "column" need not be specified. In a perfect foil the whole foil surface can be considered as the basis of a "column"; the result would in any case be independent of the choice of the size of the column and of the point (x,y) on which the column is centered. This is no longer true in deformed crystals.

It is helpful to remind at this stage that we consider the Laue case and that Bragg angles are very small in electron diffraction; moreover electron scattering is strongly peaked in the forward direction. The maximum sideways displacement that electrons can suffer on entering a foil of thickness z_0 is $\Delta = z_0\theta_n$ (Figure 1.3a)

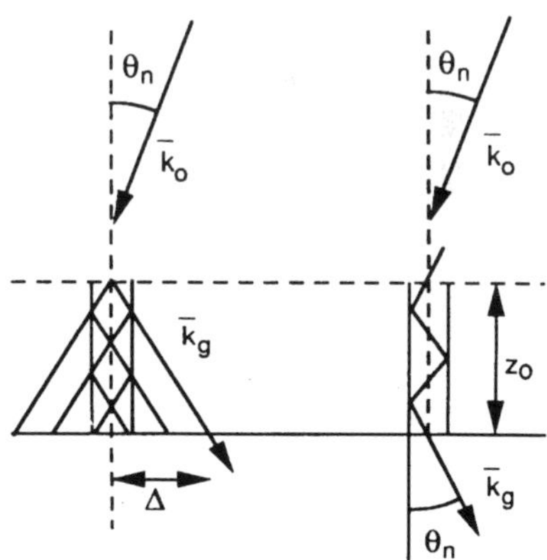

Figure 1.3. Illustrating the column approximation.
(a) Kinematical theory
(b) Dynamical theory

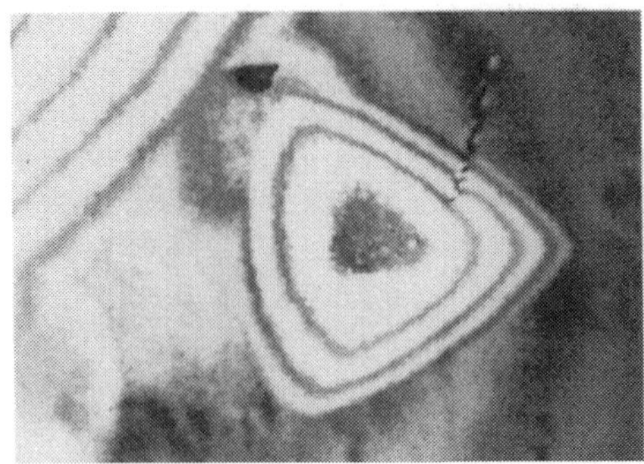

Figure 1.4. Thickness extinction contours in a silicon foil. The contours reveal the shape of an etchpit which is centered on dislocation, imaged by diffraction contrast.

which is very small since we are considering thin foils. Moreover we must realize that multiple scattering takes place; this will explicitly be discussed in the dynamical theory. This will evenmore confine the electrons to very narrow columns on traversing the foil (Figure 1.3b). It is therefore a good approximation to assume that electrons propagate along narrow columns centered on the point of impact.

The intensity of the beam, either scattered or transmitted, computed at the exit surface of the column will then be

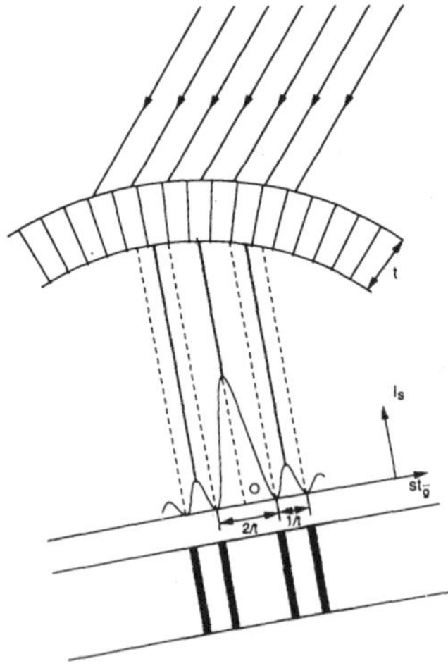

Figure 1.5. Equi-inclination contours produced by a bent foil; schematic model.

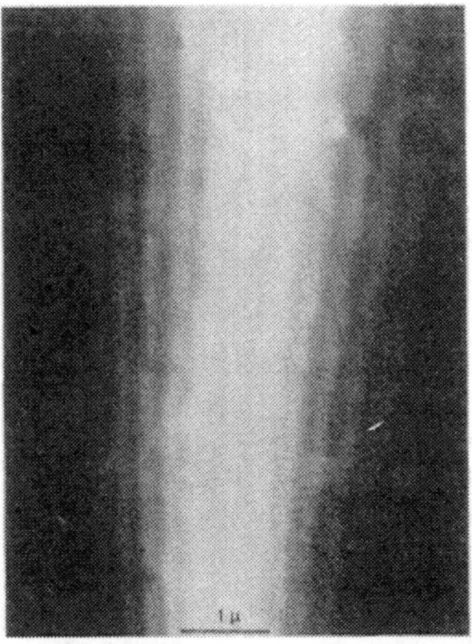

Figure 1.6. Equi-inclination or bend contours in a cylindrically bent foil of graphite; they image the rocking curve.

representative for the site on which the column is located. The columns can in a sense be considered as the picture elements (pixels) of the map of the intensity distribution in a given beam, at the back surface of the foil. For a defect-free perfectly flat foil of constant thickness this map will exhibit uniform intensity. In a wedge shaped defect-free crystal we shall observe thickness fringes (Figure 1.4). In the dark field image the dark fringes are the geometrical loci of the columns of which the length is such that the emerging scattered intensity is a minimum.

For a cylindrically bent crystal of uniform thickness the s-value becomes spatially variable; the loci of constant s being parallel to the axis of the cylinder (Figures 1.5 and 1.6). Those columns for which the s-value is such as to produce a maximum of I_g at the exit face will produce a bright pixel. The loci of bright pixels are lines parallel to the axis of the cylinder; they are lines of equal inclination or "bend contours" [3]. The bend contours in a cylindrically deformed thin cleavage foil of graphite are shown in Figure 1.6; they image directly the rocking curve.

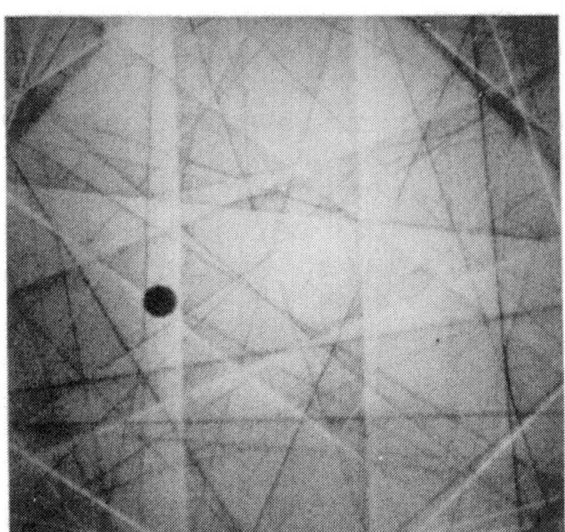

Figure 1.7. Pattern of Kikuchi lines in a rather thick silicon crystal.

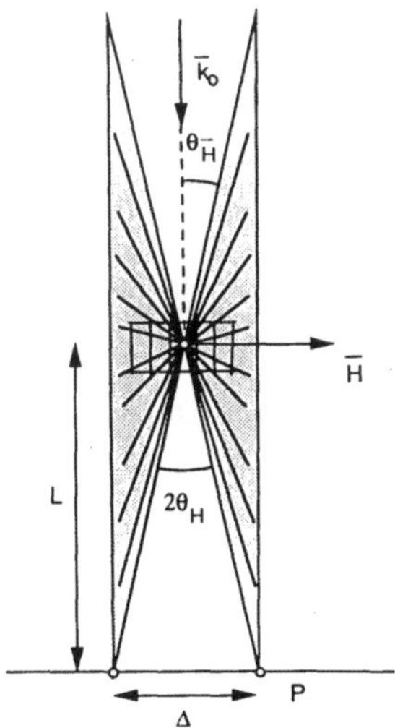

Figure 1.8. Geometry of the Kikuchi cones in the symmetrical orientation.

1.4. Kikuchi Lines; Determination of Sign and Magnitude of s

In sufficiently thick and rather perfect foils spot patterns are no longer observed; instead a diffraction phenomenon, first discovered by Kikuchi in 1928 [4], is produced. It usually consists in the occurrence of pairs of bright and dark straight lines in the diffraction pattern as shown in Figure 1.7. In foils of intermediate thickness one can observe the Kikuchi pattern superposed on the spot pattern. The geometry of the Kikuchi pattern can satisfactorily be explained by assuming that electrons are not only Bragg scattered but that a substantial fraction, especially in thick foils, is scattered inelastically and incoherently in the crystal, the energy loss being small compared to the energy of the incident electron. In this way the electron wavelength is not appreciably changed. Inside the crystal these randomly scattered electrons impinge on the lattice planes from all directions but preferentially in the forward direction and can subsequently give rise to Bragg reflection, i.e. to elastic scattering.

Let us consider for instance the symmetrical situation with respect to the set of lattice planes H, with spacing d_H, as shown in Figure 1.8. Bragg scattering out of the incident beam is then weak since the Bragg condition is not satisfied. However a fraction of the randomly scattered electrons will have the correct direction of incidence to give rise to Bragg scattering by the considered set of lattice planes. The geometrical locus of these Bragg scattered electron beams is a double cone of revolution with an opening angle $\pi/2-\theta_H$ and with its axis along H. (θ_H = Bragg angle). These cones are therefore rather "flat" and the intersection lines of the two sheets of this double cone with the photographic plate P look like two parallel straight lines, although in actual fact they are two branches of a hyperbolic conic section. It is also evident that the angular separation of these two lines, is $2\theta_H$; the separation Δ observed on the plate is thus $\Delta = 2L\theta_H$, where L is the camera length, i.e. the distance specimen to plate. This angular separation does not depend on the crystal orientation.

It should be noted that the geometry of this cone (i.e. the axis of revolution and the opening angle) is entirely fixed by the crystal lattice and independent of the incident beam direction. This implies that tilting the specimen over a small angle will lead to an equal tilt of the double cone, but would leave the geometry of the spot diffraction pattern unchanged provided the same reflections remain

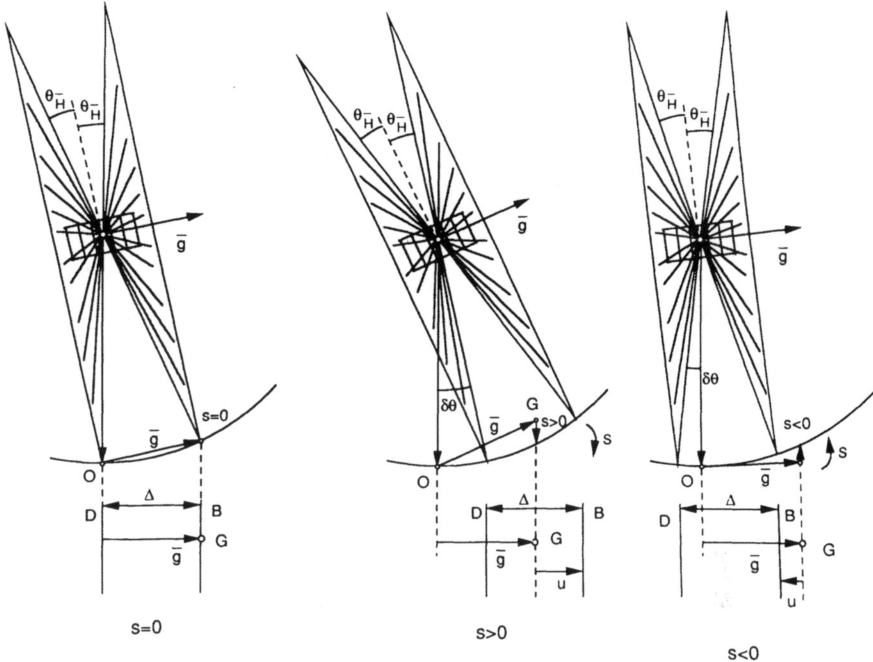

Figure 1.9 Evolution of the Kikuchi line pattern on tilting: note the relative positions of line and spot.
(a) The Kikuchi line passes through the Bragg spot (s = 0)
(b) The deviation parameter s is positive
(c) The deviation parameter s is negative

excited, i.e. as long as the same "relrods" are intersected by Ewald's sphere. The relative position of the spot pattern and of the Kikuchi line pattern is thus very orientation [illegible]sitive and as a consequence it carries very useful information which can only be obtained with difficulty otherwise, as we shall see.

We now consider the situation which arises when the specimen is tilted in such a way that the set of lattice planes **g** satisfies the Bragg condition. The situation with respect to the incident beam is then no longer symmetrical (Figure 1.9a). The elastically Bragg scattered beam, which produces the spot **G** is now one of the generators of the cone. One of the Kikuchi lines thus passes through the Bragg spot. It appears bright (B) on a positive print, i.e. it corresponds with an excess of electrons above the background. The other line (D) which appears dark, due to a deficiency of electrons, passes through the origin. These features will now be explained. The dark line passing through the origin is produced against a high background caused by the predominantly forward, inelastically scattered electrons. Among these electrons, those which satisfy the Bragg condition are scattered elastically out of this background onto the sheet of the cone which passes through the Bragg spot. Along the parallel line through the origin, which is the locus of the electrons satisfying Bragg's condition, there is as a consequence a deficiency of electrons compared to the background. On the other hand the electrons which by their absence cause the dark line through the origin, cause an excess, compared to a lower background along the part of the cone containing the coherently scattered Bragg beam. This background is somewhat smaller since the scattering angle is larger. Therefore the excess electrons produce a bright line through the Bragg spot. The angular separation of the bright - dark line pair, is clearly the same as in the symmetrical orientation; the linear separation measured on the plate may slightly depend on the tilt angle however.

The symmetrical situation is represented schematically in Figure 1.8. In this orientation the Kikuchi lines often form the limiting lines of "Kikuchi bands", the inside of which exhibits a somewhat smaller brightness than the outside. In this particular orientation the Kikuchi lines can be considered as images of the Brillouin zone boundaries belonging to the different reflections.

Starting with a foil in the exact Bragg orientation for the reflection G, the bright Kikuchi line passes through G (Figure 1.9a), whereas the dark line passes through the origin of the reciprocal lattice. Tilting the specimen over a small angle $\delta\theta$ in the clockwise sense, i.e. towards $s < 0$ about an axis in the foil plane normal to the

g-vector, the bright Kikuchi line moves towards the origin over $u = L\delta\theta$ (Figure 1.9b). The vector **g** is then rotated over the same angle $\delta\theta$ and hence *s* becomes negative and equal to $s = g\delta\theta$ the relation between u and s is thus:

$$u = (L/g)s \text{ and also } \Delta u = (L/g)\Delta s \qquad (1.13)$$

This relation allows to determine the sign and the magnitude of s from the relative position of a diffraction spot and its associated Kikuchi line (Figure 1.9). This relation also allows to determine the orientation difference between two crystal parts from the splitting of the Kikuchi lines.

The sign of s is required for a number of applications such as the determination of the sign of the Burgers vector of a dislocation, the vacancy or interstitial character of a dislocation loop, the orientation difference across a domain boundary, etc. as will be discussed below. The magnitude of s is needed when applying the weak beam method (see § 1.28.4).

1.5. Refraction of Electrons at an Interface

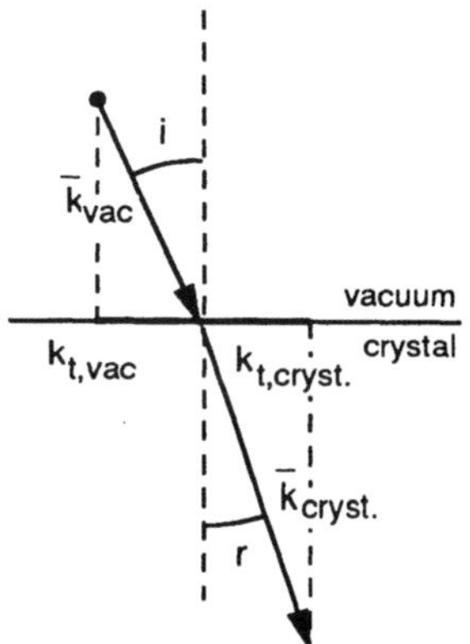

Figure 1.10 Refraction of electrons at the interface crystal-vacuum.

Refraction of the incident electron beam takes place at the interface vacuum-crystal foil because the lengths of the wavevectors are different in the two media:

$$K_{vacuum} = (2meE)^{1/2}/h \tag{1.14}$$

$$K_{crystal} = [2me(E+V_o)]^{1/2}/h \tag{1.15}$$

but the tangential components have to be conserved at the interface. Figure 1.10 shows the relation between the two wavevectors; one has

$$\begin{aligned} n &= \sin i/\sin r = (K_{t,vac}/K_{vac})/(K_{t,cryst}/K_{cryst}) \\ &= K_{cryst}/K_{vac} = [(E+V_0)/E]^{1/2} \end{aligned} \tag{1.16}$$

The refractive index is thus:

$$n \approx [1+(V_0)/E]^{1/2}$$

Since $V_0 << E$, *n* is just slightly larger than 1 and the angle of refraction is very small, especially for quasi-normal incidence as is the case in most observations. Refraction nevertheless produces an observable effect for grazing incidence. Small polyhedral particles may produce diffraction spots consisting of a number of components corresponding with the number of crystal wedges crossed by the beam.

1.6. Kinematical Theory as a Born Approximation; the Extinction Distance

Any exact diffraction theory has in fact to be based on Schrödinger's equation, which describes adequately the interaction of the imaging electrons with the periodic lattice potential of the crystal. This was done in Ref. [1], but we shall summarize a somewhat different reasoning here and present the solutions in a form as needed subsequently. The equation to be solved is

$$(h^2/8\pi^2 m)\Delta\psi + [E + V(r)]e\psi = 0 \tag{1.18}$$

where ψ is the wavefunction of the electrons, -e is the charge of the electrons, E is the accelerating potential (100 to 1000 kV) and V(**r**) the lattice potential (a few volts). The constant part V_0 of the lattice potential can be separated by introducing

$$V'(\mathbf{r}) = V(\mathbf{r}) - V_0 \text{ and setting } E + V_0 = h^2 k_0^2/2m$$

with (1.19)

$$V'(\mathbf{r}) = \sum_g V_g \exp(2\pi i\mathbf{g}.\mathbf{r})$$

where $\mathbf{k}_0$ is the wavevector of the incident beam, corrected for refraction part by the constant part V_0 of the lattice potential.

The solution, obtained by applying the Born approximation, can be formulated as follows. Let $\psi = \psi_0 + \sum_g \psi_g$ where ψ_0 is the wavefunction of the incident beam and ψ_g that of the scattered beam **g**. We can represent the transmitted beam by $\psi_0(\mathbf{r}) = C\exp 2\pi i\mathbf{k}_0.\mathbf{r}$ and the scattered beam as

$$\psi_g(\mathbf{r}) = \varphi_g(\mathbf{r})\exp[2\pi i(\mathbf{k}_0 + \mathbf{g}).\mathbf{r}] \tag{1.20}$$

The z-axis is chosen along the normal to the foil, the x and y-axis being parallel to the foil.

It can be shown [5] that the function $\varphi_g(\mathbf{r})$ has to be a solution of the following equation

$$\frac{d\varphi_g}{dz} - 2\pi i s_g \varphi_g = \frac{\pi i}{t_g} \exp(i\theta_g) \qquad (1.21)$$

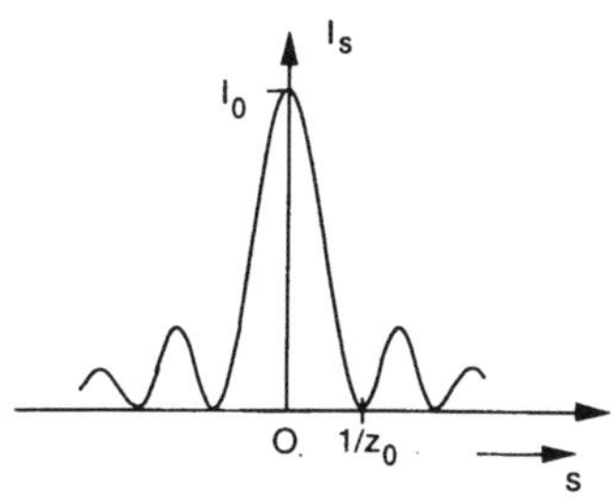

Figure 1.11 Variation of the diffracted intensity with the deviation parameter.

where θ_g is the phase of the Fourier coefficient V_g of the lattice potential $V(\mathbf{r})$. The important parameter t_g called extinction distance, is related to the other parameters:

$$t_g = h^2 k_0 \cos\theta_B / 2me|V_g| \qquad (1.22)$$

where $\cos\theta_B \approx 1$, θ_B being the Bragg angle, $V_g = |V_g| \exp(i\theta_g)$.

The extinction distance is a measure for the strength of the reflection; with a strong reflection corresponds a small extinction distance. For low order reflections in metals they are of the order of 5 - 20 nm whereas weak superstructure reflections may have extinction distance of the order of 100 - 1000 nm.

The deviation parameter s_g is introduced in terms of the wavevectors by the relations

$$s_g = \left(k_0^2 - |\mathbf{k}_0 + \mathbf{g}|^2\right) / 2|\mathbf{k}_0 + \mathbf{g}|\cos\theta_B \qquad (1.23)$$

Putting $\cos\theta_B = 1$ is justified by the smallness of the Bragg angles. In this derivation it is assumed furthermore that the column approximation is consistently applied by neglecting a term containing the partial derivative $\frac{\partial \varphi_g}{\partial x}$, replacing also $\frac{\partial \varphi_g}{\partial z}$ by $\frac{d\varphi_g}{dz}$

Integrating the equation (1.21) from front to exit surface of the foil leads to

$$\varphi_g = (\pi i / t_g)\ \exp(2\pi i s_g z_0)\ \exp(i\theta_g) \int_0^{z_0} \exp(-2\pi i s z) dz$$

$$= i \exp(\pi i s_g z_0) \exp(i\theta_g) \sin(\pi s_g z_0) / s_g t_g \qquad (1.24)$$

which is equivalent to Eq. (1.12).

For the scattered intensity per unit of incident intensity one thus finds

$$I_g = \psi\ \psi_g^* = \varphi\ \varphi_g^* = \sin^2 \pi s_g z_0 / (s_g t_g)^2 \qquad (1.25)$$

The dependence of I_g on $s_g z_0$ is represented in Figure 1.11. There is a pronounced central maximum (for $s = 0$) given by $I_g(\max) = (\pi z_0 / t_g)^2$. There are zero's for $s = n / z_0$ and maxima approximately halfway between zero's. These secondary maxima are much smaller than the central peak.

1.7. Diffraction by Deformed Crystals: Models

Deformations of the crystal can be modeled by stating that the unit cell which in the undeformed crystal was in $\mathbf{A}_L$ occupies, after deformation, the position $\mathbf{A}_L + \mathbf{R}(\mathbf{r})$. The deformation field $\mathbf{R}(\mathbf{r})$ simulates the defect. A few simple examples of deformation fields are the following:

(i) Planar translation interfaces such as stacking faults, out of-phase boundaries, discommensuration walls, ... All these planar defects have a displacement field of the type $\mathbf{R} = 0$

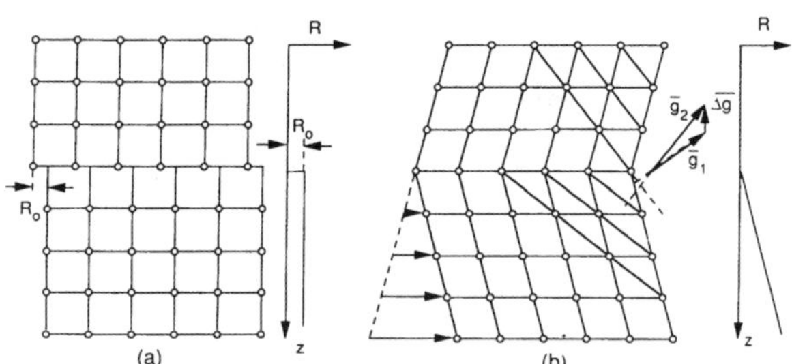

Figure 1.12 Displacement function of planar interfaces.
(a) stacking fault (b) Domain boundary

for $z \leq z_1$ and $\mathbf{R} = \mathbf{R}_0$ for $z > z_1$, where z_1 is the level at which the planar defect occurs behind the entrance face (Figure 1.12a).

(ii) Domain boundaries or twin boundaries with a small twinning vector. We now have $\mathbf{R} = 0$ for $z \leq z_1$ and $\mathbf{R} = \mathbf{k}z$ for $z > z_1$ (Figure 1.12b).

(iii) A pure screw dislocation has a displacement field described by $R = \mathbf{b}[\theta/2\pi]$, where θ is the azimuth angle, measured in the plane perpendicular to **b**. All displacements are clearly assumed to be parallel to **b.**

(iv) A spherical inclusion has a radial, spherically symmetric displacement field:

$$\begin{aligned} \mathbf{R} &= \varepsilon\, r_0^3 \mathbf{r} / r^3 \quad \text{for } r \geq r_0 \\ \mathbf{R} &= \varepsilon\, \mathbf{r} \qquad\quad\ \text{for } r < r_0 \end{aligned} \tag{1.26}$$

with $\varepsilon = (2/3)\delta$, where δ is the lattice mismtch between inclusion and matrix.

Planar interfaces which are inclined with respect to the foil surface can be considered as consisting of "steps" one column wide. Along a line perpendicular to the intersection line of the fault plane and the foil surface the columns are assumed to contain a planar fault, parallel to the foil plane, at the level where the inclined fault plane intersects the columns.

Similarly an inclined dislocation line is assumed to consist of small segments, one column long, each section being parallel to the foil plane.

1.8. Scattered Amplitude of a Deformed Foil; Kinematical Theory [7]

In the case of a deformed foil the scattered amplitude corresponding with the scattering vector $\mathbf{h} = \mathbf{g} + \mathbf{s}$ becomes, from (1.6):

$$A(\mathbf{h}) = \sum_{\mathbf{L}} F_h \exp[2\pi i(\mathbf{g} + \mathbf{s})].[\mathbf{A}_L + \mathbf{R}(\mathbf{r})] \tag{1.27}$$

or replacing the summation by an integration as in (1.11b):

$$A(\mathbf{h}) = F_{\mathbf{h}} \int_{\text{column}} \exp(2\pi i[\mathbf{g}.\mathbf{R}(\mathbf{r}) + \mathbf{s}z])dz \tag{1.28}$$

Hereby we have used the fact that $g.A_L$ is an integer and that $\mathbf{s}.\mathbf{R}(\mathbf{r})$ is very small compared to the other terms in the exponential. Putting

$$\alpha = 2\pi\mathbf{g}.\mathbf{R}(r) \tag{1.29}$$

we can write

$$A(h) = F_{\mathbf{h}} \int_0^{z_0} \exp\ i\alpha(z)\ \exp\ (2\pi isz)dz \tag{1.30}$$

1.8.1. Stacking fault contrast

Let the fault plane be parallel to the foil planes at $z=z_1$, behind the entrance face. Since R = constant in this case also α = constant and we can split the integral of (1.30) in two parts.

$$A(\mathbf{h}) / F_{\mathbf{h}} = \int_0^{z_1} \exp(2\pi isz)dz + e^{i\alpha} \int_{z_1}^{z_0} \exp(2\pi isz)dz \tag{1.31}$$

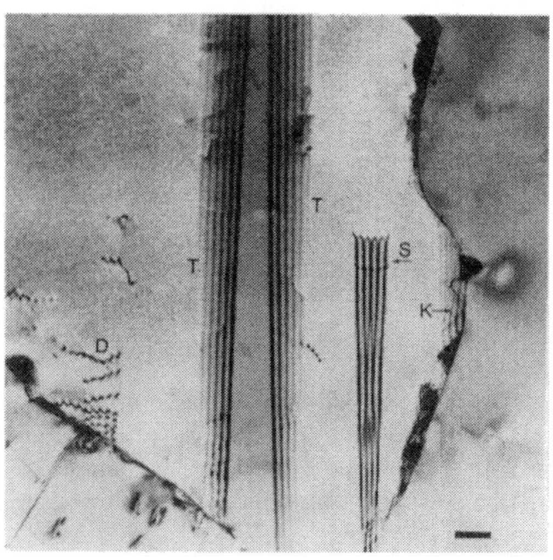

Figure 1.13 Fringe pattern due to a stackingfault in S, wedge fringes in T and dislocations in D (stainless steel).

Figure 1.14 Uniformly shaded area due to the presence of a stacking fault parallel to the foil plane (graphite).
(a) partial dislocations in contrast (b) stacking fault areas show up bright

or after evaluating the integrals one finds $I(\mathbf{h}) = A(\mathbf{h})\, A^*(\mathbf{h})$ or:

$$I(\mathbf{h}) / F_{\mathbf{h}}^2 = \{1 - \cos(\alpha + \pi s z_0)\cos \pi s z_0$$

$$+ \cos 2\pi s u\ [\cos(\alpha + \pi s z_0) - \cos \pi s z_0]\ \} / (\pi s)^2 \qquad (1.32)$$

with $u = \frac{1}{2}(2z_1 - z_0)$, i.e. u is the distance from the central plane of the foil.

The intensity $I(\mathbf{h})$ clearly depends periodically on the thickness z_0 of the foil as well as on the level of the fault in the foil, i.e. on u. For an inclined fault in a foil of constant thickness the intensity $I(\mathbf{h})$ is a periodic function of u with period 1/s; it is symmetrical in u since the cosine is an even function. An electron micrograph will produce a projection of this intensity distribution, i.e. a fringe pattern with depth period 1/s and of which the lateral extension is confined to the projected width of the fault (Figure 1.13).

If the fault plane is parallel to the foil surfaces a region of uniform shade is produced in the faulted area. This shade can be either brighter or darker than the perfect areas of the foil (Figure 1.14).

1.8.2. Domain boundary contrast

Let the domain boundary be parallel to the foil surfaces and situated at z_1 (Figure 1.15). We then have $\alpha = 0$ for $z \leq z_1$ and $\alpha = 2\pi \mathbf{g}.\mathbf{k}z$ for $z_1 < z \leq z_0$. The integral (1.30) can again be split into two parts. Calling $\mathbf{g}.\mathbf{k} = \Delta s$ we have

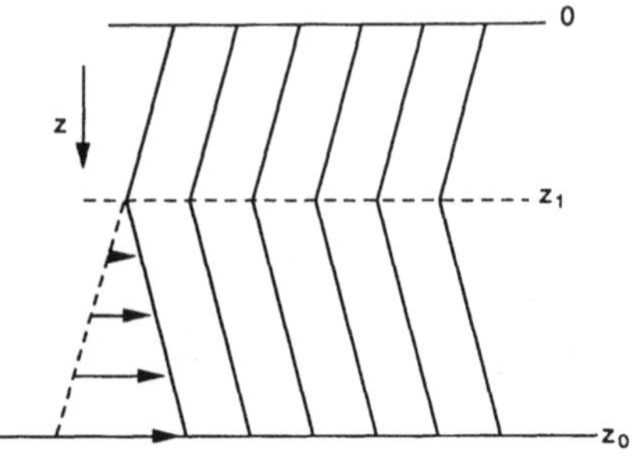

Figure 1.15 Model for a domain boundary

$$A(\mathbf{h}) / F_{\mathbf{h}} = \int_{0}^{z_1} \exp(2\pi i s z) dz + \int_{z_1}^{z_0} \exp \; [2\pi i (s + \Delta s) z] \; dz \qquad (1.33)$$

Since the two crystal parts on both sides of the surface are perfect, but slightly misoriented, s and Δs are constants. These integrals can easily be evaluated explicitly, and the corresponding expression for the intensity profile obtained.

1.8.3. Dislocation contrast

We adopt the geometry shown in Figure 1.16, i.e. the screw dislocation line is parallel to the foil plane (x.y), along the y-axis. We then have $\theta = \text{arctg}\ (z/x)$ and

$$\alpha = (\mathbf{g}.\mathbf{b}) \ \text{arctg}\ (z/x) \qquad (1.34)$$

where $\mathbf{g}.\mathbf{b} = n$ is an integer for a perfect dislocation but a fraction for a partial dislocation. The scattered amplitude is given by

$$A(\mathbf{h}) / F_{\mathbf{h}} = \int_{-z_1}^{+z_2} \exp(2\pi i s z) . \exp \ [in \ \text{arctg}\ (z/x)] \ dx \qquad (1.35)$$

The integral can be evaluated for different values of n but the results are not elementary functions. The numerical results are represented graphically in Figure 1.17 for different values of n, as a function of $\beta = 2\pi sx$. It turns out that the peak height, peak shift and peak width increase with increasing n.

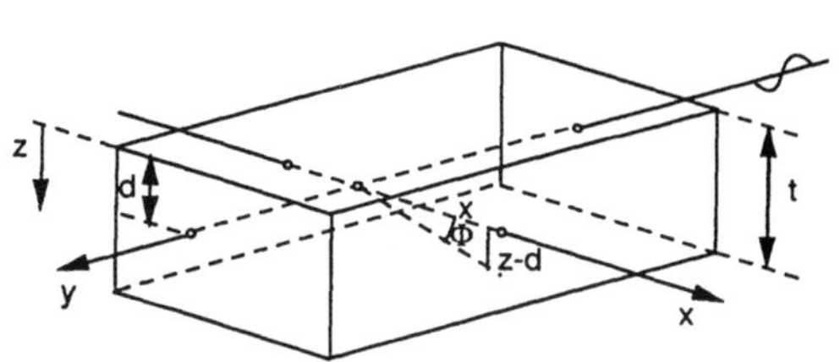

Figure 1.16 Foil containing a screw dislocation, notations used.

It is clear that no image will be produced if $n = 0$, i.e. for **g.b** = 0 since then the expression (1.35) reduces to that for a perfect foil.

The corresponding image profiles for edge dislocations are represented in Figure 1.18 [6].

1.9. Lattice Potential of Deformed Crystals

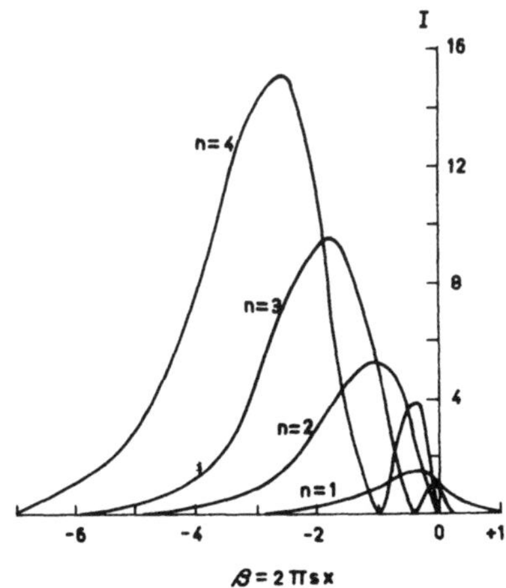

Figure 1.17 Image profiles for screw dislocations according to the kinematical theory (after Hirsch et al. [7]).

Deformation can also be introduced in the diffraction equations using the "deformable ion approximation" via the lattice potential. It is assumed that a displaced volume element of crystal carries along the lattice potential associated with it before the deformation, i.e.

$$V_{deformed}(\mathbf{r}) = V_{undeformed}(\mathbf{r} - \mathbf{R}) \tag{1.36}$$

As a result the Fourier coefficients of the lattice potential in the deformed crystal acquire phase factors since

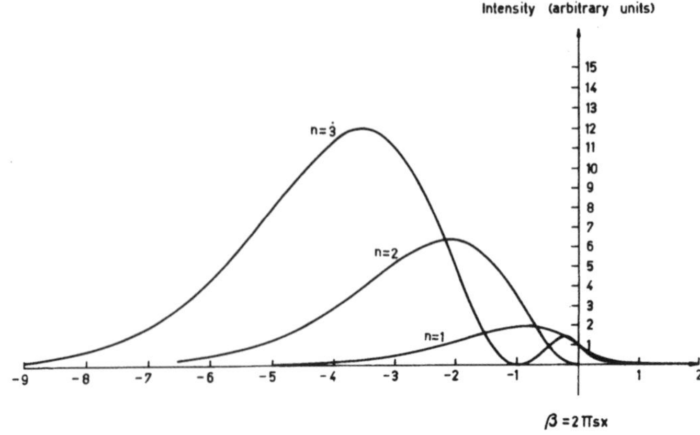

Figure 1.18 Image profiles for edge dislocations (after R. Gevers [6]).

$$V_{def.}(\mathbf{r}) = V_0 + \sum_g V_g \exp[2\pi i\mathbf{g}.(\mathbf{r}-\mathbf{R})]$$
$$= V_0 + \sum_g [\ V_g \exp(-2\pi i\mathbf{g}.\mathbf{R})\ ]\exp(2\pi i\mathbf{g}.\mathbf{r}) \qquad (1.37)$$

Strictly speaking this series is no longer a Fourier series, which is consistent with the fact that the function $V_{def}(\mathbf{r})$ is no longer a periodic function. Accepting nevertheless this approximation the presence of a deformation field can be accounted for by replacing

$$V_g \rightarrow V_g \exp[-2\pi i\mathbf{g}.\mathbf{R}(\mathbf{r})] \qquad (1.38)$$

In view of the relation between V_g and $1/t_g$ this also amounts to replacing $1/t_g$ by $(1/t_g)\exp(-i\alpha_g)$ with $\alpha_g = 2\pi\mathbf{g}.\mathbf{R}(\mathbf{r})$.

For a deformed foil the equation (1.21) thus becomes

$$\frac{d\phi_g}{dz} - 2\pi i s_g \phi_g = (\pi i / t_g)\ \exp(i\theta_g)\ \exp(-i\alpha_g) \qquad (1.39)$$

Introducing a new function $\varphi_g = \phi_g \exp(-i\alpha_g)$ which does not change the scattered intensity since $I_g = \varphi_g\varphi_g^* = \phi_g\phi_g^*$ this equation becomes

$$\frac{d\varphi_g}{dz} - 2\pi i\left(s_g + \mathbf{g}.\frac{d\mathbf{R}}{dz}\right)\varphi_g = \left(\pi i / \mathbf{t}_g\right)\exp\left(i\theta_g\right) \tag{1.40}$$

This equation is of the same form as (1.21), except that s_g has been replaced by a local value

$$s_{eff} = s_g + g.\frac{dR}{dz} \tag{1.41}$$

which is in general a function of x, y and z.

Depending on the problem to be treated one has to use (1.39) or (1.40); if α_g is a constant Eq. (1.40) cannot be used, since it reduces to that of a perfect crystal.

1.10. Matrix Formulation of the Kinematical Theory

In preparation of the subsequent matrix formulation of the dynamical theory we introduce here the use of matrices in the simpler kinematical theory [G1].

It is possible to write the normalized amplitudes of the scattered and transmitted beams, S and T, for a perfect foil, as the result of a matrix multiplication, operating on their initial values 1 and 0, at the entrance face

$$\begin{pmatrix} T \\ S \end{pmatrix} = \begin{pmatrix} \exp(-\pi isz) & 0 \\ i\ \sin(\pi sz / st_g) & \exp(\pi isz) \end{pmatrix}\begin{pmatrix} 1 \\ 0 \end{pmatrix} = M(s,z)\begin{pmatrix} 1 \\ 0 \end{pmatrix} \tag{1.42}$$

We have hereby omitted the factor exp $(i\theta_g)$ (which is equal to ± 1 in a centro-symmetrical crystal) since $\theta_g = 0$ or π. It is easy to verify that this leads to the correct expressions for T and S.

$$\begin{aligned} &T=\exp(-\pi isz) \text{ i.e. } |T|=1 \\ &S=i\ \sin(\pi sz/st_g) \end{aligned} \tag{1.43}$$

by comparing with Eq. (1.24) and noting that φ_g = S exp (πisz). The response matrices M(s,z) have the property

$$M(s,z_1)\ .\ M(s,z_2)=M(s,z_1+z_2) \tag{1.44}$$

This follows intuitively from the consideration that the result must be independent of the fact whether or not we consider the slab to be divided in two slabs. It is easy to show by matrix multiplication.

For a stacking fault with $\alpha = 2\pi\mathbf{g}.\mathbf{R}_0$ the scattered amplitude can then be obtained by means of the matrix multiplication

$$\begin{pmatrix} T \\ S \end{pmatrix} = \begin{pmatrix} \exp[-\pi is(z_0 - z_1)] & 0 \\ \dfrac{i \sin[\pi s(z_0 - z_1)]}{st_g} & \exp[\pi is(z_0 - z_1)] \end{pmatrix} \begin{pmatrix} 1 & 0 \\ 0 & \exp\ (i\alpha) \end{pmatrix}$$

$$\times \begin{pmatrix} \exp(-\pi isz_1) & 0 \\ \dfrac{i \sin \pi sz_1}{st_g} & \exp(\pi isz_1) \end{pmatrix} \begin{pmatrix} 1 \\ 0 \end{pmatrix} \qquad (1.45)$$

It is easy to verify that this leads to the result obtained above (1.32).

The matrix $\begin{pmatrix} 1 & 0 \\ 0 & e^{i\alpha} \end{pmatrix}$ is called the shift matrix and takes care of the phase shift introduced by the stacking fault.

Using the shorthand notation introduced in (1.43) we can write

$$\begin{pmatrix} T \\ S \end{pmatrix} = \begin{pmatrix} T_2 & 0 \\ S_2 & T_2^{(-)} \end{pmatrix} \begin{pmatrix} 1 & 0 \\ 0 & \exp(i\alpha) \end{pmatrix} \begin{pmatrix} T_1 & 0 \\ S_1 & T_1^{(-)} \end{pmatrix} \begin{pmatrix} 1 \\ 0 \end{pmatrix} \qquad (1.46)$$

where the minus sign indicates that the diffraction vector is -**g** and hence that $s \rightarrow -s$ in the corresponding expressions. The indices 1 and 2 mean that the corresponding expressions have to be taken for $z=z_1$ or $z=z_2$ respectively. After performing the matrix multiplications one obtains:

$$T = T_1T_2; \quad S = T_1S_2 + S_1T_2^{(-)} \exp(i\alpha) \tag{1.47}$$

which shows that the transmitted amplitude is simply that of the doubly transmitted beam. According to the kinematical theory the stacking fault remains undetected in the bright field image since α does not occur in the expression for T. This is in contradiction with the observation.

The scattered amplitude on the other hand is formed by the interference between the beam transmitted through the front part and subsequently scattered by the rear part and the beam scattered by the front part and subsequently transmitted through the rear part, taking into account the phase shift α caused by the fault. The dark field image thus reveals the fault.

We shall see below that the expressions obtained from the dynamical theory are generalizations of (1.46) and (1.47) taking into account that doubly scattered beams occur which interfere with the doubly transmitted beam. This can be taken into account by replacing the zero in the response matrices by $S_2^{(-)}$.

1.11. Limitation of the Kinematical Theory

The rocking curve representing the scattered intensity as a function of the deviation parameter (Figure 1.11) is symmetrical in s_g. This will still be the case for the scattered beam in the dynamical theory, but not for the transmitted beam, if anomalous absorption is taken into account. The experimental results show that the transmitted intensity is asymmetrical in s_g; in particular the transmitted intensity is larger for $s_g > 0$ than for $s_g < 0$, for the same absolute value of s_g(Borrmann effect [8]).

The kinematical theory does not describe what happens to the transmitted beam. At best one could assume that electrons are conserved and that the transmitted intensity I_T varies in a manner complementary to that of the scattered beam I_s i.e. $I_T = 1 - I_s$.

Under which conditions can the kinematical theory lead to useful results? Neglecting the depletion of the incident beam, as is assumed in the kinematical theory, is clearly only acceptable for thin crystals such that $I_S << 1$, i.e. the thickness should be small compared to the extinction distance. Another way to satisfy the condition $l_s << 1$ is to assume s_g large, i.e. to consider foil orientations which are far from the exact Bragg orientation.

It should be noted that for $s_g = 0$ the thickness fringes acquire an infinite period since this is inversely proportional to s_g. In actual fact thickness extinctions contours are always observed, especially for $s_g = 0$. The kinematical theory is then not valid for small s-values, not even qualitatively. It will nevertheless be useful, especially for large s-values as used in the weak beam technique, or in very thin slices in multi-slice considerations.

1.12. Basic Equations of the Two-beam Dynamical Theory for Perfect Crystals

Rather than deriving the fundamental equations of the two-beam dynamical theory directly from Schrödinger's equation as was done in Ref. 1, we shall use a very simple physically transparent method, related to the one originally used by Darwin [9]. It includes all usual approximations already in the model.

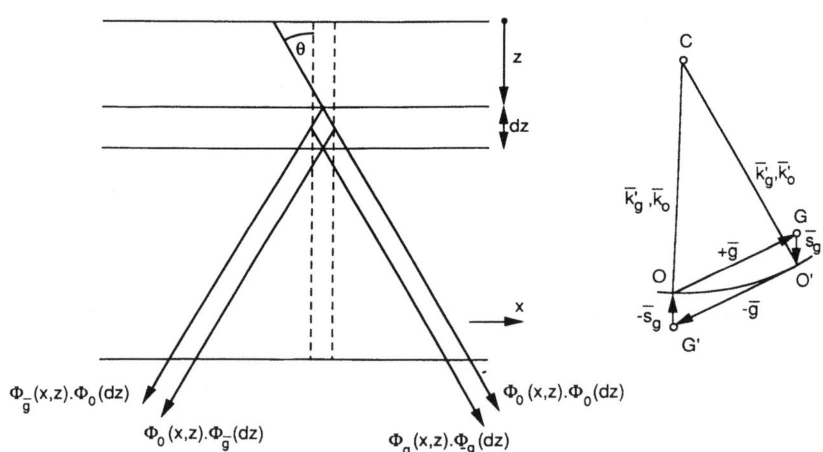

Figure 1.19 Illustrating the derivation of the fundamental equations of the dynamical theory.

The electrons travelling down a column of crystal can be represented by the wavefunction:

$$\psi(\mathbf{r}) = \phi_0(z)\exp(2\pi i\mathbf{k}_0.\mathbf{r}) + \phi_g(z)\exp(2\pi i\mathbf{k}.\mathbf{r}) \qquad (1.48)$$

where the $\mathbf{k}$ and $\mathbf{k}_0$ represent respectively the wavevectors of the scattered and of the incident beam; the amplitudes ϕ_g and ϕ_0 of these beams depend on z. The column approximation justifies to neglect the x dependence of these amplitudes. We shall write down the amplitude after interaction with a slice dz of the column, at level z behind the entrance face (Figure 1.19).

We can write

$$\phi_0(z+dz) = \phi_0(z)\phi_0(dz) + \phi_g(z)\phi_{-g}(dz) \qquad (1.49)$$

where ϕ_{-g} means that scattering with scattering vector -$\mathbf{g}$ has to be taken into account. This relation states that the transmitted beam results from the interference between the twice transmitted beam of which the amplitude is $\phi_0(z)$ $\phi_0(dz)$ and the twice scattered beam with amplitude $\phi_g(z)\phi_{-g}(dz)$. Since the slices dz are arbitrarily thin, it is justified to use the kinematical approximation for $\phi_0(dz)$ and $\phi_g(dz)$. Note that changing $\mathbf{g} \rightarrow -\mathbf{g}$ implies changing $s \rightarrow -s$, since the origin of reciprocal space is now in the node point G. From the equation (1.24) we can conclude, ignoring the phase factor of the Fourier coefficient and leaving out irrelevant phase factors:

$$\phi_g(dz) = d\phi_{\mathbf{g}} = (\pi i / t_g)\exp(-2\pi isz)dz \qquad (1.50)$$

We have furthermore $\phi_0(dz) = 1$ and hence

$$\phi_0(z+dz) = \phi_0(z) + \phi_g(z)\frac{\pi i}{t_{-g}}\exp(2\pi isz)dz \qquad (1.51)$$

or

$$\phi_0(z+dz) - \phi_0(z) = \phi_g(z)\frac{\pi i}{t_{-g}}\exp(2\pi isz)dz$$

and in the limit for $dz \rightarrow 0$:

$$(d\phi_0 / dz) = (\pi i / t_{-g}) \exp(2\pi isz)\phi_g(z) \tag{1.52}$$

Similarly we can write for the scattered beam

$$\phi_g(z+dz) = \phi_0(z)\ \phi_g(dz) + \phi_g(z)\phi_0(dz) \tag{1.53}$$

and taking into account the expressions for $\phi_g(dz)$(1.50) and $\phi_0(dz) = 1$, and making the same transformations as for ϕ_0, we obtain:

$$d\phi_g / dz = (\pi i / t_g)\exp(-2\pi isz)\phi_0(z) \tag{1.54}$$

The equations (1.52) and (1.54) form the Darwin-Howie-Whelan [10] set of coupled differential equations describing the interplay between the incident and the scattered beam in a perfect crystal. This set can be given a number of alternative forms by introducing amplitudes differing from ϕ_0 and ϕ_g only by phase factors. This does not change the resulting intensity distributions.

One can for instance substitute T′and S' for ϕ_0 and ϕ_g

$$\phi_0 = T'\exp(\pi isz) \qquad \phi_g = S'\exp(-\pi isz) \tag{1.55}$$

The system of equations then acquires a symmetrical form

$$\begin{cases} dT'/dz + \pi isT' = (\pi i / t_{-g})S' & (1.56a) \\ dS'/dz - \pi isS' = (\pi i / t_g)T' & (1.56b) \end{cases}$$

or one can also make the substitution

$$\phi_0 = T \qquad \phi_g = S\ \exp(-2\pi isz) \tag{1.57}$$

and obtain the following asymmetrical set of equations:

$$\begin{cases} dT/dz = (\pi i / t_{-g})S & (1.58a) \\ dS/dz = 2\pi is\ S + (\pi i / t_g)T & (1.58b) \end{cases}$$

Depending on the problem to be solved either the symmetrical or the asymmetrical set will have to be used. In Ref. [1] the multiple beam set of Howie-Whelan equations (2.51) was derived from the Schrödinger equation. In the two-beam case the equations (2.51) and (1.58) have the same physical content.

1.13 Dynamical Equations for Deformed Crystals [10]

We have seen previously (§8 and 9) that the deformation of the lattice can be modeled by using a local effective deviation parameter s_{eff}, which is a function of **r**, rather than a constant s. We found that s_{eff} is given by

$$s_{eff} = s + \frac{d}{dz}\,(\mathbf{g}.\mathbf{R})$$

or (1.59)

$$s_{eff} = s + d\alpha / dz$$

with $\alpha = \alpha_g / 2\pi$. One can alternatively replace $\frac{1}{t_g}$ by $\frac{1}{t_g}\exp(-i\alpha_g)$.

These substitutions allow to adapt the different sets of equations to the case of deformed crystals. The presence of a defect is now described by a variation of *s* along the integration columns, the diffraction vector **g** being the same all along the column.

It is also possible to describe the defect by focussing attention on a local diffraction vector **g**+Δ**g** where Δ**g** now varies along the integration columns. We must then define Δ**g** by the relation:

$$\mathbf{g}.[\mathbf{r} - \mathbf{R}(\mathbf{r})] = (\mathbf{g} + \Delta\mathbf{g}).\mathbf{r} \qquad (1.60)$$

i.e.

$$-\mathbf{g}.\mathbf{R}(\mathbf{r}) = \Delta\mathbf{g}.\mathbf{r} \qquad (1.61)$$

which allows to write alternatively

$$\alpha_g = 2\pi\mathbf{g}.\mathbf{R}(\mathbf{r}) = -2\pi\Delta\mathbf{g}.\mathbf{r} \qquad (1.62)$$

1.14. Solution of the Two-Beam Dynamical Equations for a Perfect Crystal

The system of equations (1.58a,b) can be solved by uncoupling the two equations, on eliminating respectively S and T between the two equations of the set. One then obtains the same second order differential equations with constant coefficients for S and T, i.e.:

$$\frac{d^2T}{dz^2} - 2\pi i s_g \frac{dT}{dz} + \left(\pi^2 / t_g^2\right)T = 0 \tag{1.63}$$

and a similar equation for S.

The equations have to be solved with the initial values T= 1, S = 0 at z= 0, and by requiring that the solutions satisfy both equations of the original set (1.58a,b).

One finds:

$$T = \left[\cos(\pi\sigma_g z) - i(s_g / \sigma_g)\sin(\pi\sigma_g z)\right]\exp(\pi i s_g z) \tag{1.64a}$$

$$S = \left(i / \sigma_g t_g\right)\sin(\pi\sigma_g z)\exp(\pi i s_g z) \tag{1.64b}$$

where

$$\sigma_g^2 = \left[1 + (s_g t_g)^2\right] / t_g^2 \tag{1.65}$$

For the scattered intensity one then obtains:

$$I_S = \sin^2(\pi\sigma_g z) / (\sigma t_g)^2 \tag{1.66}$$

and

$$I_T = 1 - I_S \tag{1.67}$$

since absorption was neglected as yet.

The depth periods of both I_S and I_T are now $1/\sigma_g$ as compared to $1/s_g$ according to the kinematical theory. For s_g=0 the depth period now remains finite and is equal to t_g. Note that for large values of s_g, σ_g reduces to s_g.

The expressions (1.66) and (1.67) as well as the equations (1.52) and (1.54) describe the periodic transfer of electrons from the transmitted beam into the scattered beam and vice versa (Pendellösung effect). The periodic variations of I_S(or I_T) with the crystal thickness give rise to the formation of thickness extinction contours in wedge-shaped crystals (Figure 1.4). The contours are the geometrical loci of equal thickness. According to the expressions (1.66) and (1.67) such fringes should be periodic; in actual fact the fringes are found to be damped with increasing thickness, showing that absorption takes place. We shall take absorption into account further below.

The expressions I_S and I_T also describe the dependence on s_g. The geometrical loci with s_g= constant are the equi-inclination or bend contours. According to the expressions for I_S and I_T this dependence should be symmetrical in *s*, i.e. $I_S(-s)=I_S(+s)$ and $I_T(-s) = I_T(+s)$. In fact one finds that $I_T(-s)<I_T(+s)$ whereas $I_S(-s)=I_S(+s)$. This effect, called the Borrmann effect [8], was shown to be a consequence of anomalous absorption in Ref. [1].

1.15. Two-beam Lattice Fringes

The total wavefunction at the exit face of a crystal foil can be written as:

$$\Psi = \exp(2\pi i\mathbf{K}.\mathbf{r}) \left[T + S \ \exp(2\pi i\mathbf{g}.\mathbf{r})\right] \qquad (1.68)$$

where **K** is the wavevector of the incident electron beam corrected for reflection by the constant part V_0 of the lattice potential. The total wavefunction Ψ can be imaged directly in the microscope by collecting the two interfering beams in the objective aperture and focussing on the exit face of the foil [11]: one then observes the intensity distribution:

$$I = \Psi\Psi^* = TT^* + SS^* + T^*S \ \exp(2\pi i\mathbf{g}.\mathbf{r}) + TS^* \exp(-2\pi i\mathbf{g}.\mathbf{r}) \qquad (1.69)$$

$$= I_T + I_S + 2\sqrt{I_T I_S}\ \sin(2\pi i\mathbf{g}.\mathbf{r} + \varphi) \qquad (1.70)$$

where

$$tg\varphi = (s_g / \sigma_g)tg(\pi\sigma_g z_0) \qquad (1.71)$$

z_0 being the foil thickness. The image consists of sinusoidal fringes with a period $1/|\mathbf{g}|$, which is equal to the interplanar distance of the lattice planes normal to **g** The fringes are parallel to the traces of these lattice planes. For s_g= 0 the expression (1.70) simply reduces to

$$I = 1 + (\sin 2\pi z_0 / t_g)\sin(2\pi g.x) \qquad (1.72)$$

where the x-axis was chosen along **g**

The contrast of the fringes clearly depends on z_0, whereas the localization of the fringes depends on s_g since this determines φ. For s=0, φ=0, the bright fringes coincide with the lattice planes gx= integer +1/2. The fringes are therefore called lattice fringes; their formation can be understood on a purely geometrical basis.

The phase φ in (1.70) will in general also depend on the exact imaging conditions. Since the beams T and S may enclose different angles with the optical axis of the microscope additional phase shifts will in general occur. As long as we consider diffraction contrast images the image transfer function of the microscope is not important since the intensity distribution in a single beam does not depend on its shape. However when considering the interference of more than one beam it plays an important role and cannot be ignored.

1.16. Absorption

Absorption can phenomenologically be described by assuming the refractive index n = c/υ (c = velocity in vacuum; υ = velocity in medium) to be complex. Representing a plane wave as

$$\Psi = \Psi_0 \exp\ i(kz - \omega t) \qquad (1.73)$$

with k = ω/υ = ωn/c and making n complex n_c = n + iμ leads to a complex wavevector

$$k_c = \frac{\omega}{c}(n + i\mu) \tag{1.74}$$

and to a wave

$$\Psi = \Psi_0 \exp[i(k_c z - \omega t)] = \Psi_0 \exp(-\omega\mu z / c)\exp[i(kz - \omega t)] \tag{1.75}$$

which is clearly damped, the amplitude absorption coefficient being $\mu\omega/c$.

Replacing V_0 by V_0 + i W_0 in the lattice potential has the desired effect since it makes the wavevector K complex:

$$K^{*2} = 2me[(E + V_0) + iW_0] / h^2 \tag{1.76}$$

$$= K^2 + i[2me\ W_0 / h^2K]K \tag{1.77}$$

We now define

$$\frac{1}{\tau_0} = 2me\ W_0 / h^2K \tag{1.78}$$

by analogy with

$$\frac{1}{t_0} = 2me\ V_0 / h^2K \tag{1.79}$$

The complex wavevector K^* then becomes

$$K^{*2} = K^2 + i(K / \tau_0) \tag{1.80}$$

and since $1/K\tau_0 << 1$ also:

$$K^* = K[1 + i / (K\tau_0)]^{1/2} \approx K(1 + \tfrac{1}{2}\ i / K\tau_0)$$

or finally

$$K^* = K + \tfrac{1}{2}\ i / \tau_0 \tag{1.81}$$

The boundary conditions now require that at the entrance face z= 0 the tangential component must be conserved, i.e. $K_t^* = K_t$. The imaginary part being zero for z < 0 this means that for z > 0 the imaginary part must be oriented along $\mathbf{e}_n$, i.e.

$$\mathbf{K}^* = \mathbf{K} + \left(\tfrac{1}{2}\, i/\tau_0\right)\mathbf{e}_n \tag{1.82}$$

The expression exp $(2\pi i\mathbf{K}.\mathbf{r})$ then becomes

$$\exp(2\pi i\mathbf{K}.\mathbf{r})\exp(-\pi z/\tau_0) \tag{1.83}$$

since $\mathbf{e}_n.\mathbf{r}= z$. The absorption coefficient for the amplitude is thus $\mu = \pi/\tau_0$.

It has been shown by Yosioka [12] that anomalous absorption can be taken into account by assuming the lattice potential to become complex: $V(\mathbf{r}) + i\, W(\mathbf{r})$. We have just shown that this applies to the constant term $V_0 + i\, W_0$, and leads to normal absorption in that case. This procedure can be generalized replacing also the other Fourier coefficients by complex quantities: $V_g \rightarrow V_g + i\, w_g$. In view of the relation between $1/t_g$ and V_g (Eq. (1.22)) this is equivalent to replacing

$$\frac{1}{t_g} \rightarrow \frac{1}{t_g} + \frac{i}{\tau_g} \tag{1.84}$$

where we have introduced the <u>absorption lengths</u> τ_g, which are related to W_g in the same way as t_g is related to V_g:

$$\frac{1}{\tau_g} = 2meW_g / h^2k \tag{1.85}$$

$\tau_g >> t_g$ since $W_g << V_g$. If $1/t_g$ becomes complex it is clear that also σ_g defined in (1.65) must become complex

$$\sigma_g = \sigma_r + i\sigma_i \tag{1.86}$$

Omitting for simplicity the index **g** at σ_r and σ_i one finds to a good approximation, noting that $\sigma_i^2 << \sigma_r^2$ and $1/\tau_g^2 << 1/t_g^2$:

$$\sigma_r = (1/t_g)\left[1 + s_g^2\, t_g^2\right]^{1/2} \; ; \; \sigma_i = \left(\sigma_r t_g \tau_g\right)^{-1} \tag{1.87}$$

We note that $\sigma_r \mathbf{r} = \sigma_i z$

1.17. Dynamical Equations including Absorption

Anomalous absorption effects can be taken into account in the basic equations (1.52) (1.54) or in the equations derived from these, by the above mentioned substitutions (1.84) or (1.86). The equations (1.52) (1.54) then become for instance

$$\frac{d\phi_0}{dz} = \pi i\left(\frac{1}{t_{-g}} + i\frac{1}{\tau_{-g}}\right)\exp(2\pi i s_g z)\phi_g(z) \tag{1.88}$$

$$\frac{d\phi_g}{dz} = \pi i\left(\frac{1}{t_g} + i\frac{1}{\tau_g}\right)\exp(-2\pi i s_g z)\phi_0(z) \tag{1.89}$$

The substitution can also simply be performed directly in the final solutions (1.64). Although the method is purely phenomenological the result allows a physically meaningful interpretation.

In order to obtain tractable analytical solutions one adopts moreover the following approximation: σ is replaced by $\sigma_r + i\sigma_i$ in the periodic terms of the expressions (1.64) but in the coefficients we replace σ by σ_r. This approximation preserves the essential features of anomalous absorption. It allows to show that the wavefunction ψ_T of the transmitted beam results from the interference of two waves with slightly different wavevectors ($\mathbf{e}$ = unit normal on foil)

$$\mathbf{K} + (1/2)(s + \sigma_r)\mathbf{e} \quad \text{and} \quad \mathbf{K} + (1/2)(s - \sigma_r)\mathbf{e} \tag{1.90}$$

For the wavefunction ψ_S of the scattered beam the two interfering waves have wavevectors

$$\mathbf{K} + \mathbf{g} + (1/2)(s + \sigma_r)\mathbf{e} \quad \text{and} \quad \mathbf{K} + \mathbf{g} + (1/2)(s - \sigma_r)\mathbf{e} \tag{1.91}$$

The beating of these two waves causes the periodic depth variation of ψ_T and ψ_S with a period $1/\sigma_r$ i.e. the "Pendellösung effect".

The four waves present in the total wavefunction $\psi = \psi_T + \psi_S$ belong to two Bloch wavefields ψ_1 and ψ_2. For the waves of the first wavefield ψ_1 the sign of σ_r is positive; it contains the two waves with wavevectors $\mathbf{K} + 1/2(s - \sigma_r)\mathbf{e}$ and $\mathbf{K} + \mathbf{g} = (1/2)(s + \sigma_r)\mathbf{e}$. The second wavefield ψ_2 corresponds to the negative sign of σ_r; it contains the two waves with wavevectors $\mathbf{K} + 1/2(s - \sigma_r)\mathbf{e}$ and $\mathbf{K} + \mathbf{g} + (1/2)(s + \sigma_r)\mathbf{e}$. In the simple case $s = 0$ the wavefield ψ_1 has a maximum amplitude in planes coinciding with the atomic planes, whereas ψ_2 has maximum amplitude along planes exactly in between the atomic planes.

The first type of wavefield which is strongly excited for $s < 0$ is strongly attenuated whereas the second wavefield, which is more strongly excited for $s > 0$ is less attenuated by anomalous absorption. These results are physically meaningful. Electrons propagating close to atomic cores, i.e. for which the wavefunction is peaked at the atomic positions, will have a larger probability to excite X-rays and hence to be "absorbed" than those passing between atomic planes. (This was also discussed in Ref. [1])

The damping of the thickness fringes in a wedge shaped crystal can be understood on the basis of these results. The depth variations of I_S and I_T are in both cases caused by the beating of two waves, one of each wavefield. Since one of these waves is much more attenuated than the other one the beating envelope, which has maximum amplitude when the two beating waves have equal amplitude decreases with depth in the crystal, even though one of the waves has still an appreciable amplitude.

Also the Borrmann effect can be understood on the basis of the same model. For $s > 0$ the amplitude of the rapidly attenuated wave in ψ_T i.e. $(\frac{1}{2})\left[1 - (s/\sigma_2)\right]$ is smaller than that of the passing wave $\left((\frac{1}{2})\left[1 + (s/\sigma_2)\right]\right)$, which is enhanced by anomalous absorption. As a result ψ_T will have a larger amplitude for $s > 0$ than for $s < 0$ for the same absolute value of s. A similar asymmetry is absent for ψ_S where the amplitudes of the two constituent waves are both $1/\left(2\sigma t_g\right)$, which only depends on s^2.

1.18. Rocking Curves for Perfect Crystals taking into account Anomalous Absorption

Explicit expressions for I_T and I_S are obtained by computing $I_T = \psi_T . \psi_T^*$ and $I_S = \psi_{S*} \psi_{S*}^*$. Since $\sigma_i << \sigma_r$ and $\tau_g >> t_g$ one can approximate the expressions by neglecting higher order terms in t_g / τ_g. One obtains after lengthy but straightforward calculations

$$I_T = \left[\cosh u + (s/\sigma_r)\sinh u\right]^2 - \left[1/(\sigma_r t_g)^2\right]\sin^2 v \qquad (1.92)$$

with $u = \pi\sigma_i z$; $v = \pi\sigma_r z$; $\sigma_i = (\sigma_r t_g \tau_g)^{-1}$ and $\sigma_r = \left[1 + (st_g)^2\right]^{1/2} / t_g$

Similarly

$$I_S = \left[\sinh^2 u + \sin^2 v\right] / (\sigma_r t_g)^2 \qquad (1.93)$$

These expressions are represented in Figure 1.20.

Note that in the limit $\sigma_i \to 0$, i.e. $u \to 0$ one obtains the corresponding expressions for the non-absorption case:

$$I_S = \sin^2(\pi\sigma_r z) / (\sigma_r t_g)^2 \quad \text{and} \quad I_T = 1 - I_S \qquad (1.94)$$

For s = 0 one obtains

$$I_S = \sinh^2 u + \sin^2 v \quad \text{and} \quad I_T = \cosh^2 u - \sin^2 v \qquad (1.95)$$

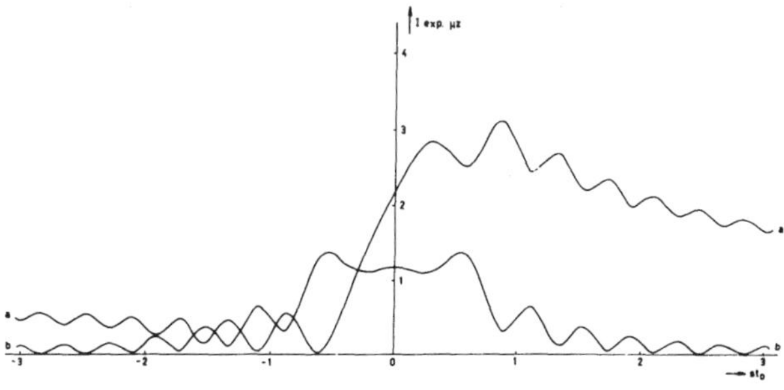

Figure 1.20 Rocking curves for perfect crystal taken into account anomalous absorption. The transmitted intensity I_T (a) is asymmetrical in $s.t_g$ whereas the scattered intensity I_s (b) is symmetrical in $s.t_g$.

Note that now $I_T + I_S = \sinh^2 u + \cosh^2 u > 1$! This apparently contradictory result is due to the fact that normal absorption, which attenuates both beams to the same extent, was neglected.

The expressions for I_S and I_T can be rewritten (s = 0) as

$$I_{T,S} = \tfrac{1}{2}[\cosh 2u \pm \cos 2v] \qquad (1.96)$$

where the + sign refers to the transmitted beam and the - sign to the scattered beam.

1.19. Dynamical Diffraction by Deformed or Faulted Crystals [10] [G2]

Quantitative studies of defects in crystals are usually performed under optimized two-beam conditions since this allows the most straightforward and detailed interpretation. The system of two-beam equations which lends itself most conveniently to describe the diffraction by non-perfect crystals is the Darwin-Howie-Whelan system of coupled differential equations.

Different formulations of this system are available; their solutions for the amplitudes of scattered (S) and transmitted (T) beams differ by phase factors only and hence lead to the same intensity distribution, i.e. to the same image. We shall use two different forms of this system depending on the problem to be treated.

The asymmetric form (from (1.58)):

$$\begin{cases} dT/dz + (\pi i/t_{-g})S & (1.97a) \\ dS/dz = 2\pi i s_g S + (\pi i/t_g)T & (1.97b) \end{cases}$$

and the symmetric form (from (1.56))

$$\begin{cases} dT'/dz + \pi i s_g T' = (\pi i/t_{-g})S' & (1.98a) \\ dS'/dz - \pi i s_g S' = (\pi i/t_g)T' & (1.98b) \end{cases}$$

The amplitudes T′, S′ and T, S are directly related:

$$T = T'\exp(\pi i s_g z) \tag{1.99a}$$

$$S = S'\exp(\pi i s_g z) \tag{1.99b}$$

We have shown in §9 that deformation can be introduced in the diffraction equations by applying the deformable ion model, a defect being modeled by means of a vector field **R**(**r**), called its displacement field. It was also shown that the Fourier coefficients of the lattice potential then become functions of **r** and acquire a phase factor $V_g\exp(-i\alpha_g)$ with $\alpha_g = 2\pi\mathbf{g.R}$. Strictly speaking the series (1.19) is then no longer a Fourier series in agreement with the fact that the lattice potential is no longer periodic. It nevertheless turns out to be a good approximation especially for small gradients of **R**(**r**). The presence of the defect can then be taken into account by substituting in the diffraction equations $V_g \to V_g\exp(-i\alpha_g)$. In view of the direct proportionality of $1/t_g$ and V_g (Eq. 1.22) this implies that $1/t_g$ has to be replaced by $(1/t_g)\exp(-i\alpha_g)$. The equations (1.97) then take the form:

$$\begin{cases} dT/dz = (\pi i/t_{-g})\exp(i\alpha_g)S & (1.100a) \\ dS/dz = 2\pi i s_g S + (\pi i/t_g)\exp(i\alpha_g)T & (1.100b) \end{cases}$$

and the equations (1.98) become:

$$\begin{cases} dT'/dz + \pi i s_g T' = (\pi i/t_{-g})\exp(i\alpha_g)S' & (1.101a) \\ dS'/dz - \pi i s_g S' = (\pi i/t_g)\exp(i\alpha_g)T' & (1.101b) \end{cases}$$

Both systems of equations can be transformed so as to involve only the gradient of the displacement field.

Putting $T = T''$ and $S = S''\exp(-i\alpha_g)$ the first system of equations (1.100) is transformed into the set

$$\begin{cases} dT''/dz = (\pi i/t_{-g})S'' & (1.102a) \\ dS''/dz = 2\pi i(s_g + d\alpha'_g/dz)S'' + (\pi i/t_g)T'' & (1.102b) \end{cases}$$

with $\alpha'_g = \alpha_g / 2\pi$
Performing the substitution:

$$T'' = T''' \exp(\pi i \alpha_g); \; S'' = S''' \exp(\pi i \alpha_g) \tag{1.103}$$

leads to the system

$$\begin{cases} dT'''/dz + \pi i(s_g + d\alpha'_g / dz)T''' = (\pi i / t_{-g})S''' & (1.104a) \\ dS'''/dz - \pi i(s_g + d\alpha'_g / dz)S''' = (\pi i / t_g)T''' & (1.104b) \end{cases}$$

The sets of equations (1.101) and (1.104) are of the same form as the corresponding sets for the perfect crystal except for the s-value. They suggest a simple interpretation of the diffraction phenomena. They show that the presence of the defect causes the local s-value $s_{eff} \equiv s_g + \mathbf{g}.d\mathbf{R}/dz$ to be in general different from its value in the perfect part of the foil. Locally, close to the defect, the Bragg condition is thus better or less well satisfied than in the rest of the foil and hence the locally diffracted beam has a larger or smaller amplitude than in the perfect part.

Bearing in mind the relation (1.61):

$$\mathbf{g}.\mathbf{R}(\mathbf{r}) = -\Delta\mathbf{g}.\mathbf{r} \tag{1.105}$$

one can alternatively interpret the equations as meaning that the diffraction vector **g** changes locally in orientation and length in the vicinity of the defect. This is consistent with the local lattice geometry for instance in the vicinity of a dislocation, where the lattice planes are inclined with respect to those in the perfect part.

If $\alpha = 2\pi\mathbf{g}.\mathbf{R}_0$ = constant, which is the case for a translation interface, with displacement vector $\mathbf{R}_0$, we have $d\alpha / dz = 0$ and the sets of equations (1.102) and (1.104) reduce to those for a perfect crystal. This is consistent with the fact that the two parts of the crystal on either side of the interface remain perfect. We have therefore to use the sets (1.101) or (1.100) to treat this problem, as we shall demonstrate below, rather than the sets (1.102) or (1.104).

If the interface is a domain boundary described by $\mathbf{R} = kz\,\mathbf{e}_\tau$ (§ 8) we have $d\mathbf{R}/dz = k\mathbf{e}_\tau$ and now the sets (1.102) and (1.104) can be used. They show that the crystal can be described as an assembly of two juxtaposed crystal parts separated by the interface and having different s values: s and $s+\Delta s$ with $\Delta s = k\mathbf{e}_\tau$.

For the displacement field of a dislocation neither the factor exp $i\alpha_g$ nor the gradient $d\mathbf{R}/dz$ disappear (except for those **g** for which extinction occurs). The sets of eqs. (1.102) and (1.104) as well as (1.100) and (1.101) are thus suitable descriptions in this case.

We remind at this stage that in the various systems of equations the following approximations were introduced:

(i) The column approximation. In the case of a perfect crystal this is not an approximation since all columns are identical, i.e. S and T are independent of x and y, chosen parallel with the foil plane, but only depend on z, the depth in the foil. For a foil containing defects the column approximation consists in fact in replacing the deformed crystal by a succession of thin, perfect slices, parallel to the foil planes and which are only displaced one with respect to the next by an amount given by $R(x_0,y_0,z)$ for the slice at level z and for a column at $x_0.y_0$. The shifts in successive slices vary with z in a way which depends on the chosen column, i.e. on $x_0.y_0$. By integration along z for a grid of columns a two-dimensional intensity distribution is obtained, which is the image of the defect. The columns are in fact the "pixels" of the image. The lateral dimensions of the columns, i.e. the mesh size of the grid of columns determines the resolution of the computed image, but it is unimportant for the computation since each slice is considered to be perfect and only displaced. For a dislocation line parallel to the foil plane the image can be described by a single profile, but for an inclined dislocation line a two-dimensional intensity map is required. For a foil containing defects it is not obvious how good the column approximation still is. However images computed numerically with and without the column approximation have shown very little difference for the image profiles in cases where the foil is close to the symmetrical Laue orientation [13].

(ii) The high energy approximation implies neglecting back scattered electrons. This is not appreciably influenced by the presence of defects and the approximation is therefore believed still to hold in defected crystals.

(iii) The two-beam approximation. Although in practice weak beams are usually excited next to the two most intense beams, the images obtained in carefully oriented foils of materials with not too large unit cells, are well described, at least qualitatively, by the computed two-beam images.

(iv) The anomalous absorption has not explicitly been introduced as yet in the different sets of equations for faulted crystals. Phenomenologically this can easily be done by replacing $1/t_g$ by $1/t_g + i/\tau_g$ in the sets of equations. Alternatively it can be done by making the same substitution directly in the final results. We shall follow the latter approach for a discussion of the fringe patterns at inclined planar interfaces.

1.20. Matrix Formulation for the Amplitudes of Transmitted and Scattered Beams

For a systematic discussion of the contrast at planar interfaces we shall make use of a matrix formulation for the amplitudes of the transmitted and scattered beams by a perfect crystal slab, which we now derive first (see for instance [14]).

Let T and S represent the transmitted and the scattered amplitudes for an incident wave with unit amplitude. The initial values at the entrance face of the slab are represented by the column vector $\binom{1}{0}$; at level z the amplitudes of transmitted and scattered beams are represented by the column vector $\binom{T}{S}$. From §14 we know that, ignoring a common irrelevant phase factor:

$$T \equiv T(z, s_g) = \cos(\pi\sigma_g z) - i(s_g/\sigma_g)\sin(\pi\sigma_g z) \qquad (1.106)$$

$$S \equiv S(z, s_g) = (i/\sigma_g t_g)\sin(\pi\sigma_g z) \qquad (1.107)$$

are the solutions of the set of eq. (1.56) or (1.58) with initial values $\binom{1}{0}$. Anomalous absorption is taken into account by assuming $\sigma_g = \sigma_r + i\sigma_i$ with

$$\sigma_r = (1/t_g)\left[1+(s_g t_g)^2\right]^{1/2} \text{ and } \sigma_i = (\sigma_r t_g \tau_g)^{-1} \tag{1.108}$$

For an incoming wave with an arbitrary amplitude we can write in view of the linear character of the system of differential equations:

$$\begin{pmatrix}T\\S\end{pmatrix}_{out} = \begin{pmatrix}AC\\BD\end{pmatrix}\begin{pmatrix}T\\S\end{pmatrix}_{in} \tag{1.109}$$

where the elements A, B, C and D of the 2x2 matrix remain to be determined. From

$$\begin{pmatrix}T(z,s_g)\\S(z,s_g)\end{pmatrix} = \begin{pmatrix}AC\\BD\end{pmatrix}\begin{pmatrix}1\\0\end{pmatrix} \equiv \begin{pmatrix}A\\B\end{pmatrix} \tag{1.110}$$

we can conclude that

$$A \equiv T \quad \text{and} \quad B \equiv S$$

We now make use of the symmetry of the system of equations (1.56). We note that this system is mapped on itself by the substitution $T \to S, S \to T$, $s_g \to -s_g$ since $t_g = t_{-g}$ in a centro-symmetric crystal. This means that the solution for initial values $\binom{0}{1}$ is given by

$$\begin{pmatrix}S^-\\T^-\end{pmatrix} = \begin{pmatrix}AC\\BD\end{pmatrix}\begin{pmatrix}0\\1\end{pmatrix} \equiv \begin{pmatrix}C\\D\end{pmatrix} \tag{1.111}$$

where the minus sign means: $T^- = T(z,-s_g)$ and $S^- = S(z,-s_g)$. We conclude that $C = S^{(-)}$ and $D = T^{(-)}$. The response matrix is thus completely defined for arbitrary initial values:

$$\begin{pmatrix}T\\S\end{pmatrix}_{out} = \begin{pmatrix}T & S^{(-)}\\S & T^{(-)}\end{pmatrix}\begin{pmatrix}T\\S\end{pmatrix}_{in} \tag{1.112}$$

We shall represent the response matrix M of a perfect crystal slab as

$$M(z,s) \equiv \begin{pmatrix} T & S^{(-)} \\ S & T^{(-)} \end{pmatrix} \quad (1.113)$$

Imagining a slab of perfect crystal with total thickness z_0 to be sliced in perfect slabs with thicknesses $z_1, z_2, \ldots, z_{n-1}, z_n$ such that $z_1 + z_2 + \ldots + z_n = z_0$ should clearly not influence the final result. We must therefore have

$$M(z_1 + z_2 + \ldots + z_n, s_g) = M(z_n, s_g) \,.\, M(z_{n-1}, s_g) \ldots M(z_1, s_g) \quad (1.114)$$

This property of the response matrix can be verified in a straightforward manner by multiplying the matrices.

The relation (1.114) can formally be generalized to include also the subtraction of a lamella, i.e.

$$M(z_1 - z_2, s_g) = M(-z_2, s_g)\, M(z_1, s_g) \quad (1.115)$$

where

$$M(-z, s_g) = \begin{pmatrix} T(-z) & S^{(-)}(-z) \\ S(-z) & T^{(-)}(-z) \end{pmatrix} \quad (1.116)$$

1.21. Matrix Formulation for a Foil containing a Translation Interface

A description of the diffraction effects associated with translation interfaces can be based on the set of equations (1.101). The vector $\mathbf{R}_0$ describes the displacement of the exit part with respect to the entrance part of the foil; it determines the sign of $\alpha_g = 2\pi \mathbf{g}.\mathbf{R}_0$. In the front part $\alpha_g = 0$ whereas in the exit part $\alpha_g \neq 0$. The translation interface at $z = z_1$ is assumed to be parallel to the foil surfaces. The total foil thickness is $z_1 + z_2 = z_0$. The front part being perfect and undisplaced; its response matrix is $M(z_1, s_g)$.
Let the response matrix of the second part be represented by

$$\begin{pmatrix} U & X \\ V & Y \end{pmatrix} \quad (1.117)$$

where X, Y, U and V must be determined from the set of equations (1.101). We note that this set of equations reduces to that for a perfect undisplaced slab by means of the substitution $T' = T^s, S' = S^s \exp(-i\alpha_g)$. The solution of this set of equations is thus $T^s = T(z_2, s_g)$ and $S^s = S(z_2, s_g)$ since front and exit part have the same orientation. For the original set the solution is thus

$$T' = T(z_2, s_g) \quad \text{and} \quad S' = S(z_2, s_g)\ \exp(-i\alpha_g) \tag{1.118}$$

i.e.

$$\begin{pmatrix} T(z, s_g) \\ S(z, s_g)\exp(-i\alpha_g) \end{pmatrix} = \begin{pmatrix} U & X \\ V & Y \end{pmatrix} \begin{pmatrix} 1 \\ 0 \end{pmatrix} \equiv \begin{pmatrix} U \\ V \end{pmatrix} \tag{1.119}$$

and hence

$$U = T(z_2, s_g); \quad V = S(z_2, s_g)\exp(-i\alpha_g) \tag{1.120}$$

We note that the system (1.101) is mapped onto itself by the substitution $T' \to S', S' \to T'$, $s_g \to -s_g$ and $\alpha_g \to -\alpha_{-g}$. The solutions of this new set then also remain the same as those of the original set, except that the interchange of S' and T' has caused the initial values to become $\binom{0}{1}$ and that $s_g \to -s_g$. We thus find

$$\begin{pmatrix} S(z_2, -s_g)\exp(i\alpha_g) \\ T(z_2, -s_g) \end{pmatrix} = \begin{pmatrix} U & X \\ V & Y \end{pmatrix} \begin{pmatrix} 0 \\ 1 \end{pmatrix} \equiv \begin{pmatrix} X \\ Y \end{pmatrix} \tag{1.121}$$

and

$$X = S(z_2, -s_g)\exp(i\alpha_g) \text{ and } Y = T(z_2, -s_g) \tag{1.122}$$

The response matrix of the exit part is thus, in a more concise notation

$$\begin{pmatrix} T_2 & S_2^{(-)} \exp(i\alpha_g) \\ S_2\ \exp(-i\alpha_g) & T_2^{(-)} \end{pmatrix} \tag{1.123}$$

The response matrix of the faulted slab can thus be formulated as

$$\begin{pmatrix} T \\ S \end{pmatrix} = \begin{pmatrix} T_2 & S_2^{(-)} \exp(i\alpha_g) \\ S_2 \exp(-i\alpha_g) & T_2^{(-)} \end{pmatrix} \begin{pmatrix} T_1 & S_1^{-} \\ S_1 & T_1^{-} \end{pmatrix} \begin{pmatrix} 1 \\ 0 \end{pmatrix} \tag{1.124}$$

The matrix (1.123) can conveniently be written as the product of three matrices:

$$\begin{pmatrix} T_2 & S_2^{(-)} \exp(i\alpha_g) \\ S_2 \exp -i\alpha_g & T_2^{(-)} \end{pmatrix} = \begin{pmatrix} 1 & 0 \\ 0 & \exp(-i\alpha_g) \end{pmatrix} \begin{pmatrix} T_2 & S_2^{(-)} \\ S_2 & T_2^{(-)} \end{pmatrix} \begin{pmatrix} 1 & 0 \\ 0 & \exp i\alpha_g \end{pmatrix} \tag{1.125}$$

This suggests to introduce as a shorthand, the <u>shift matrix</u>, already defined in §10.

$$\$(\alpha_g) \equiv \begin{pmatrix} 1 & 0 \\ 0 & \exp(i\alpha_g) \end{pmatrix} \tag{1.126}$$

The final result for the response of the faulted slab can then be written as:

$$\begin{pmatrix} T \\ S \end{pmatrix} = \$(-\alpha_g) M_2 \; \$(\alpha_g) M_1 \begin{pmatrix} 1 \\ 0 \end{pmatrix} \tag{1.127}$$

with

$$M_j \equiv M_j(z_j, s_j) = \begin{pmatrix} T_j & S_j^{(-)} \\ S_j & T_j^{(-)} \end{pmatrix} \tag{1.128}$$

The shift matrices have the property

$$\begin{pmatrix} 1 & 0 \\ 0 & \exp(i\alpha_1) \end{pmatrix} \begin{pmatrix} 1 & 0 \\ 0 & \exp(i\alpha_2) \end{pmatrix} = \begin{pmatrix} 1 & 0 \\ 0 & \exp i(\alpha_1 + \alpha_2) \end{pmatrix} \tag{1.29}$$

i.e. $\$(\alpha_1)\$(\alpha_2) = \$(\alpha_1 + \alpha_2)$. The shift matrices commute.

The result can directly be generalized to a succession of overlapping translation interfaces characterized by phase angles α_j, all referred to the front slab:

$$\begin{pmatrix} T \\ S \end{pmatrix} = \ldots \$(-\alpha_2')M_3\$(\alpha_2')\$(-\alpha_1')M_2\$(\alpha_1')M_1 \begin{pmatrix} 1 \\ 0 \end{pmatrix} \tag{1.130}$$

Introducing the phase angles $\alpha_j = \alpha_j' - \alpha_{j-1}'$, which now describe the <u>relative</u> displacements of successive lamellae, the rear part being displaced with respect to the front one, we obtain

$$\begin{pmatrix} T \\ S \end{pmatrix} = \ldots M_3\$(\alpha_2)\; M_2\$(\alpha_1)\; M_1 \begin{pmatrix} 1 \\ 0 \end{pmatrix} \tag{1.131}$$

1.22. Matrix Formulation for a Foil containing a Domain Boundary

It is possible to generalize further by assuming that in successive lamellae the s-values may be different, as in the case of domain boundaries.

For a pure domain boundary the transmitted and scattered amplitudes are given by

$$\begin{pmatrix} T \\ S \end{pmatrix} = M(z_2, s_{g,2})\; M(z_1, s_{g,1}) \begin{pmatrix} 1 \\ 0 \end{pmatrix} \tag{1.132}$$

where now $s_{g,1} \neq s_{g,2}$.

Formula (1.131) describes also the most general case of overlapping mixed boundaries, i.e. boundaries containing a translation component as well as exhibiting a difference in deviation parameters, provided the s-values in the successive matrices M_j are assumed to be different.

1.23. Matrix Formulation for a Crystal containing a Non-reflecting Part: the Vacuum Matrix [14]

A foil may contain lamellae which are very far from any reflecting orientation under the diffraction conditions prevailing in the rest of the foil; except for absorption they behave as if no material was present in that lamellae. This is for instance the case for a microtwin lamella in a face centered structure if a non-common reflection is excited in the matrix. It also applies to a cavity and to a precipitate lamella with a lattice different from that of the reflecting matrix. Even though such parts of the foil do not contribute to the diffraction phenomena their presence influences the relative phases of the waves diffracted by the foil parts in front and behind these inactive lamellae. This can be accounted for by including the appropriate matrix in the matrix product, describing such a non-diffracting part.

In a non-reflecting part the extinction distance t_g is infinite and the system (1.101) thus reduces to:

$$dT/dz + \pi i s_g T = 0; \quad dS/dz - \pi i s_g S = 0 \tag{1.133}$$

which integrates to

$$T = T_0 \exp(-\pi i s_g z); \quad S = S_0 \exp(\pi i s_g z) \tag{1.134}$$

where T_0 and S_0 are the amplitudes at the entrance face of the non-reflecting "vacuum" lamella. One can thus write

$$\begin{pmatrix} T \\ S \end{pmatrix}_{out} = \begin{pmatrix} \exp(-\pi i s_g z) & 0 \\ 0 & \exp(\pi i s_g z) \end{pmatrix} \begin{pmatrix} T \\ S \end{pmatrix}_{in} \tag{1.135}$$

The "vacuum" matrix is thus

$$V(z, s_g) = \begin{pmatrix} \exp(-\pi i s_g z) & 0 \\ 0 & \exp(\pi i s_g z) \end{pmatrix} \tag{1.136}$$

where z is the thickness of the non-reflecting part as measured along the beam path and s_g is the deviation parameter of the crystal part preceding the "vacuum" lamella.

1.24. Fringe Profiles at Planar Interfaces

1.24.1. General formulae

Multiplication of the matrices in Eq. (1.127) results in the following expressions for T and S.

$$T = T_1 T_2 + S_1 S_2^{(-)} \exp(i\alpha); \quad S = T_1 S_2 \exp(-i\alpha) + S_1 T_2^{(-)} \tag{1.137}$$

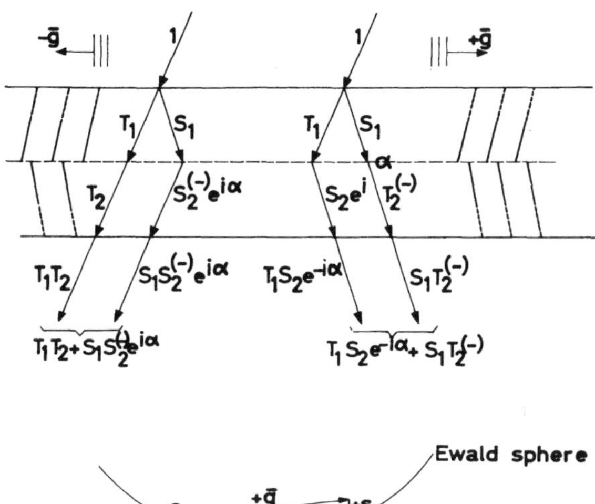

Figure1.21 Transmitted and scattered amplitude for crystal containing a planar interface.

The minus signs in $S_2^{(-)}$ and $T_2^{(-)}$ indicates that the expressions S_2 and T_2 have to be modified by changing s into -s. The relation (1.137) expresses the fact that the transmitted amplitude results from the interference between the doubly transmitted beam T_1T_2 and the doubly scattered beam $S_1S_2^{(-)} \exp i\alpha$. The minus sign in $S_2^{(-)}$ indicates that the scattering, by the second part of the foil of the beam which has already been scattered by the first part takes place from the **-g** side of the lattice planes. This implies that the deviation parameter has to be changed from s to -s since for this second scattering event the node G acts as the origin of reciprocal space (Figure 1.21). This second scattering process is accompanied by a phase shift $\alpha = 2\pi \mathbf{g}.\mathbf{R}$ due to the translation over **R** of the lattice planes in part II with respect to those in part I. This is taken into account by the phase factor $\exp i\alpha$. A similar interpretation can be given to the expression for S. The phase factor is now $\exp -i\alpha$ rather than $\exp i\alpha$ because the phase shifts due to the displacement of part II have opposite signs for $S_2^{(-)}$ and for S_2 since **g** has opposite sign for the two processes.

Introducing the explicit expressions for T_j and $S_j (j = 1,2)$ given by (1.106 and 1.107), in the relations (1.137) and noting that

$\sigma = \sigma_r + i\sigma_i$ since anomalous absorption is taken into account, leads to explicit expressions for $I_T = TT^*$ and $I_S = SS^*$. It turns out that it is possible to cast these explicit expressions for I_T and I_S in a form which allows a detailed analytical discussion of the fringe profiles. Detailed calculations are given in Ref. 15 also for the general case, where the deviation parameters s_1 and s_2 as well as the extinction distances t_{g_1} and t_{g_2} in parts I and II respectively, are assumed to be different. We shall not give the details of these straightforward but tedious calculations; we summarize the significant results.

The expressions for I_T and I_S can be written as sums of three terms

$$I_{T,S} = I^{(1)}_{T,S} + I^{(2)}_{T,S} + I^{(3)}_{T,S} \tag{1.138}$$

1.24.2. Translation interfaces

We consider first pure translation interfaces; thus $s_1 = s_2 \equiv s$ and limiting ourselves to the case $S = 0$, the expressions become:

$$I^{(1)}_{T,S} = \tfrac{1}{2}\cos^2(\alpha/2)\left[\cosh(2\pi\sigma_i z_0) \pm \cos(2\pi\sigma_r z_0)\right] \tag{1.139}$$

$$I^{(2)}_{T,S} = \tfrac{1}{2}\sin^2(\alpha/2)\left[\cosh(4\pi\sigma_i u) \pm \cos(4\pi\sigma_r u)\right] \tag{1.140a}$$

$$I^{(1)}_{T,S} = \tfrac{1}{2}\sin\alpha\left[\sin(2\pi\sigma_r z_1)\sinh(2\pi\sigma_i z_2)\right]$$

$$\pm \sin(2\pi\sigma_r z_2)\sinh(2\pi\sigma_i z_1) \tag{1.141}$$

where the upper sign corresponds to I_T and the lower sign to I_S. The total thickness is $z_0 = z_1 + z_2$, where z_1 is the thickness of the front part and z_2 that of the rear part. We further have $u = (\frac{1}{2})(z_1 - z_2)$, i.e. u is the distance of the interface from the midplane of the foil. Along a planar interface intersecting the foil surfaces, as is often the case, z_1 and z_2 vary along the foil in such a way that $z_1 + z_2$ remains constant and equal to z_0. In the projected area of the interface fringes will be formed, which according to the column approximation, can be considered as being due to the intersection of the depth variation of I_T (or I_S) with the inclined interface.

If $\alpha = n.2\pi$ (n = integer), which is the case if there is <u>no</u> stacking fault, $\sin\alpha = 0$ and $\sin(\alpha/2) = 0$. The only remaining term is then $I^{(1)}_{T,S}$, which as a result must represent the contribution due to the perfect crystal; this is consistent with the fact that this term only depends on the total thickness z_0. This term describes a background on which the fringes represented by the other terms are superimposed.

The second term $I^{(2)}_{T,S}$ depends on u and not on z_1 and z_2 separately. It represents a function which is periodic in u, with a depth period $\frac{1}{2}\sigma_r$. The center of the pattern, at u = 0, exhibits an extremum; it is a minimum for $I^{(2)}_S$ and a maximum for $I^{(2)}_T$. This fringe pattern consists of fringes which are parallel to the central line of the pattern. We shall see that the amplitude of these is only large enough to be visible in the central part of the pattern.

The dominant features of the pattern, in sufficiently thick foils, are described by $I^{(3)}_{T,S}$. Where the interface is close to the entrance face z_1 is small and $z_2 \approx z_0$ and the factor $\sinh 2\pi\sigma_i z_2$ is then large. The term $\frac{1}{2}\sin\alpha \sinh(2\pi\sigma_i z_2)\sin(2\pi\sigma_r z_1)$ represents a damped sinusoid with depth period $1/\sigma_r$. This term disappears at the rear surface where $z_2 = 0$. If $\sin\alpha > 0$ the first extremum is a maximum; as a result the first fringe will be bright at the entrance face. For $\sin\alpha < 0$ the first fringe will be dark.

Where the interface is close to the exit face $z_1 \approx z_0$ and $z_2 \approx 0$; the term $\pm\frac{1}{2}\sin\alpha \sinh 2\pi\sigma_i z_1 \sin 2\pi\sigma_r z_1$ is now dominant. It represents again a damped sinusoid. The first extremum, which now refers to the last fringe, is either a maximum or a minimum depending on the sign of $\sin\alpha$. Note that it is different for I_T and I_S since in the first case the + sign applies, whereas for I_S the minus sign applies.

Figure 1.22 and Table I summarize this discussion. Note that these results, in particular as to the nature of the edge fringes, imply that anomalous absorption must be sufficiently large to make sure that the dominant behavior is described by the term $I^{(3)}_{T,S}$

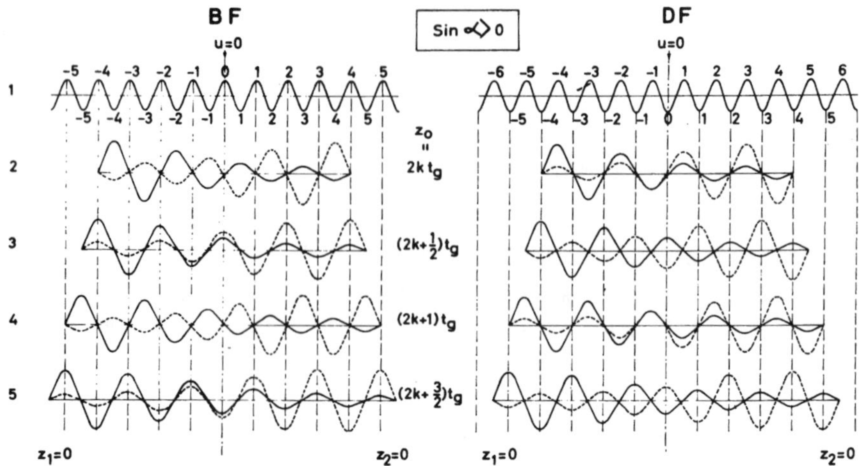

Figure 1.22 Schematic representation of the different terms occurring in the expressions for I_T (left) and I_S (right) for a foil containing a stacking fault.

For a discussion of the behavior in the central part of the pattern the term $I^{(2)}_{T,S}$ may become important since the two terms of $I^{(3)}_{T,S}$ compensate for certain thicknesses and $I^{(2)}_{T,S}$ may then become the dominant term.

For the important case $\alpha = \pm\, 2\pi/3$, which occurs in cubic close-packed structures, the stacking fault fringes have the following properties, provided the foil is sufficiently thick so as to make anomalous absorption a dominant feature.

The B.F. fringe pattern is symmetrical with respect to the line u = 0. This can be deduced quite generally from the implicit expressions (1.137) for $s_1 = s_2 \equiv s$ which shows that $I_T = TT^*$ has the symmetry property

$$I_T(z_1 z_2, s, \alpha) = I_T(z_2 z_1, s, \alpha) \tag{1.142}$$

On the other hand $I_S = SS^*$ has the property

$$I_S(z_1 z_2, s, \alpha) = I_S(z_2 z_1, -s, -\alpha) \tag{1.143}$$

i.e. the dark field fringe pattern is anti-symmetrical with respect to the foil center since interchanging z_1 and z_2 also requires changing the signs of s and α, which changes the nature of the edge fringes.

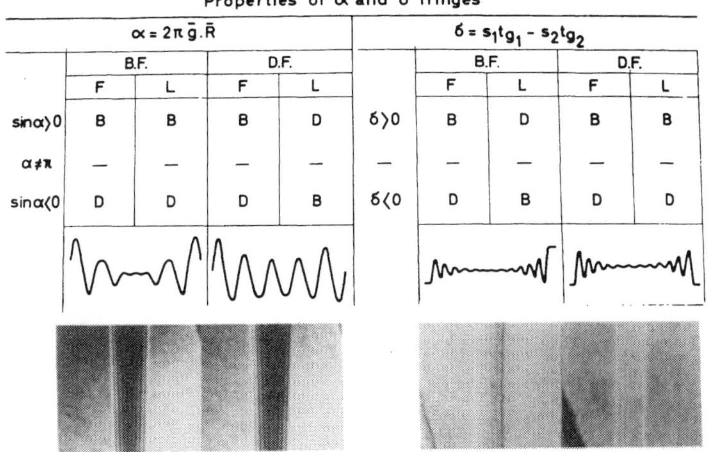

Properties of α and δ fringes

	$\alpha = 2\pi \bar{g}.\bar{R}$					$\delta = s_1 t_{g_1} - s_2 t_{g_2}$			
	B.F.		D.F.			B.F.		D.F.	
	F	L	F	L		F	L	F	L
$\sin\alpha > 0$	B	B	B	D	$\delta > 0$	B	D	B	B
$\alpha \neq \pi$	—	—	—	—	—	—	—	—	—
$\sin\alpha < 0$	D	D	D	B	$\delta < 0$	D	B	D	D

Table I Survey of the properties of fringe profiles due to planar interfaces: Nature of the edge fringes F (first) L (last) for different signs of sin α and δ. A schematic profile and an observed pattern are given for the two types of fringes.

The fringes are parallel to the closest surface; as a result new fringes, caused by an increase in foil thickness are generated by fringe splitting close to the center of the pattern. This result can be understood by noting the relative shift, with increasing thickness, of the curves representing the two terms in $I^{(3)}_{T,S}$ in the central part of the foil where they overlap.

Close to the entrance face of the foil the fringe patterns are similar, but close to the exit part they are complementary. This property is generally true for diffraction contrast images of defects in sufficiently thick foils. We shall in particular show further below that this is also true for dislocation images. The computed profiles of Figure 1.23, which can be compared with the observed fringes in a silicon wedge allow to verify most of these properties.

It is clear from the foregoing discussion that the nature of the edge fringes depends on the sign of sin α. The case $\alpha = \pi$ is singular since now $\sin \alpha = 0$ and the term $I^{(3)}_{T,S}$ which is mostly dominant is now absent as well as the background term $I^{(1)}_{T,S}$ since $\cos \alpha/2 = 0$. The complete fringe pattern is then represented by

(s = 0!)

$$I^{(2)}_{T,S} = (1/2)\left[\cosh(4\pi\sigma_i u) \pm \cos(4\pi\sigma_r u)\right] \qquad (1.140b)$$

since $\sin \alpha/2 = 1$.

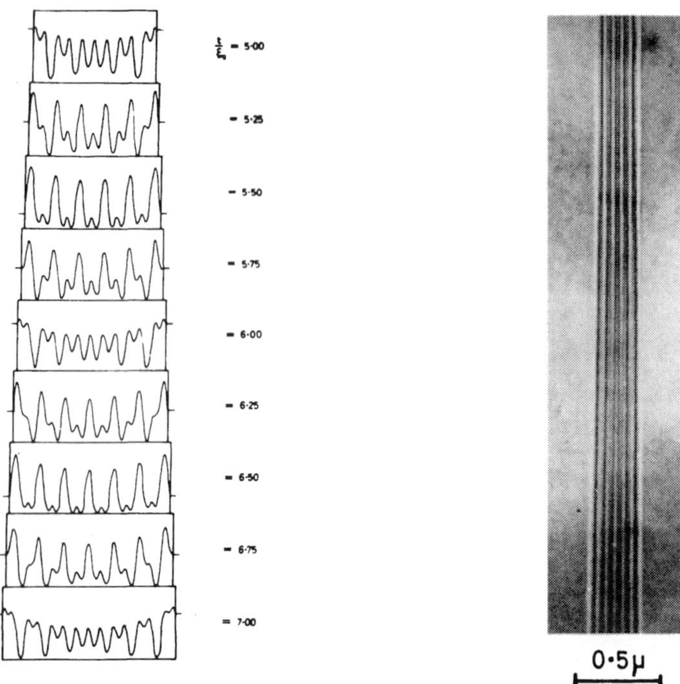

Figure 1.23 Computed profiles for a stacking fault with increasing thickness of the specimen (left).
Fringe pattern due to a stacking fault in a wedge shaped foil of silicon (right) (after Booker in G3). ($s = 0$; $g = 220$; $\alpha = 2\pi/3$; thickness 5-7 t_g)

This expression shows that the B.F. and D.F. image are complementary with respect to the non-periodic background which is now described by cosh $4\pi\sigma_i u$. This background exhibits a minimum for $u = 0$ i.e. in the central part of the pattern. The fringes with a depth period $1/(2\sigma_r)$ are parallel to the central line $u = 0$; they are superimposed on this background. As the thickness increases new fringes are added at the surfaces.

1.24.3. Domain boundary fringes

We now consider interfaces which can be modeled as the boundaries separating two juxtaposed crystal parts in which the deviation parameters for homologous, simultaneously excited reflections are slightly different. Such boundaries occur for instance in the microstructure that results from a phase transition in which rotation symmetry elements are lost. Often the interface is a coherent twin with a small twinning vector. It is easy to show that under these conditions the simultaneously excited diffraction vectors $\mathbf{g}_1$ and $\mathbf{g}_2$ differ by $\Delta\mathbf{g} = \mathbf{g}_2 - \mathbf{g}_1$, where $\Delta\mathbf{g}$ is perpendicular to the coherent twin interface (Figure 1.24). The difference in deviation parameter $\Delta\mathbf{s} = \mathbf{s}_2 - \mathbf{s}_1$ is the projection of $\Delta\mathbf{g}$ along the normal to the foil plane.

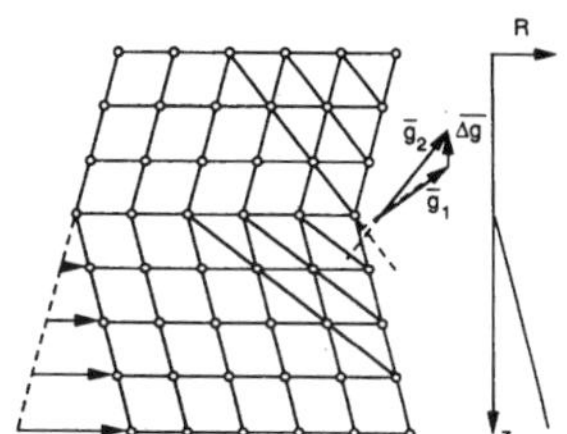

Figure 1.24 Schematic representation of a coherent domain boundary.

In the "symmetrical" case one has $\mathbf{s}_2 = 1/2\Delta\mathbf{s}$ and $\mathbf{s}_1 = -1/2\Delta\mathbf{s}$ i.e. $\mathbf{s}_1 = -\mathbf{s}_2$.

The expressions for the transmitted and scattered amplitudes for a pure domain boundary (i.e. without a translation component) are given by

$$T = T_1 T_2 + S_1 S_2^-; \quad S = TS_2 + T_2^{(-)} S_1 \tag{1.144}$$

of which the interpretation has been discussed in detail for the case of a stacking fault.

The general expressions for the intensities I_T and I_S also assuming the extinction distances in the two parts to be different, can again be written as the sum of three terms, and a discussion similar to the one given for translation interfaces is possible. For sufficiently thick foils the behavior is again dominated by the terms $I_{T,S}^{(3)}$ which we will now discuss.

The general features of the fringe pattern are adequately exhibited by the "symmetrical" case defined above. The terms $I_{T,S}^{(3)}$ are now given by

$$\begin{aligned} w^4 I_{T,S}^{(3)} = & -(1/2)\delta\{\cos(2\pi\sigma_{r,1}z_1)\sinh[2(\pi\sigma_{i,2}z_2 \pm \varphi_2)] \\ & \mp\cos(2\pi\sigma_{r,2}z_2)\sinh[2(\pi\sigma_{i,1}z_1 + \varphi_1)]\ \} \end{aligned} \tag{1.145}$$

with

$$w^2 = 1 + \left(st_g\right)^2; \quad \delta = s_1 t_{g,1} - s_2 t_{g,2}; \quad 2\varphi_j = \operatorname{arghsinh}\left(st_{gj}\right)$$

The upper sign applies to I_T and the lower to I_S. The nature of the fringes is visibly determined by the parameter δ, which is assumed to be sufficiently small so that the same **g**-vector is excited in both crystal parts.

Close to the front surface the first term of $I^{(3)}_{T,S}$ is determining the behavior since sinh $(\pi\sigma_{i,2} z_2 \pm \varphi_2)$ is large for $z_2 \approx z_0$ and $z_1 \approx 0$. Close to the exit face the second term is dominant. The dependence of the nature of the edge fringes on the sign of δ is summarized in Table I and in Figure 1.25.

The most striking and useful feature is the difference in symmetry with the fringe patterns due to translation interfaces. Whereas the bright field pattern for a translation interface is symmetrical with respect to the central line it is roughly anti-symmetrical for a domain boundary, the edge fringes having opposite nature. On the other hand for the special case $s_1 = -s_2$ and $t_{g_1} = t_{g_2}$ the dark field image is symmetrical for domain boundary fringe patterns, but it is anti-symmetrical for translation interfaces. If t_{g_1} and t_{g_2} are significantly different, the depth periods close to front and rear surfaces may be different. Like for translation interfaces the fringes are parallel to the closest surface; this is a consequence of anomalous absorption and it is therefore only true in sufficiently thick foils.

A characteristic feature of domain boundary images is that the domain contrast on either side of the interface (i.e. on either side of the fringe pattern) may be different, which is never the case for translation interfaces. However for $s_1 = -s_2$, i.e. in the symmetrical situation the domain contrast is the same in both domains in the dark field image, but not in the bright field image. This is a consequence of the symmetry of the rocking curve for the scattered beam on the one hand and the asymmetry of the rocking curve for the transmitted beam (the Borrmann effect) on the other hand.

Along certain interfaces there may be simultaneously a phase shift as well as a difference in orientation or length of the excited diffraction vector in the two crystal parts. The fringes produced along such interfaces have properties which are intermediate between those of pure α-fringes and δ-fringes [15].

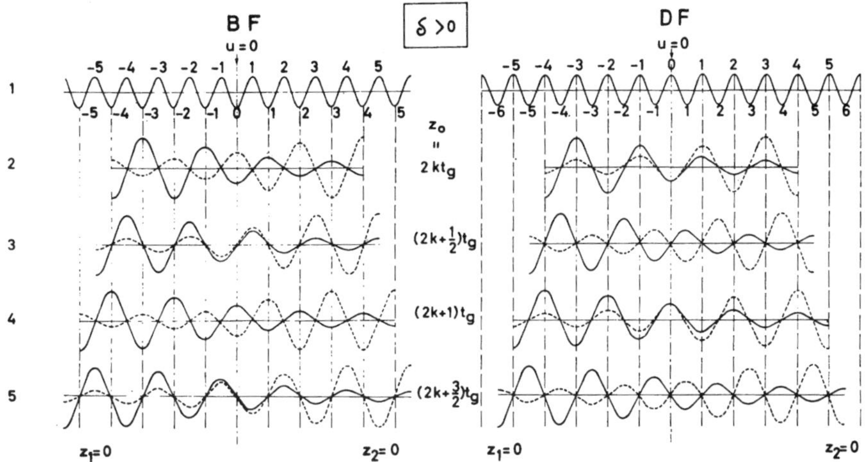

Figure 1.25 Schematic representation of the fringe profile due to a coherent domain boundary. I_T (left); I_S (right)

1.24.4. Extinction criteria

It is clear that no α-fringes are produced if **g.R**= integer. In the expressions (1.139), (1.140) and (1.141) for $I_{T,S}$ the terms $I^{(2)}_{T,S}$ and $I^{(3)}_{T,S}$ become zero. Only $I^{(1)}_{T,S}$ is different from zero, however this term represents thickness fringes since it only depends on z_0. In fact it is easy to verify that for a perfect crystal eqs. (1.140b) and (1.96) are identical (for $s = 0$).

If an image is made using a diffraction vector which is common to the two crystal parts, i.e. if a diffraction spot belonging to the unsplit row or the unsplit plane is selected the δ component of a mixed interface becomes inoperative and only a possible translation component may produce α-fringes. It is also possible to eliminate selectively the translation component from the images of mixed boundaries. In this way it is for instance possible to image the lattice relaxation along anti-phase boundaries or stacking faults with a displacement vector $\mathbf{R}_0+\boldsymbol{\varepsilon}$ by exciting only a systematic row of reflections - ... -2g,−g,0, +g,+2g for which $\mathbf{g}.\mathbf{R}_0$ = integer. The presence of relaxation is then revealed by the occurrence of weak residual fringes, due to the additional displacement ε, for which $\mathbf{g}.\boldsymbol{\varepsilon} \approx$ integer [16]. Using a number of different reflections for which $\mathbf{g}.\mathbf{R}_0$ = integer but for which $\mathbf{g}.\boldsymbol{\varepsilon} \neq$ integer one can obtain a fair idea of the direction, sense and magnitude of **ε** from observations of the nature and the contrast of the edge fringes in the residual fringe patterns.

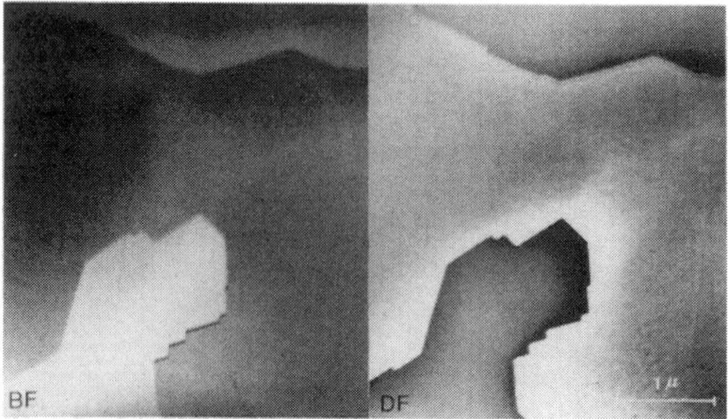

Figure 1.26 Dauphiné twin domains in α-quartz.

1.25. Domain Fragmented Crystals: Microtextures

Many phase transformations lead to a decrease in space group symmetry, the space group of the low temperature phase being a subgroup of that of the high temperature phase. As a result a single crystal of the high temperature phase becomes usually fragmented in domains after transformation into the low temperature, low symmetry phase. The structures within these domains are then related by symmetry operations lost during the transformation. The lost symmetry elements can be either translations or rotations [17]. The interfaces resulting from lost translation symmetry are translation interfaces such as anti-phase boundaries, stacking faults, discommensurations, crystallographic shear planes, etc. ... Lost rotation symmetry elements give rise to twins or domain boundaries. The use of the term "domain boundary" will be reserved for those cases where the lattices of the two domains are only slightly different. The reciprocal lattice nodes belonging to the two domains are then sufficiently close to each other to be excited simultaneously, albeit with different deviation parameters, and produce δ-fringes.

On the other extreme, if the diffraction spots in a diffraction pattern made across the interface are sufficiently split so as to be able to make a dark field image in one of the components separately, we call the interface a "twin". The image so obtained then exhibits wedge fringes in the selected domain. It is clear that the distinction between twins and domains boundaries is not very strictly defined in this way and intermediate situations are possible.

Figure 1.27 Inversion domains in the χ-phase of Fe-Cr-Mo-Ti.

In some cases the lattices of the two domains separated by the interface are the same but the structures may be different. This is for instance the case in non-centrosymmetrical crystals where the structures in the domains may be related by an inversion operation, or by a two-fold axis, the lattice being unperturbed by the interface. The domain structure in α-quartz provides an example of the latter type. The high temperature β-form of quartz has point-group symmetry 6 2 2, whereas the low temperature α-form belongs to the point group 3 2 i.e. the sixfold axis of the β-phase becomes a threefold axis in the α-phase. On cooling to below the $\beta \rightarrow \alpha$ transition temperature (~573°C) the β-phase breaks up into Dauphiné twins, α_1 and α_2, of the α-phase. The structures of α_1 and α_2 are related by the lost 180° rotation about the threefold axis,

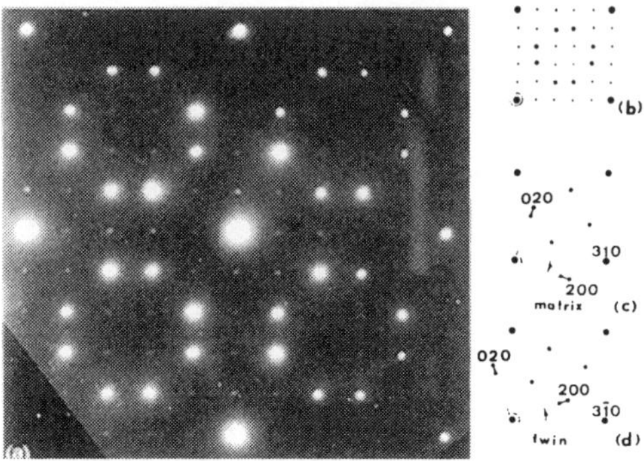

Figure 1.28 Diffraction pattern of Ni_4Mo exhibiting weak double diffraction spots.

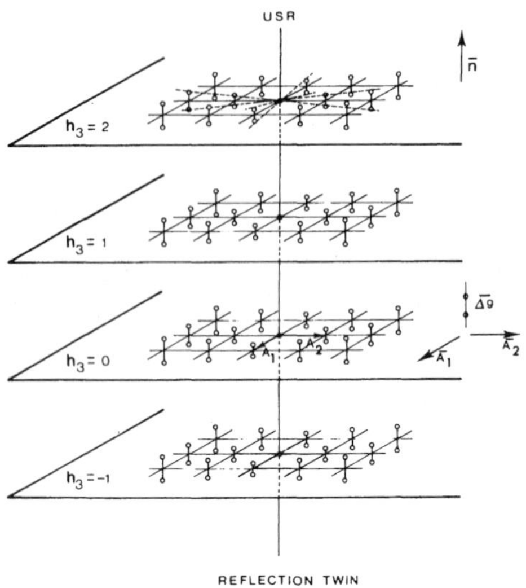

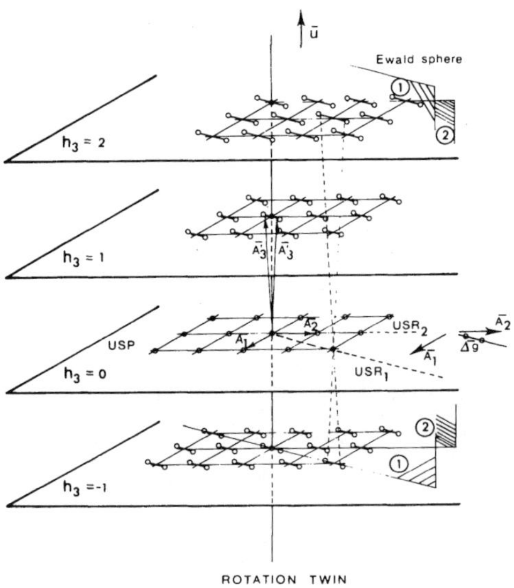

Figure 1.29 Reciprocal latitude of a reflection twin (a) and of 180° rotation twin (b).

whereas the lattice of α_1 and α_2 remains common (Figure 1.26) [18].

Inversion boundaries occur in many non-centrosymmetric crystals and for instance in the cubic χ-phase alloy Fe-Cu-Mo-Ti (Figure 1.27) [19].

Domain textures can conveniently be studied by a combination of diffraction and imaging techniques exploiting different diffraction contrast phenomena.

1.26. Diffraction Patterns of Domain Textures

Domain textures produce a composite diffraction pattern which is the superposition of the diffraction patterns of the separate domains. This usually affects the geometry of the diffraction pattern by the occurrence of spot splitting; in some cases only the intensities are changed as compared to those of a monodomain pattern.

Where differently oriented domains overlap when viewed along the zone axis double diffraction may occur. This may sometimes complicate the interpretation by simulating a diffraction pattern with lower translation symmetry than that of the separate domains, as a result of the double diffraction spots (Figure 1.28).

If several orientation variants are present in the selected area, the diffraction pattern may become quite complicated and difficult to "unscramble". The interpretation can be simplified by first making mono-domain diffraction patterns of the domains on both sides of the interface and subsequently from an area across the interface separating the two domains. However this is only possible if the domains are sufficiently large.

The diffraction patterns across twins have characteristic features which allow to determine the twinning elements. The reciprocal space of a reflection twin is represented in Figure 1.29a; it exhibits a central row of unsplit nodes, perpendicular to the mirror plane in real space. This is a general feature of the relationship between direct and reciprocal space. A common lattice plane in real space (the coherent mirror plane) is represented in reciprocal space as a common lattice row perpendicular to the mirror plane. The reverse is also true. A common lattice row in direct space, as is the case for the lattice row along a 180° rotation twin axis, is represented in reciprocal space as a common reciprocal lattice plane perpendicular to the twinning axis. All other spots are split (Figure 1.29b) [20].

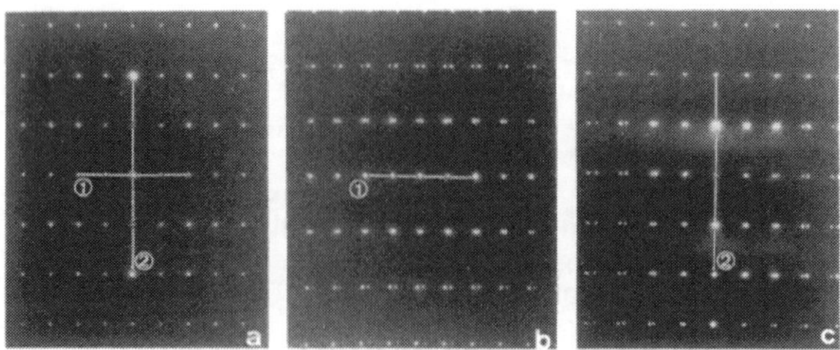

Figure 1.30 Tilting experiment in $MoTe_2$ allowing to show the presence of 180° rotation twins [21].

In the case of a reflection twin the spot splitting is parallel to the unsplit row and its magnitude is proportional to the distance from the unsplit row. The magnitude of the spot splitting is a direct measure for the twinning vector.

For a 180° rotation twin the spots are all split along a direction parallel to the unsplit plane, the magnitude of splitting is proportional to the distance from the unsplit plane.

It is not always obvious how to distinguish the two cases since many sections of reciprocal space will look very similar. Tilting experiments exploring the relevant parts of reciprocal space, are required to differentiate between the two cases. An example of the type of experiment to be performed is shown in Figure 1.30.

The presence of higher order symmetry elements relating the structures in the different domains is reflected in the symmetry of the diffraction pattern. Figure 1.31 shows for instance the presence of three orthorhombic orientation variants related by 120° rotations along the zone axis. Such microstructures can usually be analyzed in terms of reflection or 180° rotation twins, by considering pairs of domains.

Figure 1.31 Composite diffraction pattern of a foil containing three orthorhombic orientation variants of Ni_3 related by 120° rotations.

The distinction between a diffraction pattern produced by a quasi-crystal along a non-crystallographic zone (e.g. a fivefold or tenfold symmetry axis) and a diffraction pattern due to multiply-twinned "classical" crystals is not always obvious and has given rise to much debate.

The presence of domains which are built on a common lattice is not reflected in the geometry of the diffraction pattern since it causes neither spot splitting nor additional spots compared with a monodomain pattern. The relative intensities of the spots are affected but this is not easily detected in electron diffraction patterns. We shall see that imaging techniques are of considerable help in the study of such textures.

1.27. Imaging of Microtextures

Microtextures can be imaged either by means of domain contrast, by interface contrast or by both.

1.27.1. Domain contrast of orientation variants

Domain contrast usually finds its origin in a small difference in the deviation parameters in adjacent domains leading to a significant difference in brightness in either the bright field image or the dark field image, made in a split reflection. More pronounced contrast arises if the dark field image is made in one of the components of a split reflection. However this is only possible if the spot splitting is large enough.

The difference in brightness in the bright field image can be understood with reference to the asymmetric rocking curve for I_T. In the vicinity of s = 0 the s-dependence of I_T is quite pronounced and a small difference in deviation parameter leads to a pronounced difference in transmitted intensity. Optimum domain contrast is thus obtained if the average deviation parameter is close to s = 0.

In the dark field image optimum contrast is achieved if a single spot can be isolated. If this is not the case the symmetry of the rocking curve for I_s shows that now the optimum contrast is obtained for an average s which is different from zero.

Domain contrast can also arise because the moduli of the structure factors, and hence the extinction distances are different in adjacent domains. This is for instance the case for Dauphiné twins in quartz [18]. The lattices of α_1 and α_2 coincide, no spot splitting occurs and the above mentioned contrast phenomena are inoperative. However a number of coinciding reflections have structure amplitudes of different magnitudes. Dark field images made in such reflections will give rise to domain contrast, often called "structure factor contrast" (Figure 1.26).

It is clear that translation variants cannot give rise to domain contrast since the lattices, as well as the structures, are strictly parallel in the two domains.

1.27.2. Interface contrast

It is also possible to image the interfaces rather than the domains. This is the only possibility for translation interfaces. For orientation variants domain contrast and interface contrast are often produced simultaneously.

The interfaces separating translation variants such as out-of-phase boundaries, crystallographic shear planes and stacking faults are imaged as α-type fringes in reflections for which $\mathbf{g}.\mathbf{R}_0 \neq$ integer. This is also the case for interfaces separating structural variants built on a common lattice, but having different structure amplitudes. We have seen above that domain contrast arises as a result of structure factor contrast when the moduli of the structure factors are different. However it often happens that the structure factors have the same modulus in the two domains, but have a different phase. This is the case for certain reflections in domain fragmented α-quartz. A dark field image in such a reflection will not exhibit domain contrast but will reveal the interfaces as α-type fringe patterns.

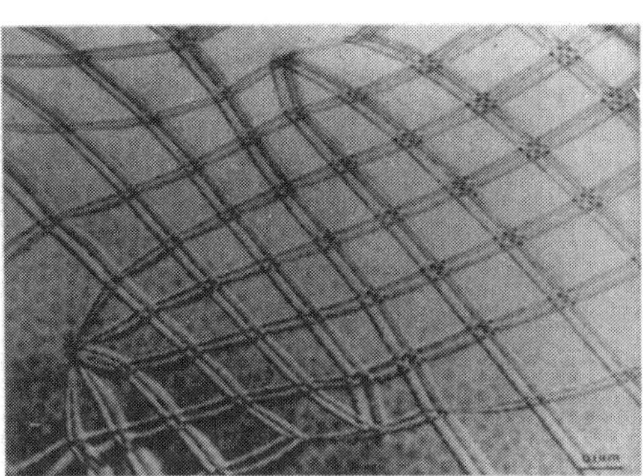

Figure 1.32 inversion domains in the χ-phase of Fe-Cr-Mo-Ti as revealed by interface contrast.

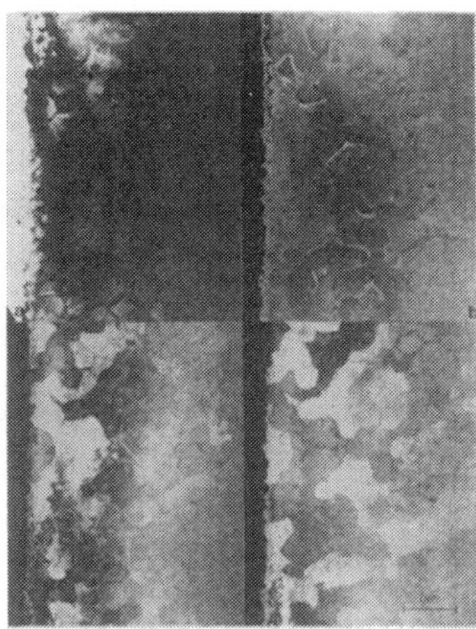

Figure 1.33 Interfaces in Ni_4Mo revealed by different imaging modes: (a)BF image (b) in double diffraction spot; (c) D F image in (020)of one variant.

The phase angle α can be deduced as follows. The structure factors for the structures in the two domains are written with respect to a common origin. The two structure factors are then related as follows:

$$F_H^{(2)} = F_H^{(1)} \exp(i\alpha) \quad (1.146)$$

where α is the phase angle characterizing the fringe pattern, $F_H^{(2)}$ and $F_H^{(1)}$ are the structure factors with indices **H** in the exit and front part respectively.

Inversion boundaries revealed by means of interface contrast in the c-phase of the alloy Fe-Cr-Mo-Ti are visible in Figure 1.32.

A particular type of interface contrast arises in dark field images made in double diffraction spots caused by overlapping orientation domains. The interfacial region will now appear bright since double diffraction is only produced in the regions of overlap along the interfaces (Figure 1.33).

1.27.3. Inversion boundaries

The first observations of inversion boundaries were made on the cubic non-centro-symmetric χ-phase in the alloy system Fe-Cr-Mo-Ti [22,23] (Figure 1.27). The contrast at this type of boundaries requires some specific discussion. It was found experimentally that under the appropriate diffraction conditions the domain structure can be revealed by domain contrast as well as by interface contrast. Inversion domains have a common lattice and hence there is no spot splitting. The structures are related by an inversion operation, i.e. the reflections **H** in one domain and -**H** in the other domain are always excited simultaneously and to the same extent. The moduli of the structure factors of simultaneously excited reflections **H** and -**H** are always the same according to Friedel's law $I_H = I_{-H}$. The phases α_H and α_{-H} are different for most reflections since the structure is non-centro-symmetric. For a non-centro-symmetric crystal the phases associated with the Fourier coefficients of the imaginary part of the lattice potential need not be equal to those associated with the Fourier coefficients of the real part.

It was shown in Ref. 22 that domain contrast arises as a result of the violation of Friedel's law in dark field images in non-centro-symmetrical crystals under multiple beam conditions, along a zone which does not produce centro-symmetry in projection. This means that the zone axis cannot be a symmetry axis of even order. In the χ-

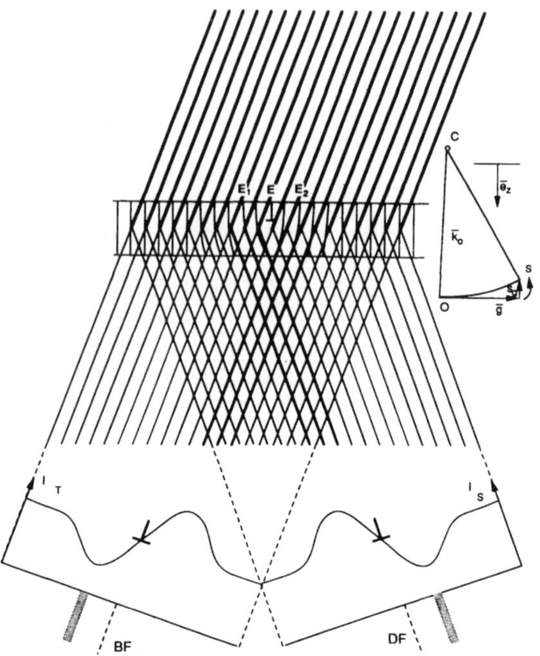

Figure 1.34 Intuitive model for the origin of contrast at an edge dislocation. The thickness of the lines is a measure for the intensity of the electron beams.

phase domain, contrast is produced for instance when the zone axis is along the threefold rotation axis.

Interface contrast arises as a result of the difference in phase of the structure factors associated with the different domains. The interfaces are imaged as α-fringes, the lattices being parallel in the domains. The difference in phase between the Fourier coefficients related to real and imaginary part of the lattice potential leads to weak interface contrast, even under two-beam conditions.

1.28. Dislocation Contrast

1.28.1. Intuitive considerations

Dislocations are usually visible as dark lines in two-beam diffraction contrast images, made with small values of the deviation parameter. When applying the weak beam method, i.e. for large values of s, they appear as bright lines on a darker background.

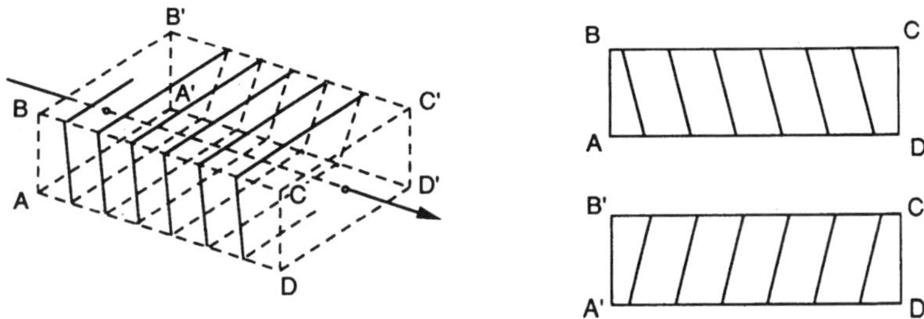

Figure 1.35 Geometry of lattice planes in the vicinity of a screw dislocation, leading to image contrast.

The two-beam image formation at dislocations can easily be understood on intuitive grounds [G1]. The foil represented in Figure 1.34 contains an edge dislocation in E. The lattice planes used for imaging are indicated schematically. Due to the presence of the dislocation the lattice planes in the vicinity of E are slightly curved and inclined in opposite sense left and right of E Since the specimen is a thin foil the Bragg condition is relaxed; the reciprocal lattice nodes have become "relrods". We can therefore assume that diffraction occurs even though the Bragg condition is only approximately satisfied with s< 0 in the part of the foil which is not affected by the presence of the dislocations. On the left of the dislocation, at E_1, the rotation of the lattice planes is then such that locally the Bragg condition is better satisfied, i.e. *s* is smaller, and hence the diffracted beam will be more intense than in the perfect parts of the foil. On the right of the dislocation in E_2 the lattice rotation is in the opposite sense and hence the diffracted beam will locally be weaker than in the perfect part of the foil. The relative intensities of the diffracted beams are indicated schematically by lines of different widths in Figure 1.34. Since no electrons are lost the transmitted beam will be depleted where the scattered beam is enhanced.

Selecting the diffracted beam by means of an aperture and magnifying the corresponding diffraction spot will produce a map of the intensity distribution in this beam. This map will reveal a lack of intensity, i.e. a dark line, to the right of the dislocations in E_2 and an excess of intensity over the background in E_1. The dislocation will thus be imaged as a bright-dark line pair. This image is called a dark field image.

When selecting the transmitted beam a similar intensity map can be produced by magnifying the intensity distribution in the direct beam. Such an image is called a bright field image; in this

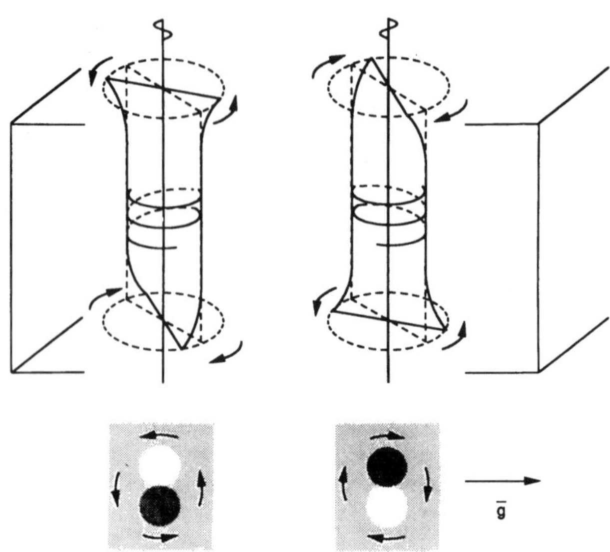

Figure 1.36 Schematic illustration of the surface relaxation around the emergence points of screw dislocations in the foil surfaces. This surface relaxation produces image contrast for **g**.**b** = 0.

approximation it is complementary to the dark field image; bright and dark lines being interchanged. Bright and dark field images are in fact only small parts of strongly magnified diffraction spots, the intensity distribution being the image.

The possibility of forming such images is a consequence of the "local" character of electron diffraction. Electrons are only sensing a narrow column of material because the Bragg angles are small, electron diffraction is strongly peaked forward and the foil is thin. The columns form in a sense the "pixels" of the image. The assumption that electrons travel in narrow columns is the basis of the "column approximation" introduced before.

The same type of reasoning can be used to demonstrate that also screw dislocations produce a line image. As a consequence of the presence of the screw dislocation the families of lattice planes intersecting the dislocation line are transformed into helical surfaces. Left and right of the dislocation the lattice planes are slightly inclined in opposite senses and hence the local diffraction conditions are different left and right. Again a bright-dark line is produced (Figure 1.35).

We note that in both cases, edges and screws, the dark line image is not produced at the dislocation core, but in a slightly displaced position called the image side. Changing the diffraction conditions so as to make $s > 0$ in the foil part which is far away from the dislocations, changes the image side as can be demonstrated by the same reasoning as used above. Also changing **g** into **-g** changes the image side since now reflection takes place from the other side of the lattice planes. Finally, changing the sign of the Burgers vector changes the sense of inclination of the lattice planes on a given side of the dislocation, and hence also changes the image side. Summarizing we can say that the image side depends on the sign of p = (**g**.**b**)*s* as we shall see further below.

This rule becomes undetermined if **g**.**b** = 0. The relation **g**.**b** = 0 is in fact the criterion for the absence of contrast. It expresses the fact that no image is produced when diffraction occurs by the lattice planes which are left undeformed by the presence of the dislocation. To a first approximation all displacements around a dislocation are parallel to the Burgers vector and they are thus parallel to the lattice planes for which **g**.**b** = 0. This extinction criterion is strictly valid for screw dislocations in an elastically isotropic medium, for which all displacements are parallel to **b**, but it is only a first approximation for edge dislocations. Deviations occur even for screws in strongly anisotropic media, the reason being that the actual extinction criterion is **g**.**R** = 0. The displacement field of an edge dislocation contains a component perpendicular to the glide plane which causes some residual contrast even if **g**.**b** = 0, as we shall discuss below.

Some contrast may also result, even though **g**.**b**= 0, from the fact that the specimen is a thin foil. The presence of dislocations in a thin foil modifies the displacement field as a result of surface relaxation effects and this may produce contrast. For instance a pure screw dislocation parallel to the incident beam and perpendicular to the foil surfaces is not expected to produce any contrast since **g**.**b** = 0 for all active **g**-vectors. However it was found that such dislocations produce a dark-bright dot contrast which was attributed to the lattice twist. It was shown by Eshelby and Stroh [24] that close to the emergence point of a screw dislocation in the foil surfaces significant elastic relaxation takes place which transforms the lattice planes parallel to the dislocation line into helical surfaces, the sense of the helical twist being determined by the sign of the screw dislocation. This helical twist produces a bright-dark dot pair because on one side of the emergence point the lattice planes are tilted into the Bragg condition and on the other side out of the Bragg condition. The line joining the bright-dark dot pair is perpendicular to **g**; (Figure 1.36). Depending on the sense of the helical twist (i.e. on the sign of the screw dislocation), the dot pair is bright-dark or dark-bright. Applying this effect the sign of the screw dislocation can thus be determined from such images.

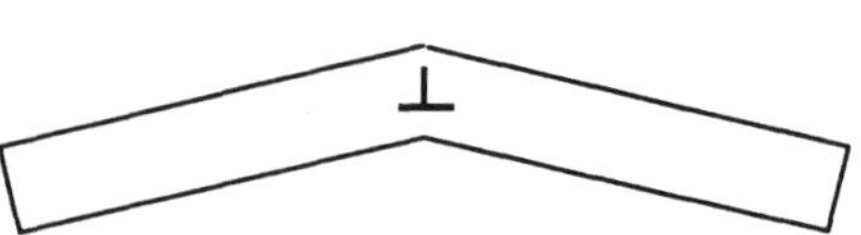

Figure 1.37 Buckling of thin foil due to the presence of an edge dislocation parallel to the foil surfaces.

An edge dislocation parallel to the foil surfaces and with its glide plane also parallel to the foil causes a slight misorientation of the two crystal parts separated by the dislocation. The tilt angle θ depends on the foil thickness and on the position of the dislocation within the foil, being a maximum $\theta_{max} = b/t$ (t = foil thickness) if the dislocation is in the central plane. As a result of this slight "buckling" of the foil a brightness difference is produced between the two crystal parts, separated by the dislocation. The tilt angle θ can be measured by the displacement of the Kikuchi lines; its sense depends on the sign of the dislocation, and therefore this effect allows to determine the sign of the dislocation (Figure 1.37) [25].

An edge dislocation viewed end-on along the beam direction, produces contrast because in the vicinity of the dislocation the interplanar spacing is slightly modified, **g** changes in length and in orientation and consequently also the diffraction conditions change. Along a column parallel to the dislocation, i.e. along z the s-value remains constant, but s becomes a function of x and y chosen in the foil plane. As a result the scattered and transmitted intensities depend on the column positions, i.e. an image is produced. The contours of equal s-value i.e. of equal brightness are shown in Figure 1.38, they image the strain field around the edge dislocation [26].

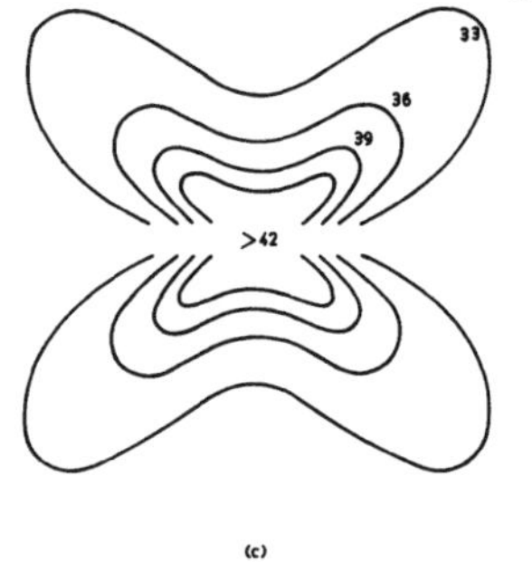

Figure 1.38 Contours of equal deviation parameter s in the vicinity of an edge dislocation viewed end on [26].

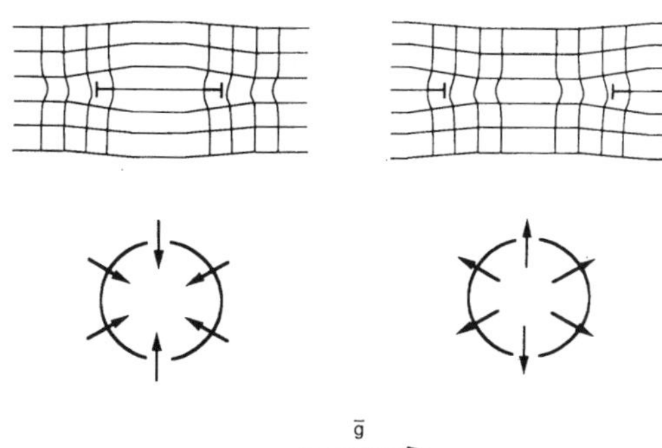

Figure 1.39 Displacement field around prismatic loops.

According to the simple **g.b** = 0 criterion an edge dislocation with its supplementary half plane parallel to the foil plane, or stated otherwise, with its Burgers vector **b** along the incident beam, would not produce any contrast. Due to the presence of the "bump" in the glide plane, i.e. the component of the displacement field, towards the supplementary half-plane, perpendicular to the glide plane, **g.R** is not zero for all **g** vectors perpendicular to **b** and some contrast is produced. Prismatic dislocation loops in planes parallel to the foil plane consist of dislocations having this configuration. The displacement field of such loops now contains a radial component $\mathbf{R}_r$ which is inward or outward respectively for vacancy and interstitial loops (Figure 1.39) as well as a normal component $\mathbf{R}_n = \mathbf{b}$.

For a diffraction vector **g** parallel to the foil plane the dot product with the normal component $\mathbf{g.R}_n = \mathbf{g.b}$ will be zero everywhere along the loop. However $\mathbf{g.R}_r$ varies along the loop and vanishes only along the two diametrically opposite segments where **g** is perpendicular to $\mathbf{R}_r$ as represented in Figure 1.39. As a result there will be two short segments only along which complete extinction occurs; the "line of no contrast" joining these two segments is perpendicular to the active **g** vector.

Somewhat against intuition one finds that parallel dislocation lines with the same Burgers vector do not necessarily exhibit the same contrast especially when they are close one to the other as in a ribbon. One of the lines is usually imaged as a darker line than the other(s); it depends on the sign of s and on the sense of **g** which one will exhibit the strongest line contrast. The effect is especially striking in triple ribbons in face centered cubic, low stacking fault energy alloys and in graphite. An analytical theory, based on the kinematical diffraction theory, allows to account satisfactorily for the observations, on noting that the total strain field of a triple ribbon is different from

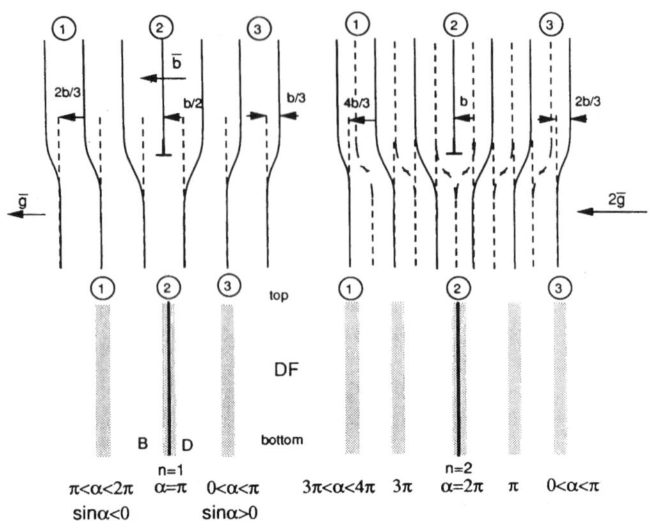

Figure 1.40 Schematic illustration of the displacement field around an edge dislocation for the cases n = 1 (a) and n = 2 (b) illustrating how the qualitative features of the image can be deduced from the profiles of stacking fault fringes.

that resulting from the mere superposition of the strain fields of three isolated dislocations [27] (see also § 1.29.2.4).

1.28.2. Semi-quantitative considerations

It is often useful to be able to predict semi-quantitatively or even qualitatively only, the two-beam image characteristics to be expected for a given defect. For dislocation lines this is possible within the framework of the dynamical theory, including anomalous absorption, by referring to the analytically soluble case of the stacking fault.

Consider for example an inclined edge dislocation with its Burgers vector parallel to the foil plane and an active reflection such that **g.b** = 1. A section of the displacement field of the dislocation is represented schematically in Figure 1.40. The bright field and dark field image profiles can be obtained by considering a row of columns along a line perpendicular to the dislocation line, and computing for each of these columns the amplitude of the transmitted and scattered beams.

We first note that these amplitudes are to a large extent determined by the phase relation between top and bottom end of the columns and not so much by the details of the variation of this phase along the column. In any case the phase varies rapidly only in the close vicinity of the dislocation core. We therefore accept as a reasonable approximation that the amplitudes emerging from such a column will be the same as those emerging from a column which intersects a stacking fault at the level of the dislocation core and which introduces abruptly the same shift between top and bottom as the dislocation. Consider as a simple example the columns passing through the dislocation core for the case $n \equiv \mathbf{g.b} = 1$. The phase shift between top and bottom of these columns is then π as is immediately evident from the geometry of Figure 1.40a. The inclined dislocation will then exhibit along its core the same contrast variation with depth as an inclined stacking fault with $\alpha = \pi$ situated everywhere at the same level as the dislocation core. If for the same dislocation $\mathbf{g.b} = 2$ the brightness along the core will be the same as that of a perfect crystal with the same thickness since now $\alpha = 2\pi$ along the central strip (Figure 1.40b).

The image of an inclined dislocation is a two-dimensional brightness map and requires the knowledge of a large number of section profiles or alternatively of a number of longitudinal profiles, parallel to the dislocation. Profiles of the latter type can be obtained by considering strips of stacking fault, all parallel to the dislocation core and at the same level, but at increasing distances from this core. For the case $n = 1$ the central strip corresponds, as mentioned above, to $\alpha = \pi$. The corresponding α-values for successive strips on moving to the right away from the dislocation, vary from $\alpha = \pi$ to 2π, far to the right. As a result on the right of the dislocation $\sin \alpha < 0$. On the left of the dislocation the α-values vary from 0 at the extreme left to $\alpha = \pi$ at the dislocation position; on the left of the dislocation $\sin \alpha > 0$. The image profiles of stacking faults (for $s = 0$) show that the contrast will oscillate. Along the strips where $\sin \alpha > 0$ (i.e. on the left of the dislocation) the first extremum behind the entrance face will be a maximum in the bright field image whereas on the left of the dislocation, where $\sin \alpha < 0$, the first extremum will be a minimum. Near the exit face the last fringe in the bright field image will be the same as the first, on the left as well as on the right, of the dislocation. The dislocation contrast is thus oscillating, as a function of depth since the maxima and minima in brightness on the two sides of the dislocation (i.e. for $\sin \alpha > 0$ and $\sin \alpha < 0$) are in anti-phase. The oscillations will be most

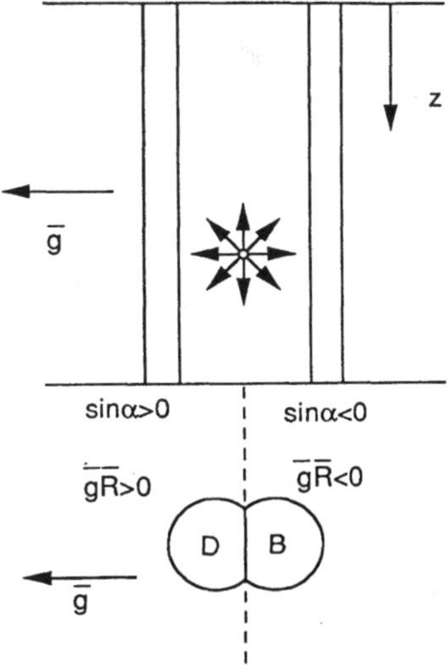

Figure 1.41 Strain field associated with a spherical inclusion.

pronounced for foil thicknesses equal to $(2k+1/2)t_g$ or $(2k+3/2)t_g$ since for these thicknesses the stacking fault fringes in the bright field image vary strongly in brightness (Figure 1.22). In the dark field image the brightness variation is less pronounced for the same thickness; on the other hand the variation is now more pronounced for foil thicknesses of the form $2k\ t_g$ and $(2k+1)t_g$. At the same time the difference between the brightness for $\sin\alpha > 0$ and $\sin\alpha < 0$ is more pronounced in the bright field for thicknesses of the form $(2k+1/2)t_g$ and $(2k+3/2)t_g$ than for thicknesses of the form $2k\ t_g$ or $(2k+1)t_g$. In the former case the contrast is "oscillating", whereas in the latter case it is "dotted". The inverse is true for the dark field image. The stacking fault fringe profiles suggest in the same manner that the bright field image will be similar to the dark field image close to the front surface but quasi-complementary close to the exit surface.

As a second example we consider the image associated with the displacement field around a spherical inclusion with $\varepsilon > 0$, as described by eq. (1.26) [28]. Also in this case we can deduce from intuitive considerations in which areas $\sin\alpha > 0$ and in which $\sin\alpha < 0$ and hence conclude for a defect close to the surface (within the first extinction distance) which area will be bright and which will be dark. We consider in particular the spherically symmetrical displacement or strain field represented in Figure 1.41. A line (or plane!) of no contrast, along which $\mathbf{g.R} = 0$ separates two regions, one in which $\mathbf{g.R} > 0$ and one in which $\mathbf{g.R} < 0$. Since $|R|$ is small as compared to a lattice vector we shall have $\sin\alpha > 0$ if $\mathbf{g.R} > 0$ and $\sin\alpha < 0$ if $\mathbf{g.R} < 0$. That this is so can be deduced from the consideration that if $\mathbf{g.R}$ is positive for all z-values along the column the integrated phase difference between top and bottom of the column is positive but smaller than π and hence $\sin\alpha > 0$, the fastest phase change occurring at the level of the inclusion. The brightness at the exit end of the column is then the same as that of a stacking fault, assuming the effective phase shift α_{eff} to occur at the level of the defect. The value of α_{eff} decreases with increasing distance from the inclusion and changes sign along the line of no contrast. The image characteristics of stacking faults in sufficiently thick foils, close to the surface, allows to deduce the dark field image of this kind of defect when close to the back surface. The last fringe in the dark field image of a stacking fault for which $\sin\alpha > 0$ is dark. We can thus conclude that **g** points towards the dark lobe for an inclusion with $\varepsilon > 0$ situated close to the back surface. Black and dark are reversed for $\varepsilon < 0$. The model also accounts for the periodic interchange with period t_g of bright and dark lobes with the depth position of the spherical inclusion.

1.28.3. Kinematical theory of dislocation contrast [7]

In the framework of the kinematical diffraction theory image profiles of dislocations are obtained by inserting the adequate expression for the displacement field $\mathbf{R}(\mathbf{r})$ in eq. (1.30) and integrating along columns situated on lines normal to the dislocation line. Due to the symmetry of the displacement field of a dislocation the profile so obtained is independent of the chosen line of columns for dislocations parallel to the surfaces of the foil.

For example, for a screw dislocation oriented along the *y*-axis parallel to the foil plane and situated at a depth d behind the entrance face the displacement field is described, according to the isotropic linear elasticity theory by the expressions

$$R_x = 0; \quad R_y = b\phi / 2\pi; \quad R_z = 0 \tag{1.147}$$

with $\phi = \text{arctg}[(z-d)/x]$ i.e. all the displacements are parallel to **b**. The image profile is then obtained by performing the integration

$$A(\mathbf{g}) = F_g \int_0^{z_0} \exp(2\pi i s_g z).\exp\{in \ \text{arctg}[(z-d)/x]\}dz \tag{1.148}$$

where $n \equiv \mathbf{g}.\mathbf{b}$ (z_0 = foil thickness), for various values of the parameter x. After a number of approximations the integrals can be obtained analytically in terms of Bessel functions.

In their discussion of image profiles of dislocations Hirsch et al [7] and Gevers [6] made extensive use of amplitude-phase (A-P) diagrams. We shall follow the same type of reasoning since this allows to identify more clearly the approximations and limitations of the theory. The integration along a column is represented graphically by the vector sum of the elementary contributions due to the slices dz along the column. In a perfect crystal we have seen that the vectors representing the amplitudes scattered by successive slices enclose a constant angle $d\theta = 2\pi s \, dz$, as a result of the constant phase difference between successive slices dz These small vectors form a regular polygon which in the limit for $dz \to 0$ becomes an arc of a circle with radius $1/2\pi s$. The length of the circular arc is equal to the column length and the amplitude scattered by the column is given by the length of the vector joining the two end points of the circular arc (see § 2.2).

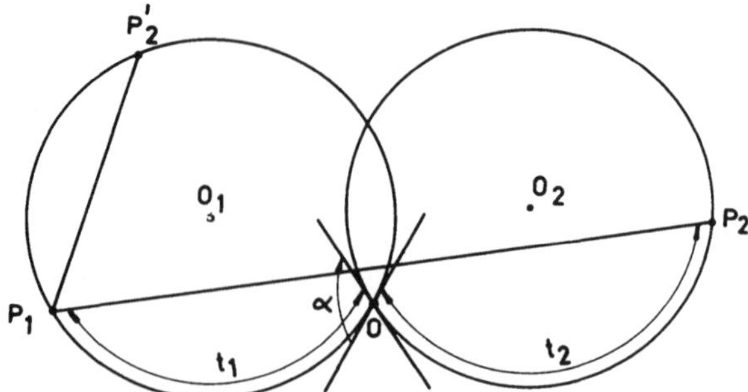

Figure 1.42 Amplitude-phase diagram for a column intersecting a stacking fault. The thickness of the two crystal parts are $t_1 \equiv z_1$ and $t_2 \equiv z_2$.

When a stacking fault is present a discontinuous phase change $\alpha = 2\pi\mathbf{g.R}$ occurs in each column at the level of the stacking fault. This is reflected in the A-P diagram by a relative rotation over an angle α of the two circles representing the A-P diagrams of the perfect parts, the tangents enclosing an angle α (Figure 1.42). The amplitude diffracted by the faulted crystal is then given by the vector joining the end points P_1 and P_2. If we choose the origin of the diagram at the position of the stacking fault the two circular arcs have lengths equal to the front (z_1) and rear part (z_2) of the foil: $z_1 + z_2 = z_0$ (z_0 = foil thickness). For an inclined stacking fault the endpoints P_1 and P_2 corresponding with successive columns along a profile, shift continuously over the same arc length in the same sense. The resulting amplitude thus varies periodically with a depth period $1/s$, describing the stacking fault fringes, represented analytically by the relation (1.31).

The A-P diagram for a column intersecting a domain boundary also consists of two circular arcs with lengths equal to the thicknesses of front and rear part; they join smoothly with a common tangent at the level of the boundary, but they have different radii $\frac{1}{2}\pi s_1$ and $\frac{1}{2}\pi s_2$, since the s-values are different in the two perfect parts. Also in this case an inclined domain boundary will produce a fringe pattern with a depth quasi-period, which is somewhat variable over the width of the fringe pattern between $1/s_1$ and $1/s_2$. The A-P diagram is the geometrical representation of the analytical expression (1.33).

In the A-P diagram for a foil containing a dislocation the phase difference between successive slices of the column at levels z and z + dz is no longer a constant $d\phi = 2\pi s\, dz$ since a supplementary phase difference results from the displacements described by $\alpha(x,y,z)$. Depending on the signs of x, s and z, this additional phase shift will either be added or subtracted; its magnitude depends on x and z and is given in the simple case of the screw dislocation by $n.d\,[\mathrm{arctg}(z/x)]$. For z>>x this additional shift becomes zero and the final shape becomes again a circle with radius $\frac{1}{2}\pi s$ as for the perfect crystal. Close to the dislocation and for s and nx having the same sign, i.e. for $n\beta > 0$ (with $\beta = 2\pi sx$) the quantity n arctg(z/x) has the same sign as $2\pi sz$ and the angle between two successive vectors is now larger than $2\pi s$ dz at least near the depth position of the dislocation, which is chosen as the origin of the diagram (i.e. of the z-axis). As z becomes larger the angle approaches again $2\pi s$ dz. The resulting curve will be a wound-up spiral which gradually tends to a circle, approaching it from the interior, the circle being the limiting curve (Figure 1.43b). If on the other hand s and nx have opposite signs i.e. for $n\beta < 0$, arctg(z/x) and $2\pi sz$ have opposite signs and the resulting angle between successive vectors will be smaller than the value $2\pi s$ dz in the perfect crystal by $n.d\,[\mathrm{arctg}(z/x)] \equiv n[x/(x^2+z^2)]dz$. Again as z becomes large the additional phase difference tends to zero and the curve approaches again a circle with radius $\frac{1}{2}\pi s$. The A-P diagram is now an unwound spiral approaching the limiting circles from the outside as shown in Figure 1.43a. The scattered amplitude is again obtained by taking an arc proportional to z_1 on this curve, in the negative sense leading to P_1 and an arc proportional to z_2 in the positive sense leading to P_2. The vector $P_1 P_2$ is then proportional to the scattered amplitude for the given column, i.e. for a given *x*-value. Since *x* has different signs on

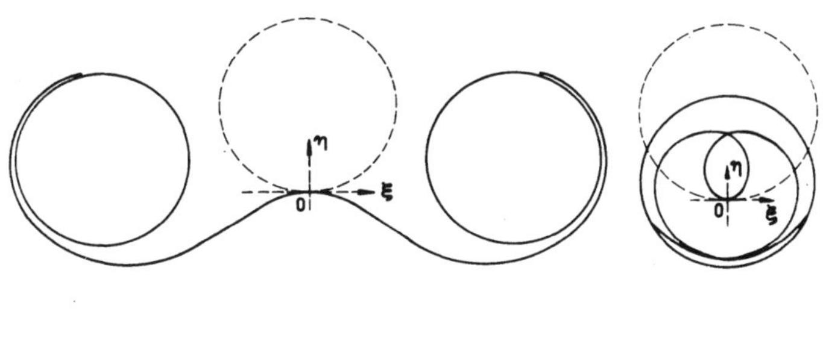

Figure 1.43 Amplitude-phase diagram for a column passing close to a dislocation core. (n = 2). (a) Unwound spiral (b) Wound-up spiral

the two sides of the dislocation the A-P diagram will be an unwound spiral on one side and a wound-up spiral on the other side. The vector representing the diffracted amplitude will clearly be larger for those columns for which the distance between the centers of the two limiting circles will be the largest, i.e. the amplitude will be largest on that side of the dislocation where the A-P diagram is an unwound spiral. This is the side where in the bright field image a dark line will be observed, called image side (see § 1.28.1). We note that the A-P diagram only depends on $n \equiv \mathbf{g.b}$ and on the product $\beta \equiv 2\pi sx$ but not on s and x separately. This is consistent with the fact that changing the sign of s changes the image side. Constructing a sufficient number of A-P diagrams allows in principle to deduce the image profiles. It is clear that for an inclined dislocation line the length of the arcs to be taken along the spiral shaped A-P diagrams will vary continuously with the position along the dislocation, one increasing, the other one decreasing. Hereby the endpoints of the A-P diagram, which determine the scattered amplitude, will in general vary periodically as the endpoints describe the limiting circles. This oscillatory behavior was suppressed in the approximation introduced by Hirsch et al. [7]. The assumption was made that the square of the separation of the centers of the limiting circles is a convenient measure for the scattered intensity. This is a reasonable assumption if *s* is sufficiently large, so that the limiting circles acquire a small radius compared to the separation of their centers. For screw dislocations Hirsch et al [7] obtained the computed profiles of Figure 1.17 for different values of n Similar calculations, using the same approximations, have been performed by Gevers [6] for perfect as well as partial dislocations of edge and mixed character. The result for pure edge dislocations are shown in Figure 1.18.

1.28.4. The weak-beam method

In § 1.11 we have discussed the limitations of the kinematical theory and its range of applicability. These limitations also apply to the results of the preceding chapter and we therefore conclude that the computed image profiles are only valid for very thin foils and for large s-values.

From the image profiles of Figure 1.17 and 1.18 we can deduce that when s is large the same $\beta(= 2\pi sx)$ value is reached for small x This implies that for large s-values the peak shift and the peak width will become small. This effect which is consistent with the observations is systematically exploited in the weak-beam method [29]. It allows to obtain very well localized and sharp images of the partial dislocations in narrow ribbons, as required for the measurement of stacking fault energies. Unfortunately with increasing s-value the image contrast decreases and long exposure

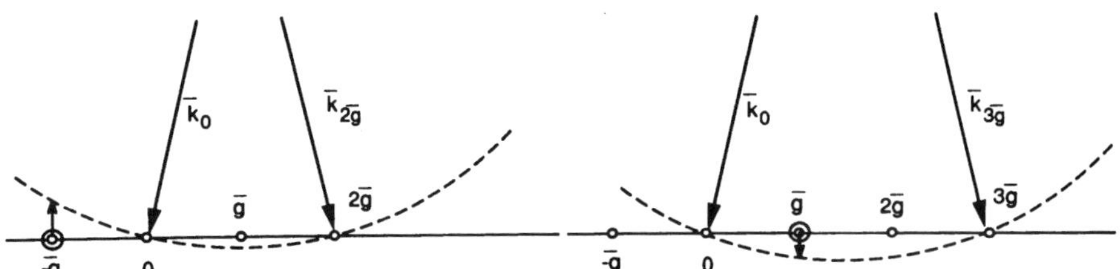

Figure 1.44 Imaging conditions used in the weak beam method; left: s = 0 for 2**g**, image is made in **g**; right: s = 0 for 3 **g**, image is made in **g**.

times are required to record the image. In practice a reasonable trade-off between image resolution and exposure time seems to be achieved for s-values of the order of 0,2 nm^{-1} for 100 kV electrons.

Usually a high order reflection 3**g** or 4**g** is brought in the exact Bragg position and a dark field image is made in the reflection **g** Alternatively a low order reflection, **g** or 2**g** may be excited and -**g** used for imaging. These imaging conditions are represented in Figure 1.44. In order to realize these diffraction conditions exactly, the Kikuchi pattern is of great help, moreover it allows to measure s. In weak beam images the depth period of extinction contours and of stacking fault fringes is given approximately by its kinematical value $1/s_g$. Using such large s-values it is possible to image for instance anti-phase boundaries in alloys as fringe patterns, even though the extinction distance of the superlattice reflection used is larger than the foil thickness.

The kinematical theory allows to derive approximate expressions for the peak width and peak positions of weak beam

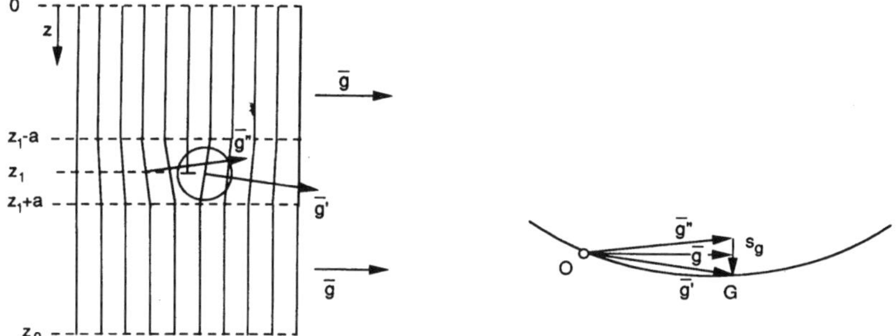

Figure 1.45 Foil containing an edge dislocation. Model used in discussing dislocation contrast according to the weak beam method. The foil is assumed to consist of three lamellae: 1 and 3 are perfect; part 2 contains the dislocation.

dislocation images [30]. The columns close to the dislocation core can be considered as consisting of three parts (Figure 1.45). The central part contains the dislocation, the two other parts are perfect. In the central part the lattice planes of interest are inclined with respect to their orientation in the perfect parts, in such a way that somewhere close to the dislocation core the local deviation parameter is much smaller than in the perfect part.

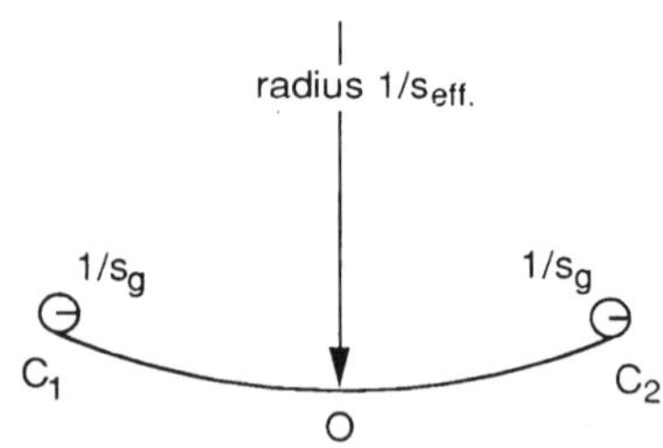

Figure 1.46 Amplitude-phase diagram for a column close to the dislocation core according the weak-beam method.

The scattered intensity will then mainly originate from this region, producing a bright peak on a darker background in the dark field, weak beam, image. The amplitude scattered by a column at x is given by:

$$A \propto \int_0^{z_0} \exp\{2\pi i[s_g z + \mathbf{g}.\mathbf{R}(x,z)]\} dz \tag{1.149}$$

This integral can be split in three parts corresponding to the three lamellae in the model of Figure 1.45.

$$A \propto \int_0^{z_1-a} \exp(2\pi i s_g z) dz + \int_{z_1-a}^{z_1+a} \exp[2\pi i(s_g z + \mathbf{g}.\mathbf{R})] dz \tag{1.150}$$

$$+ \int_{z_1+a}^{z_0} \exp(2\pi i s_g z) dz$$

The first and third integral refer to the perfect parts; they do not depend on the presence of the defect. Since s is large in these parts their contribution is small. Their A-P diagrams consist of small circles with a radius 1/2 πs. These two circles are connected by a circular arc with a much larger radius $1/s_{eff}$, which is the A-P diagram of the central part. The amplitude scattered by the column is then given to a good approximation by the length of the segment joining the centers of the two small circles. This length is well approximated by the second integral, which we shall now consider (Figure 1.46).

We can write the displacement function R(x,z) as a Taylor expansion in the vicinity of the core positions $z = z_1$

$$\mathbf{R} = \mathbf{R}(z_1) + (z - z_1)\left(\frac{\partial \mathbf{R}}{\partial z}\right)_{z_1} + 1/2(z - z_1)^2\left(\frac{\partial^2 \mathbf{R}}{\partial z^2}\right)_{z_1} + \cdots \tag{1.151}$$

Retaining only the first two terms the second integral can be written as

$$\exp\left\{2\pi i\left[\mathbf{R}(z_1) - z_1\left(\frac{\partial \mathbf{R}}{\partial z}\right)_{z_1}\right]\cdot \mathbf{g}\right\} \times$$
$$\int_{z-a}^{z+a} \exp\left[2\pi i\left(s_g + \mathbf{g}\cdot\frac{\partial \mathbf{R}}{\partial z}\right)z\right]dz \tag{1.152}$$

This expression will be a maximum if the modulus of the integrand is unity, i.e. for the value of x given by

$$s_g + \frac{\partial \mathbf{R}}{\partial z}.\mathbf{g} = 0 \tag{1.153}$$

This condition is equivalent to the statement $s_{eff} = 0$ (see Eq. (1.59)).

Introducing the displacement field for edge and screw dislocations, adopting the FS/RH convention (see § 1.29.3) leads to

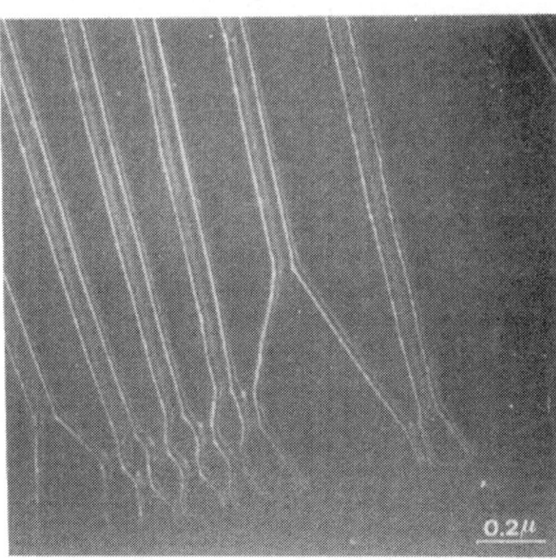

Figure 1.47 Weak beam image of fourfold dislocation ribbons in $RuSe_2$. (Courtesy J. Van Landuyt)

the peak position $x = x_m$ with

$$x_m = \{1+[K/2(1-\nu)]\}[-\mathbf{g}.\mathbf{b}/(2\pi s_g)] \quad (1.154)$$

The parameter is K = 1 for an edge dislocation and K = 0 for a screw dislocation; ν = Poisson's ratio. In this approximation the peak position does not depend on the foil thickness or on the depth position of the dislocation. The image side, i.e. the sign of x_m is clearly determined by the sign of the product $(\mathbf{g}.\mathbf{b})s_g$

Using the same model the peak width at half maximum can be deduced from the kinematical approximation. For **g.b** = 2 one finds

$$\Delta x = 0{,}28/|s_g|\ [1+K/2(1-\nu)] \quad (1.155)$$

with $\nu = 1/3$, $|s_g| = 0{,}2\ nm^{-1}$ one finds $\Delta x \approx 2{,}5$ nm for an edge dislocation.

With increasing value of **g.b** the image peak moves away from the core position. The larger the value of **g.b** the larger values of s_g are needed in order to achieve the same precision in the image position. In practice this limits the values of **g.b** ≤ 2.

An example of a weak beam image in the layered crystal $RuSe_2$ is reproduced in Figure 1.47.

The dynamical theory, neglecting anomalous absorption for simplicity, leads to essentially the same qualitative results. In terms

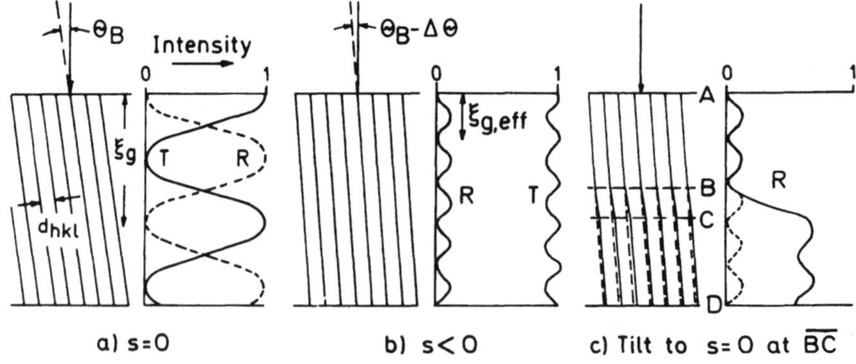

Figure 1.48 Weak-beam image formation at a dislocation according to the dynamical theory [29] ($\xi_g \equiv t_g$; T = transmitted beam amplitude; R = scattered beam amplitude).

of this theory the scattered beam, in the first part along a column close to the dislocation on the image side, oscillates with a small amplitude and with a depth period given to a good approximation by the kinematical value $1/s$ (Figure 1.48). In the second part of this column, where $s_{eff} \approx 0$, the pendellösung oscillations acquire a large amplitude and a depth period approximated by t_g. Since this second part is thin only a fraction of an oscillation can develop, and in part three the amplitude of the oscillation as well as its depth period become again the same as in part one. However the average intensity level has now become larger in part three, in particular for the column along which s_{eff} becomes zero at the level of the dislocation core. Hence the observed intensity at the exit face of that column will be larger than that for columns which are further away from the dislocation, which will thus show up as a bright line.

1.29. Dislocation Contrast: Dynamical Theory

1.29.1. Image simulation

One-dimensional profiles and two-dimensional maps which describe quantitatively the experimentally observed images are only obtained by applying the dynamical theory including anomalous absorption. The set of equations (1.104) or (1.102) has to be integrated with α being in general a function of x and z. For a screw dislocation located at a distance d behind the entrance face this function becomes for instance

$$\alpha = n \operatorname{arctg}[(z-d)/x] \tag{1.156}$$

with $n = \mathbf{g}.\mathbf{b}$: n is an integer for perfect dislocations, but it may be a fraction for partial dislocations. If d is considered to be a constant a profile along x is sufficient to describe the image. For inclined dislocations d becomes an additional parameter and a two-dimensional map is desirable for comparison with experimental images.

Analytical solutions are difficult, if not impossible to obtain in most cases. Numerically computed image profiles are available for a number of representative dislocation configurations and will be reviewed below. A semi-quantitative analytical discussion of the most striking image properties is possible.

In principle the computation procedure for profiles is a multi-slice method. It consists in considering a row of columns situated along the x-axis. The integration is performed along a column i.e. for a fixed x-value, by further dividing this column in thin slices dz, each slice being considered as perfect with an s-value: $s_{eff} = s_g + [g \cdot dR / dz]_{x_0}$, which depends on z. The amplitudes of scattered and transmitted beams can be obtained by the multiplication of a succession of response matrices of the type $M(dz, s_{eff})$ (see Eq. (1.113)). This procedure is the implementation of the "column approximation".

The linear character of the Howie-Whelan system of equations [(10) in G2] and the fact that the displacement field of a dislocation is invariant for a translation along lines parallel to the dislocation line have been exploited by Head [31] and Humble [32] to speed up the computation procedures in order to make it possible to generate rapidly two-dimensional maps which can be compared directly with observed images (Figure 1.49).

The different computer programs and subroutines required to generate two-dimensional intensity maps representing bright and dark field images for a wide variety of single and complex defects are described in full detail in Ref. 32. The displacement fields of the defects are numerically computed using anisotropic linear elasticity theory. Subroutines allow to determine the geometry of the foil, the diffraction conditions, etc., using the Kikuchi line pattern as input data. The defect identification procedure is essentially a "trial and error" method based on the inspired guess of a model based on

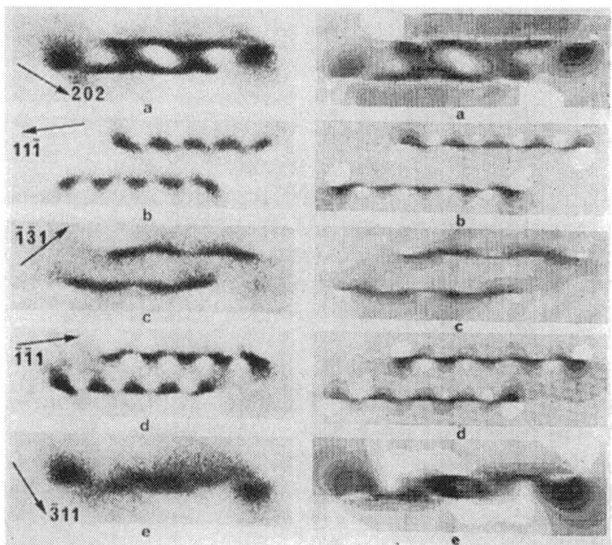

Figure 1.49 Examples of the quantitative agreement that can be achieved between observed and computed dislocation images. Left: observed images for different diffraction vectors Right: corresponding computer generated images.

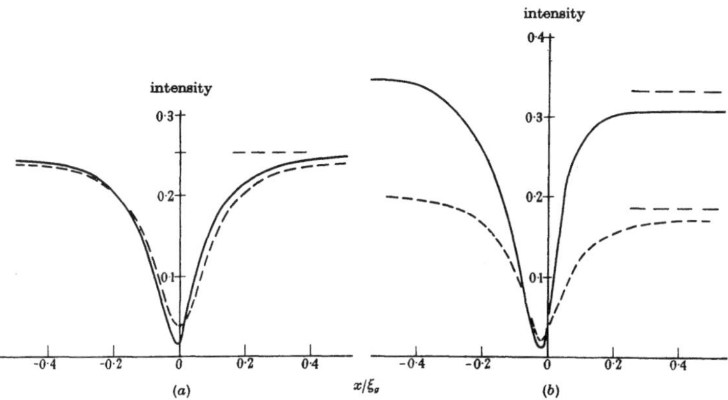

Figure 1.50 Image for screw dislocations in the central plane of the foil with thickness 8 t_g; n = 1, s = 0 (left); s = 0,3 (right). Solid line is the bright field image; dotted line is the dark field image.

symmetry considerations and on qualitative geometrical image characteristics of the type described in previous sections of this chapter. The model is then tested and where necessary further refined by the quantitative comparison of observed and computer generated images in which a small number of parameters is varied. Since the computer time is short a comparison exercise does not require an excessive computer effort. Complete listings of the software statements are provided in Ref. 32.

Remarkable agreement between computed and observed images can be achieved, even for complex defect configurations such as the one illustrated in Figure 1.49. One of the important conclusions that emerged from such simulations is that the extinction criterion for dislocations **g.b** = 0 is only a first approximation and can lead to wrong conclusions, especially in strongly anisotropic materials where the displacements around a dislocation are in general not parallel to **b** as implied in the analytical expressions based on isotropic linear elasticity.

1.29.2. Survey of results of the two-beam dynamical theory [10]

Images of screw dislocations The images for n = 1 and s = 0 for a screw dislocation parallel to the foil surfaces and located in the central plane of the foil, exhibit a single dark peak very close to the position of the dislocation core, in the bright field image as well as in the dark field image (Figure 1.50a). This is clearly in contradiction

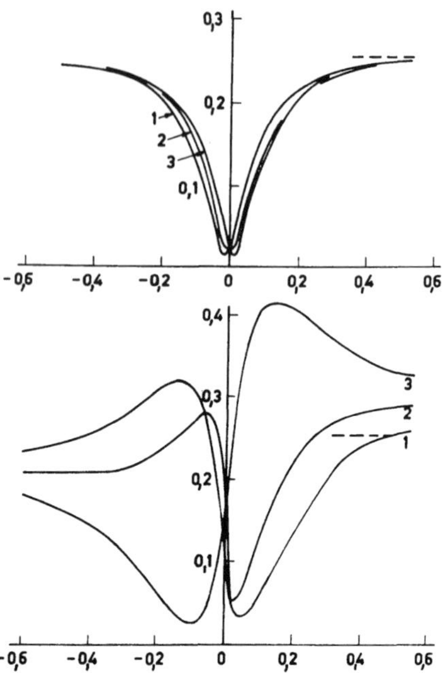

Figure 1.51 Bright field images for screw dislocations in foil with thickness 8 t_g at a distance d from entrance face. n = 1; $z_0 = 8\ t_g$; s = 0 .
top: 1,2,3: d = 4, 4.25, 4.50 t_g bottom: 1,2,3: d = 7.25, 7.50, 7.75 t_g

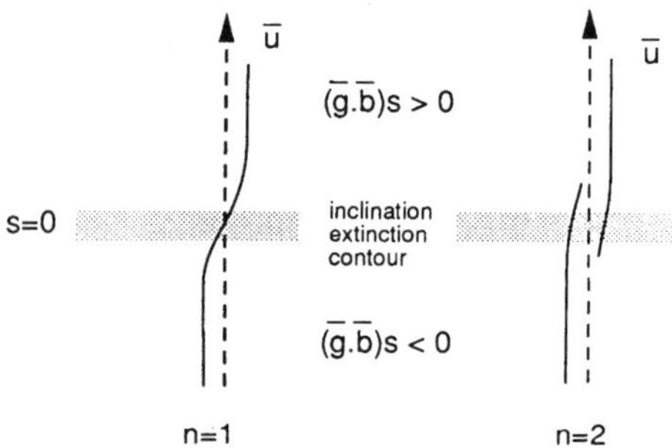

Figure 1.52 Image behavior on crossing an inclination extinction contour. (a) n = 1 (b) n = 2

with the results of the kinematical theory which predict complementary images; it is a result of anomalous absorption in thick crystals (5-10 t_g). The peak width is of the order of 0.3 - 0.4 t_g.

For screw dislocations close to the surface (Figure 1.51b) the image becomes clearly one-sided, the image side changing periodically with depth in the crystal.

For s≠0 and sufficiently large the dark line is displaced away from the core position in the sense predicted by the intuitive reasoning of § 1.28.1 for the bright field image. The sense of the image shift does not depend on the depth position of the dislocation, but changes with the sign of s. As a result the image will shift continuously sideways on crossing an inclination extinction contour in the manner represented in Figure 1.52.

For s = 0 as well as for small values of s, the image shift does depend on the depth positions z_0. Close to the surfaces the sense of

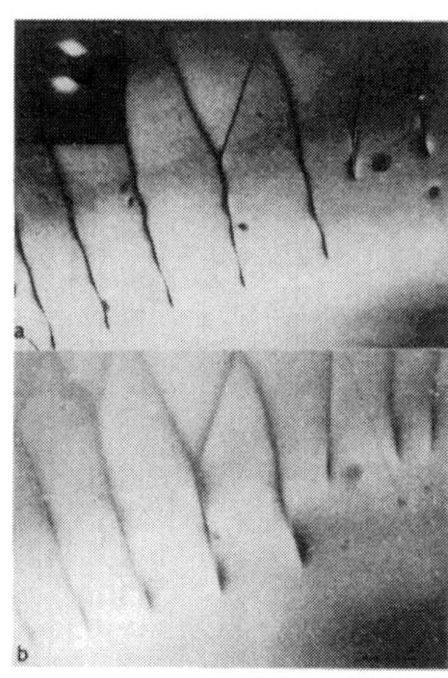

Figure 1.53 Dislocations in SnS_2 exhibiting oscillating contrast on approaching the surface; (a) bright and (b) dark field images.

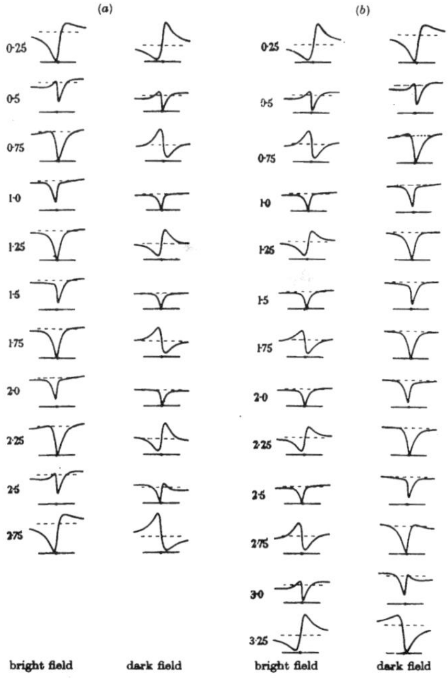

Figure 1.54 Image profiles of a screw dislocation (s = 0; n = 1) at different depths below the entrance face of the foil in two foils of different total thickness; left bright field; right dark field
(a) thickness 3 t_g; (b) thickness 3.5 t_g

the image shift alternates with a depth period t_g but in the central part of a thick foil the image shift is small. Inclined dislocations in thick foils will thus exhibit oscillating or alternating contrast close to the foil surfaces, but not in the central part (Figure 1.53). The oscillations will be in "phase" close to the entrance face in bright and dark field, but in "anti-phase" close to the exit face. This effect is a consequence of anomalous absorption and it applies to all defect images. In thick foils the bright and dark field images of defects are similar when situated close to the entrance face but quasi-complementary close to the exit face. This was in fact already pointed out for planar interfaces (see § 1.24).

This oscillating contrast can be exploited to provide a depth scale since for $s = 0$ the oscillation period is exactly t_g. In particular it allows to determine the foil thickness in units of t_g and it makes it possible to determine which end of the dislocation image corresponds to the vicinity of which foil surface front or rear.

The occurrence of "dotted" images at inclined dislocations can be understood intuitively, as pointed out above (§ 1.28.2) by noting that top and bottom parts of a column passing through the dislocation core are related by a phase jump of π which occurs at the level of the core. Along such columns the intensity profile for $n = 1$ will be the same as that for a stacking fault with $\alpha = \pi$ Whether predominantly "dotted" or "alternating" contrast occurs depends on the thickness of the foil, as does the contrast for a fault with $\alpha = \pi$. This is illustrated in Figure 1.54 which shows computed bright and dark field profiles for a screw dislocation with $n = 1$ and $s = 0$ in foils respectively with a thickness 3 t_g and 3.5 t_g. In the foil with a thickness of 3 t_g the bright field image is dotted and the dark field image alternating, whereas for a foil with a thickness of 3.5 t_g the reverse is true.

In case $n = 2$, $s = 0$ the image exhibits two dark peaks, one on each side of the dislocation core. The two peaks are different in strength, their relative strength alternates with a period t_g with the depth in the foil. These features are illustrated in Figure 1.55a, they again lead to "oscillating" contrast at inclined dislocations for $s \approx 0$. If $s \neq 0$ the two peaks become strongly asymmetrical as shown in Figure 1.55b, the sense of the asymmetry depending on the sign of *s*. Except for $s = 0$ usually only one dark peak is observed as a consequence of the asymmetry. On intersecting an equi-inclination contour with $s = 0$ the dislocation image will therefore behave as represented in Figure 1.52. It is thus possible to deduce the value of n from the behavior of the image on intersecting an inclination contour. The value of n gives the projection of **b** on the active diffraction vector **g**, and hence allows to determine the length of **b** once its direction is known. For columns passing through the

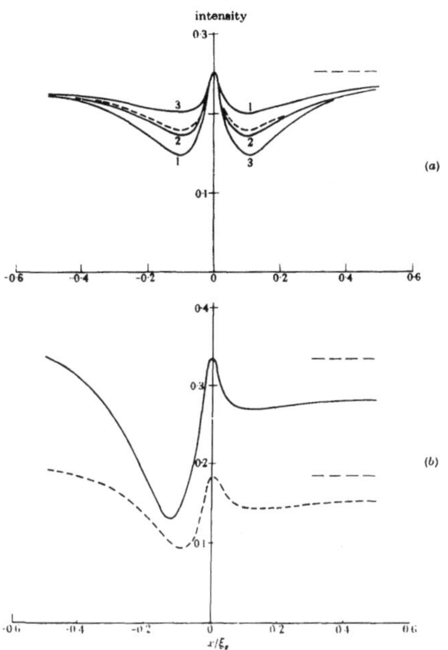

Figure 1.55 Image profile of a screw dislocation: foil thickness 8 t_g
(a) for s = 0, n = 2; 1, 2, 3 correspond to different depths in the foil: 4, 4.25 and 4.5 t_g behind the entrance face
(b) for s ≠ 0, n = 2 (st_g = 0.3). Full lines: bright field, dotted lines: dark field

dislocation core the phase shift at the level of the core in the case n = 2 is now 2π i.e. such columns will exhibit the same intensities as the perfect crystal.

Images of edge and mixed dislocations The displacement field of a mixed dislocation with a direction defined by its unit vector **u** parallel with the foil plane (Figure 1.56) is given according to isotropic linear elasticity theory by the expression

$$\mathbf{R} = (1/2\pi)\{\ \mathbf{b}\varphi + [\mathbf{b_e}/4(1-\nu)]\sin 2\varphi + [\ \langle(1-2\nu)/2(1-\nu)\ \rangle \ln|r| + (\cos 2\varphi)/4(1-\nu)\ ](\mathbf{bxu})\ \} \tag{1.157}$$

where ν is Poisson's ratio (ν= 1/3), **b** is the Burgers vector of which $\mathbf{b_e}$ is the edge component; $\varphi = \alpha\text{-}\gamma$.

For a pure screw **b** x u = 0 and $\mathbf{b_e}$ = 0 the expression reduces to **R** = (**b**/2π)φ. The term in **b** x **u** describes a displacement perpendicular to the slip plane towards the supplementary half plane. The slip plane is determined by **b** and **u**; it forms an angle γ with the foil plane. The character of the dislocation can be quantified by the parameter $p = (\mathbf{g}\,\mathbf{b_e})/(\mathbf{g}.\mathbf{b})$, which is 0 for a pure screw and 1 for a pure edge.

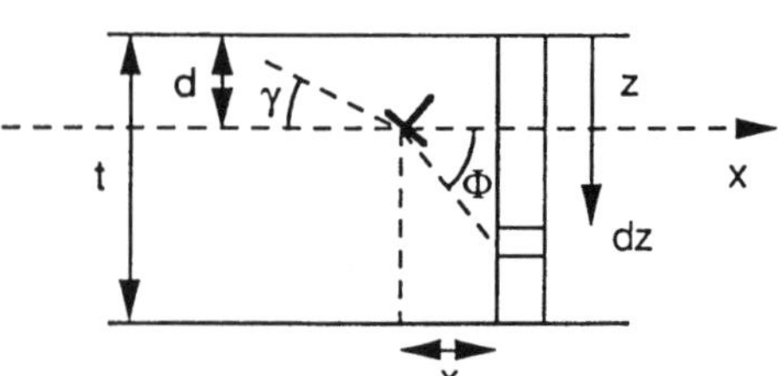

Figure 1.56 Reference system used to describe the displacement field due to mixed dislocations.

Computed images (Figure 1.57) for a mixed dislocation with its slip plane parallel to the foil plane ($\gamma = 0$), for the following values of the parameters $n = 1$, $s = 0$, $t = 8t_g$, $z_0 = 4t_g$ and for a number of values for p show that the image of a pure edge (p= 1) is wider than that of a pure screw (p = 0). The narrowest image is obtained for $p = -1/2$ and the widest for $p = 1/2$, i.e. for 45° mixed dislocations. The full width varies between 0.3 and 0.8 t_g.

Even for **g.b**= 0 a pure edge dislocation may produce contrast because of the term **b** x **u** in the displacement function. For a closed prismatic Frank loop parallel to the foil plane and for the imaging g-vectors parallel to the loop plane **g.b** = 0 but nevertheless complete extinction only occurs if moreover (**b** x **u**).**g** = 0, which is only the case if **u** is parallel to the **g**-vector. As a result only those dislocation segments, which are parallel to the acting **g**-vector, will be out of contrast. The line connecting these two segments, called "line of no

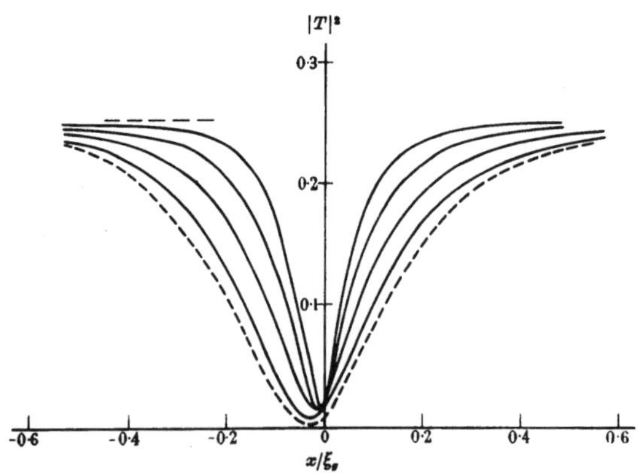

Figure 1.57 Computed images for mixed dislocations with slip plane parallel to the foil plane (g = 0) and for various values of the parameter p = $(\mathbf{g}.\mathbf{b_e})/(\mathbf{g}.\mathbf{b})$.

contrast" is perpendicular to the **g** vector (Figure 1.58). In a pure Frank loop, i.e. with its Burgers vector perpendicular to the loop plane, a line of no contrast will thus form for all **g** vectors parallel to the loop plane.

If the Burgers vector **b** is inclined with respect to the loop plane there will only be one vector **g** (as well as -**g**) parallel to the loop plane, for which a line of no contrast occurs; this is the **g**-vector perpendicular to the projection of **b** on the loop plane. The argument can be reversed; if among all **g**-vectors parallel to the loop plane.

If the Burgers vector **b** is inclined with respect to the loop plane there will only be one vector **g** (as well as -**g**) parallel to the loop plane, for which a line of no contrast occurs; this is the g-vector perpendicular to the projection of **b** on the loop plane. The argument can be reversed; if among all **g**-vectors parallel to the loop plane only one produces a line of no contrast the loop cannot be a pure Frank loop.

Changing the sign of x in the expression (1.157) for the displacement field changes the sign of φ, but since all terms in (1.157) are even functions of either φ or x we conclude that the image profile must be symmetrical in x. In the computed profiles of Figure 1.59, only the half corresponding with $x > 0$ is represented; they show that for certain depth positions the image may exhibit two broad dark lines, as for instance in Figure 1.58.

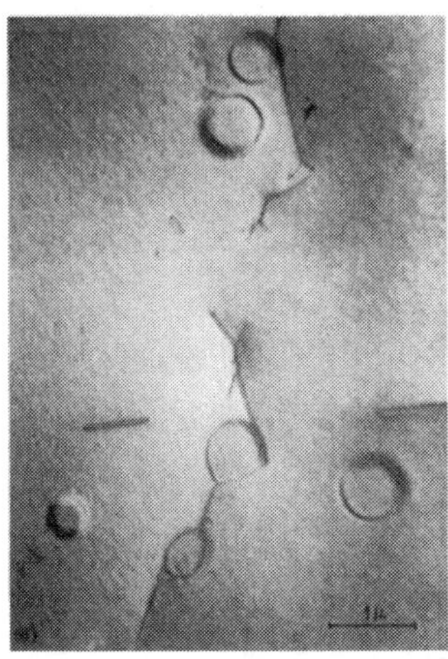

Figure 1.58 Images of prismatic dislocation loops with their Burgers vector parallel to the incident beam. Note the lines of no contrast perpendicular to the g-vector.

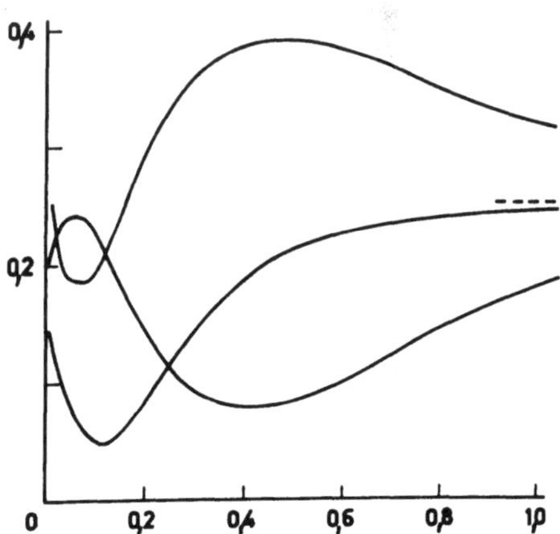

Figure 1.59 Computed image profiles for edge dislocations with their Burgers vector parallel to the incident beam. Only half of the profile is shown; it is symmetrical in x.

Images of partial dislocations Since the Burgers vectors of partial dislocations are not lattice vectors the image order $n \equiv \mathbf{g.b}$ may become fractional. For instance, for Shockley partials in F.C.C. crystals the Burgers vector is 1/6 [112] and the value of n becomes a multiple of 1/3.

Partial dislocations form the border of stacking faults. The image profile is therefore complicated by the fact that it separates two areas of which one has the brightness of a perfect region and the other one has the contrast of a faulted area at the depth level of the partial dislocation; these brightnesses are in general different.

Image profiles have been computed for $n = \pm 1/3$, $\pm 2/3$, and $\pm 4/3$. For $n = \pm 1/3$ no visible line image is formed since the profile constitutes a continuous transition between the two brightness levels. For $n = \pm 2/3$ the image consists of a dark line with a small visibility however. Images with $n = \pm 4/3$ are expected to consist of a dark line comparable to that of an ordinary dislocation (Figure 1.60).

Partial dislocations of the Frank type have pure edge character; their behavior was discussed in § 1.29.2.

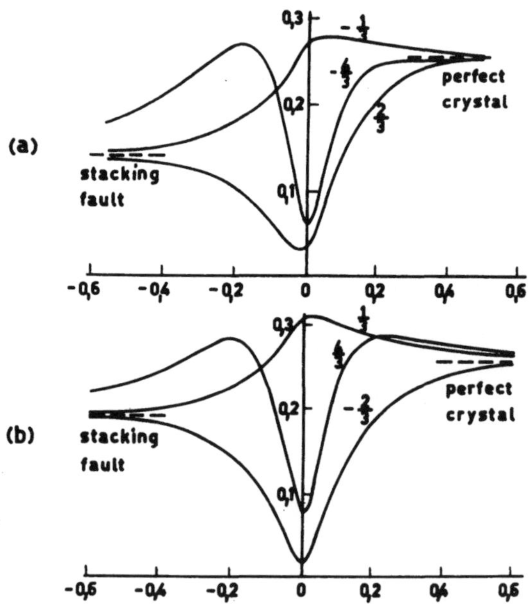

Figure 1.60 Image profiles for partial dislocations. The n-values are indicated.

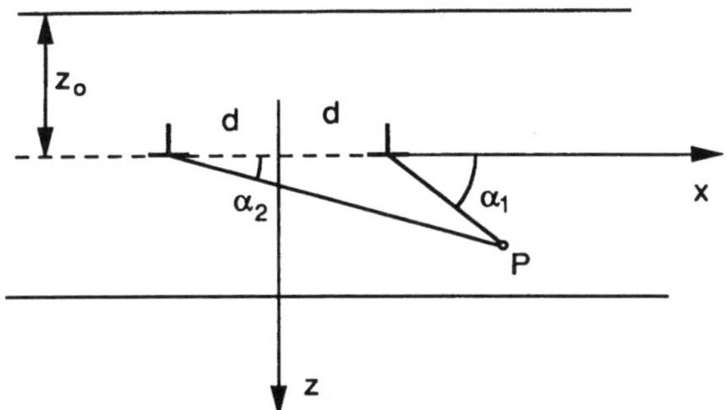

Figure 1.61 Reference system used in describing the displacement field of a dislocation ribbon.

Images of dislocation ribbons [27] The images of ribbons of partial dislocations deserve some special consideration. Since the ribbon width is a measure for the energy of the enclosed stacking fault the exact separation of the partials needs to be known accurately in order to make possible precise measurements of the stacking fault energy.

The image of a ribbon is <u>not</u> the superposition of the images of the separate partial dislocations; it is caused by the strain field of the ribbon, which is obtained in the framework of linear elasticity as the superposition of the strain fields of the two partial dislocations. For a screw ribbon consisting of two Shockley partials, enclosing an angle of 60°, and for an active diffraction vector oriented along the bisector of the acute angle between Burgers vectors $\mathbf{b}_1$ and $\mathbf{b}_2$, the ribbon behaves to a good approximation, as far as the contrast is concerned, as if it consisted of two screws with n-values ($n = \mathbf{b.g}$) which are either both +1 or -1. This is due to the fact that the edge components are perpendicular to **g** and therefore produce residual contrast only.

For the geometry shown in Figure 1.61 the phase shift α caused by the ribbon can be formulated as

$$\alpha = n_1\alpha_1 + n_2\alpha_2 \tag{1.158}$$

with

$$\alpha_1 = \operatorname{arctg}\left[(z - z_0)/(x - d)\right]$$
$$\alpha_2 = \operatorname{arctg}\left[(z - z_0)/(x + d)\right] \tag{1.159a}$$

and

$$n_1 = \mathbf{g.b}_1 \text{ and } n_2 = \mathbf{g.b}_2 \tag{1.159b}$$

Such a ribbon will produce a symmetrical image for the case $n_1 = \pm 1$ and $n_2 = \mp 1$ since changing x into -x leads to changing α_1 into $-\alpha_2$ and vice versa and thus $\alpha(x)$ into $\alpha(-x)$. Integrating along columns at -x and at +x leads to the same result and the image is thus symmetrical.

If we change simultaneously $n_1 = n_2 = +1$ into $n_1 = n_2 = -1$ and x into -x the expression for α remains unchanged. We conclude from this that the profile for $n_1 = n_2 = -1$ is the mirror image of that for $n_1 = n_2 = +1$. It is thus sufficient to discuss one of these two cases.

Bright and dark field image profiles for screw ribbons are reproduced in Figure 1.62a for various sign combinations of $n_1 = \mathbf{g.b}_1$

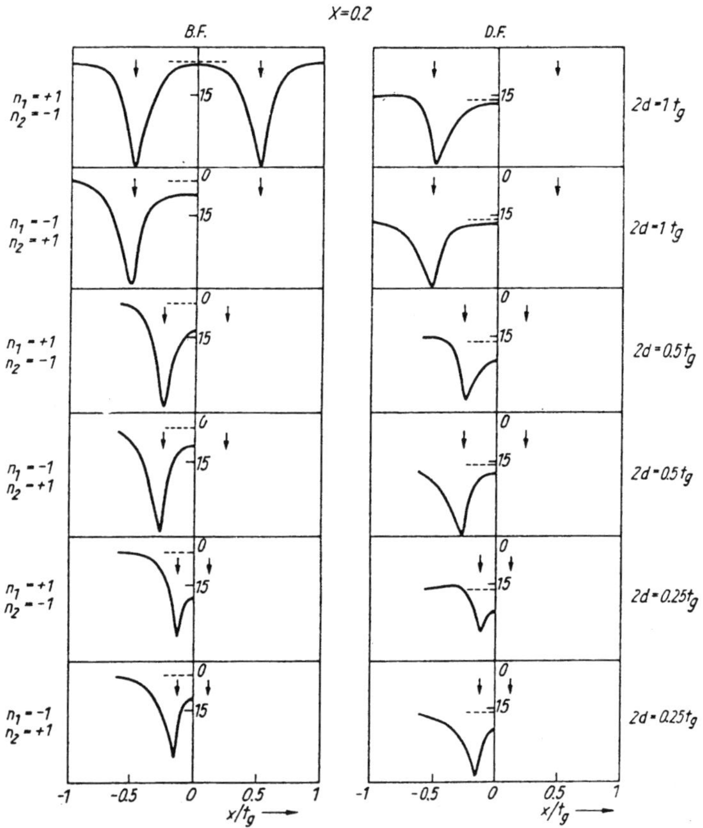

Figure 1.62a Image profiles for ribbons of varying widths ($s_g t_g = 0.2$) and varying combinations of $n_1 = \pm 1$, $n_2 = \mp 1$. The profiles are symmetrical, only one half is shown; BF and DF profiles are represented. The positions of the partials are indicated by arrows.

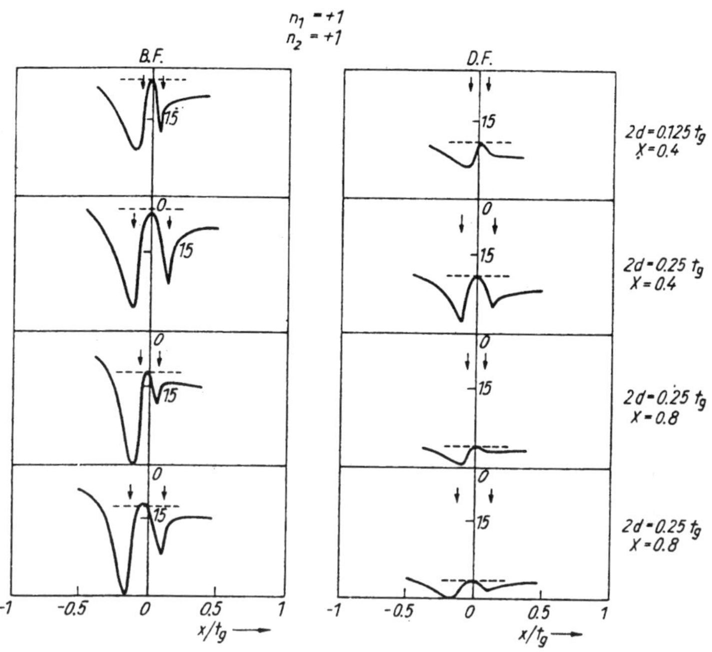

Figure 1.62b Bright and dark field image profiles for dislocation ribbons. Foil thickness tg ; x = stg; z0 = 3.5 tg; n1= n2 = ±1.

= ±1 and n_2 = **g.b**$_2$ = ±1. In all cases the foil had a thickness of 8 t_g; the ribbon was at a depth 3.5 t_g behind the entrance foil and st_g= 0,2.

When n_1 and n_2 are opposite in signs as in Figure 1.62a the profiles are symmetrical; only one half is therefore shown, the complete profile is obtained by a mirror operation. The positions of the dislocations are indicated by arrows. The ribbon widths are clearly different from the peak separations, due to the fact that the image displacements are in opposite sense for the two dislocations as a result of the sign difference of n. For the sign combination n_1 = +1 and n_2 = -1 the apparent (observed) width of the ribbon is smaller than their real width; the opposite is true for n_1 = -1 and n_2 = +1. This width difference is also found to increase with increasing s_g, since the image displacements increase with s_g. The effect of a decreasing separation 2d on the contrast in the central part of the ribbon is clearly visible; the background intensity being represented by a horizontal dotted line. The center of the image corresponds with a column for which the phase shift is given by

$$\alpha_0 = (n_1 + n_2)\phi - k\pi \tag{1.160}$$

where ϕ =arctg$[(z - z_0)/d]$. The brightness in the center is thus the same as for a stacking fault at $z = z_0$ with α_0 as a phase shift.

If on the other hand n_1 and n_2 have the same sign the image sides are the same for the two partials and the peak separation in the image is more representative of the real width of the ribbon (Figure 1.62b). However the two partials are now imaged as lines of different width and brightness for $s \neq 0$. This can be understood by noting that the displacements associated with the two partials are additive outside of the ribbon but subtractive in the region between the two partials, i.e. inside the ribbon. The strongest line image is formed outside of the ribbon on the image side of the first partial. For the second partial the image side is the same as for the first but this is now inside the ribbon where the displacements are subtractive and hence the peak is smaller. Changing the sign of s changes the image side for both partials. The strongest image will be again outside the ribbon but on the other side since this is now where the displacements are again additive albeit in the opposite sense. The brightness in the central column is now the same as the background since $\alpha(x = 0) = k\pi$. Some of these features can be observed on the computed profiles of Figure 1.62b.

The symmetrical triple ribbons in graphite are formed by three partial dislocations with the same Burgers vector, separating two fault ribbons (Figure 1.63). As a result the n-values are always the same and hence also the image sides for the three partials which are determined by the sign of ns. If the displacements are for instance additive outside of the threefold ribbon and to the left of it, they are partly additive inside the left fault ribbon, partly subtractive inside the right fault ribbon and completely subtractive (i.e. additive in the opposite sense!) outside the triple ribbon and on the right of it. As a result the peaks marking the three partials will decrease in magnitude from left to right if the image side is left. Changing the sign of *s* or changing the image side for the three partials will invert the sense in which the magnitudes of the peaks decrease. This feature can be observed in the sequence of images of Figure 1.63.

Dislocation dipoles [32] Dislocation dipoles consist of two parallel dislocations with opposite Burgers vector. If the dislocations are restrained to remain in their respective glide planes they take up a stable configuration which minimizes the elastic energy. For two edge dislocations the regions of expansion in one dislocation tend to overlap the regions of compression in the other one. The configuration is such that the plane formed by the two parallel dislocations encloses an angle ϕ with their slip planes. In the case of two pure edges $\phi = 45^\circ$.

As in the case of dislocation ribbons, the image of a dipole is not the superposition of the images of the two separate dislocations. The superposition must be carried out at the level of the strain fields in the framework of the linear elasticity theory. It was shown in Ref. 32 that the bright field image of an inclined dipole has a center of symmetry. This symmetry property allows to distinguish the images of a narrow dipole from that of a single dislocation.

Fission tracks On passing through a solid a very energetic particle causes damage along its track; this is in particular the case for the heavy and strongly charged fission fragments. It is reasonable to assume that such a track will consist of a straight cylindrical region of strongly disordered material which may lead to either an expansion or a contraction of this cylindrical region. In view of its symmetry it is reasonable to postulate a displacement field of the type

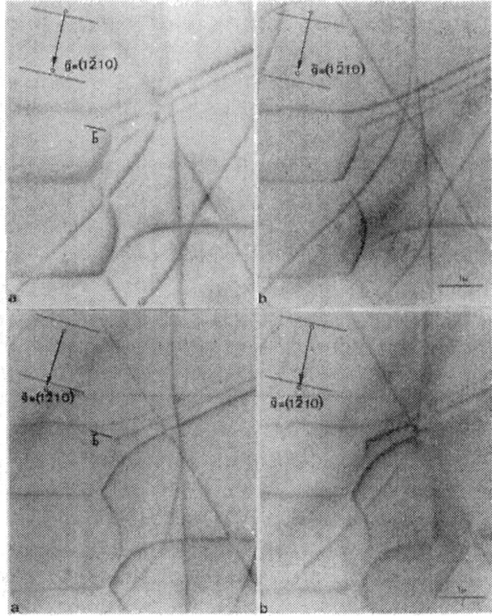

Figure 1.63 Images of triple ribbons in graphite under different diffraction conditions, i.e. for different s-values. top: $s > 0$; bottom: $s < 0$.

$$\mathbf{R}_y = 0 \quad \mathbf{R}_r = \varepsilon\, r_0 \mathbf{r} / r^2 \qquad (1.161)$$

for a fission track along the y-direction.

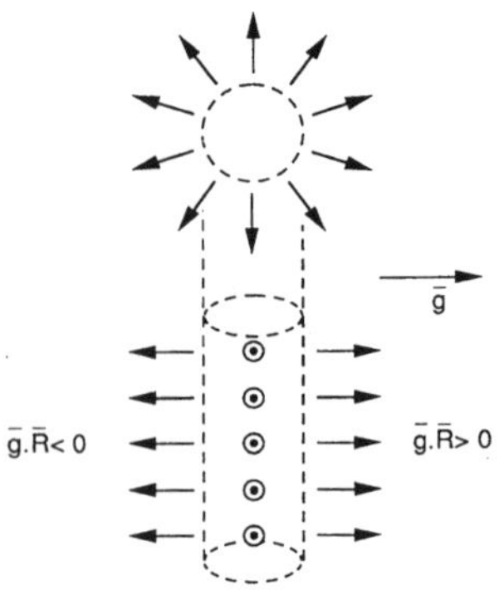

The fission track is an interesting object because it can be modeled rather simply and comparison of computed and observed image may provide some information on the sign of ε. For ε > 0 the radial displacements are outward whereas for ε < 0 they are inward.

Figure 1.64 Model used to discussthe contrast at a fission track.

It is possible to deduce some of the image characteristics intuitively for inclined tracks. The most obvious one is that for **g** parallel to the fission track no contrast will be produced according to the model since **g.R** = 0. If only the projections of **g** and y along the beam direction are parallel some contrast may remain.

An image profile along a line parallel to the inclined track will show oscillating contrast with a depth period t_g if the diffraction vector is close to perpendicular to the track. It can again be deduced with reference to the contrast at stacking faults. Let us assume that **g** is perpendicular to the track and points to the right (Figure 1.64). We then have that **g.R** > 0 and sin α > 0 along a strip to the right and parallel to the track. Columns along such a strip behave as if they contained a stacking fault with phase angle α, at the level of the track. Since on the right **g.R** > 0 (sin α > 0) whereas for a strip on the left of the track **g.R**< 0 (sin α < 0), it is clear that the contrast oscillations will be in anti-phase left and right of the track. Changing **g** into **-g** changes sin α → -sin α and vice versa and thus interchanges the phase of the oscillations. The two images are mirror images with respect to the plane determined by the track line and the foil normal. It can further be shown quite generally that the bright field image has an inversion center.

1.29.3. The image side of dislocations

From the dynamical image simulations discussed in § 1.29.1 and § 1.29.2 we concluded that the black line image of a dislocation in the bright field image is systematically one-sided provided s is large enough. This is true for **g.b** = 1 as well as for **g.b** = 2. This behavior is different for n = 1 and n = 2 if s ≈ 0 as discussed in § 1.29.2.

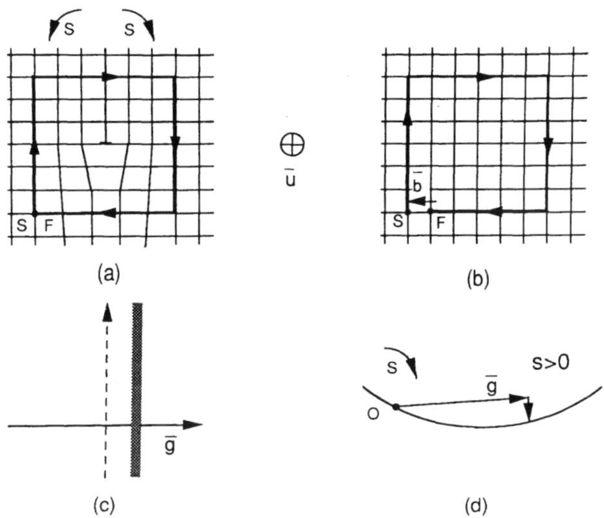

Figure 1.65 Illustrating the FS/RH convention for defining the sense of the Burgers vector of a dislocation.
(a) Real crystal
(b) Perfect reference crystal
(c) Relative position of image and dislocation
(d) Diffraction conditions

The image side, i.e. the position of the black line in the bright field image on a positive print with respect to the dislocation position is correctly given by the intuitive kinematical considerations of § 1.28.1. According to this theory the image side is on that side of the dislocation core where locally the lattice planes normal to **g** are rotated towards the exact Bragg orientation; in Figure 1.65 this is in the sense S. Finding S requires the knowledge of the sign of s; this can be determined by means of the Kikuchi line pattern as shown in § 1.4. For the edge dislocation in Figure 1.65 the positive sense, i.e. the unit vector **u** was chosen as entering the plane of the drawing. The Burgers vector **b** is determined according to the FS/RH convention. A right handed closed Burgers circuit when looking along **u** is constructed in the real crystal. In the perfect reference crystal the corresponding circuit is constructed and the Burgers vector **b** is found as the closure failure of this circuit, joining the final point to the starting point **b** = **FS**. For the concrete situation of Figure 1.65 s > 0 and (**g**.**b**)s < 0; the image is indicated by a solid line and the dislocation line by a dotted line. The rule can be formulated as follows; the image side is to the right looking along the positive sense if (**g**.**b**)s < 0. Changing the sign of one of the three parameters **g**, **b** or s changes the image side.

It should be noted that the descriptions given in different reviews are sometimes confusing and do not always agree because some authors refer to the image as seen along the incident electron beam, whereas other formulations refer to the image as seen from below. The sense of u depends on whether the first or the second viewpoint is adopted but this changes the sign of p = (**g**.**b**)s. The most direct way is to apply the intuitive reasoning, correctly taking into account possible electron optical image rotations.

1.29.4. Characterizing dislocations [G1]

The full description of a dislocation line implies the determination of its core geometry and its Burgers vector, i.e. direction, magnitude and sense of **b.** Methods are available to obtain all these elements. The precise position of the dislocation can be found by making two images leading to opposite image sides either for active diffraction vectors +**g** and -**g** for the same sign of s, or for +s and -s for the same g-vector. The true position of the dislocation is then between the two images.

The direction of the Burgers vector is determined by looking for two diffraction vectors $\mathbf{g}_1$ and $\mathbf{g}_2$ for which the dislocation is out of contrast or for which a residual contrast characteristic of **g**.**b** = 0, is

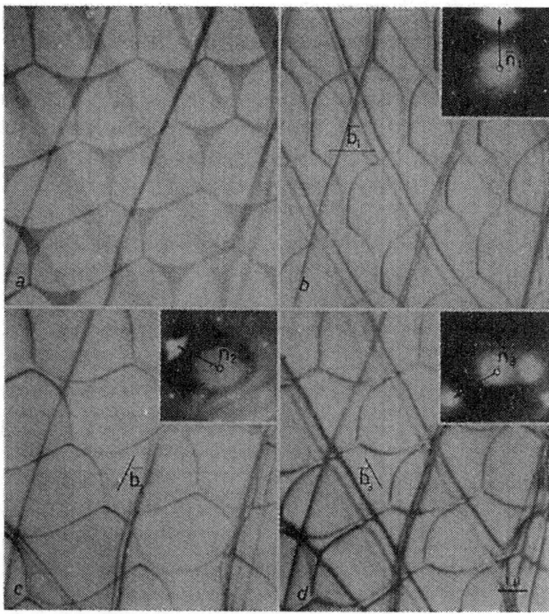

Figure 1.66 Dislocation network in graphite imaged under four different diffraction conditions leading to stacking fault contrast in (a) and to extinctions in (b), (c) and (d). Note that the triple ribbon looses contrast completely in (c).

produced. The Burgers vector then has a direction parallel with $\mathbf{g}_1 \times \mathbf{g}_2$. An example of the application of this method to a hexagonal network of dislocations in graphite is shown in Figure 1.66. In this particular case a single "extinction" is in fact sufficient since the dislocations were known to be glide dislocations and thus have their Burgers vector in the c-plane. The foil being prepared by cleavage it is also limited by c-planes. The three families of partial dislocations are seen to be successively brought to extinction using the indicated **g**-vectors. Note also the simultaneous extinction of the three partials in the triple ribbon, showing that they have the same Burgers vector. Their contrast is nevertheless different for the different partials as discussed in § 1.29.2.

It should be reminded that in highly anisotropic materials the simple extinction criterion $\mathbf{g.b} = 0$ is no longer strictly valid, as discussed in § 1.29.1. In the case of graphite, just mentioned, the extinctions can unambiguously be observed, even though graphite is highly anisotropic. However, due to the presence of the sixfold symmetry axis along c, the c-plane behaves effectively as elastically isotropic. This is also the case for dislocations in the (111) planes of face centered cubic crystals due to the threefold symmetry.

If complete extinction cannot be achieved one should look for the weakest contrast conditions and deduce from these a plausible Burgers vector, taking the crystal structure into account. Image simulations for various g-vectors, based on this Burgers vector, can then be compared with the observed images.

The magnitude of the Burgers vector for a perfect dislocation can be determined once its direction is known, by looking for diffraction vectors for which $\mathbf{g.b} = 2$. Hereby use is made of the typical contrast effect that occurs where the dislocation crosses a bend extinction contour (see § 1.29.2.1). If such a diffraction vector is

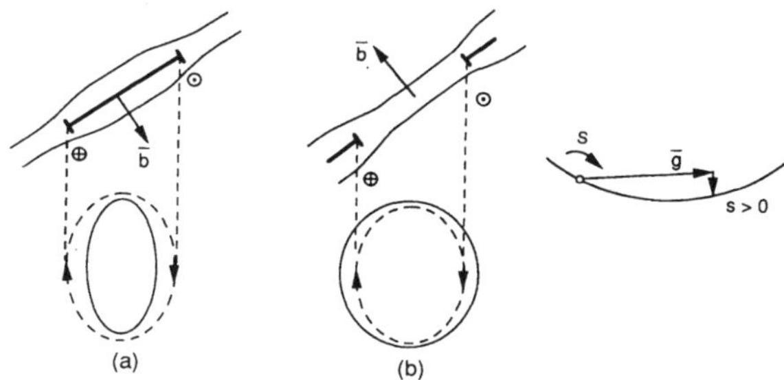

Figure 1.67 Image contrast of dislocation loops. In (a) the image is inside; in (b) it is outside the loop.

identified we know the length of the projection of **b** on **g** With the knowledge of the direction of **b** and of its projected length on **g**, the length of **b** can be found.

Finally the sense of **b** is found from the image side which defines the sign of (**g**.**b**)s. Knowing the sign of s from the Kikuchi pattern, the image side allows to find the sign of **g.b**. The knowledge of **g** then leads to the sense of **b**.

An important application of the sign determination of the Burgers vector [33,34,35] consists in determining whether a Frank loop is due to the precipitation of vacancies or of interstitials, i.e. whether **b** is either +1/3[111] or -1/3[111]. Applying the relation determining the image side to the loop represented in Figure 1.67 it follows that for a loop the image is either completely inside or completely outside the dislocation ring, depending on the sign of (**g**,**b**)s and since **b** is different for a vacancy loop and an interstitial loop so will be the image side for the same **g** and s. The type of contrast experiment required for an analysis of the nature of loops is illustrated in Figure 1.68. A difficulty with the application of this method consists in the necessity to know the sense of inclination of the loop plane. If the loops are known to be parallel to the foil

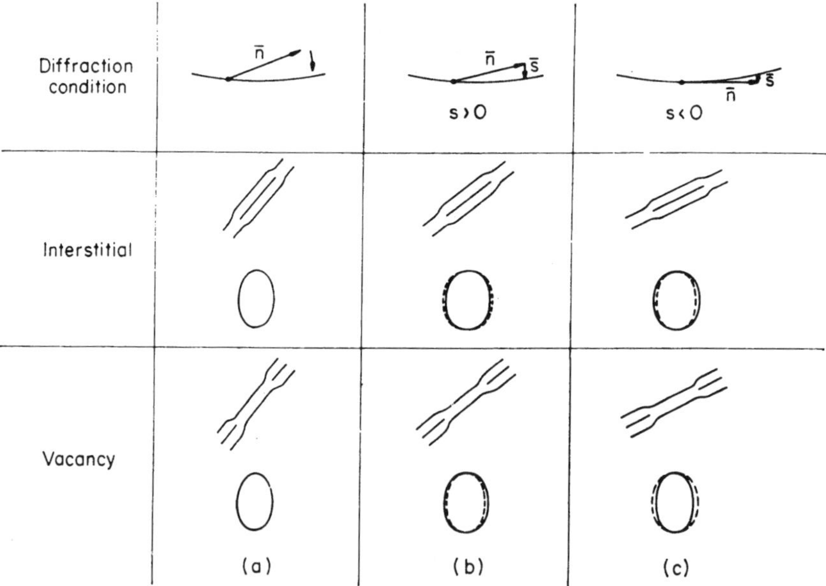

Figure 1.68 Contrast experiment allowing to determine the nature of dislocation loops. Diffraction vector n. The foil is tilted in the indicated sense.

surface, a known slope can be imposed by mounting the sample on a wedge. However this method is not always applicable.

Assuming the sense of the slope to be as represented in Figure 1.67 and **g** and s as shown, it is evident that the image is inside for interstitial loops, whereas it is outside for vacancy loops. Changing the sign of s by tilting allows to find the image side and hence to distinguish the two cases.

An alternative application of the same principle consists in rotating the specimen through the exact Bragg orientation from $s > 0$ to $s < 0$ for a given **g** (Figure 1.68). It is then found that an interstitial loop will grow in size because of two effects: (i) the projected size increases and (ii) the image goes from inside for $s > 0$ to outside for $s < 0$. A vacancy loop will grow as long as $s > 0$ because of the geometrical effect; but beyond $s = 0$ the image side changes and the image size shrinks. The experiment must clearly be performed starting with loops which are steeply inclined. One can also make use of the asymmetrical image contrast, consisting of a line of no contrast; separating a bright and a dark lobe (or crescent), characteristic of Frank loops seen end on, which moreover are close to the surface. In the dark field image the asymmetry is the same at top and bottom of the foil as a result of anomalous absorption. If the diffraction vector **g** is parallel with **b**, and points from the bright to the dark lobe in the image the loop has interstitial character. If **g** points from the dark to the bright lobe the loop is a vacancy loop. A restriction is that the loop must be close to the surface, i.e. within $1/2 t_g$ To demonstrate that the latter condition is satisfied, stereo images are required.

1.30. Properties of Dislocation Images

1.30.1. Analytical description [36]

A number of two-beam image properties of dislocations found empirically (see § 1.29) by numerical integration of the system of eq. (1.102 or 1.104) can be derived analytically for the important case

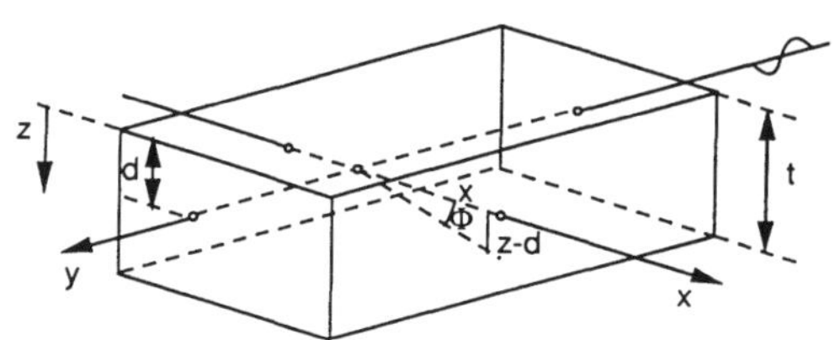

Figure 1.69 Model used in discussing the image profiles of dislocations, illustrating the notations used.

$s_g = 0$. These properties are in actual fact only approximately true but the model idealizes in a sense the situation somewhat so as to make these properties exact and mathematically tractable, without actually solving the system of eq. (1.102 or 1.104).

As a model we consider a foil containing a screw dislocation along the y-axis with a displacement field $\mathbf{R} = \mathbf{b}(\theta / 2\pi)$ with $\mathrm{tg}\theta = z/x$ (Figure 1.69); the reasoning can easily be extended to a general dislocation however. The origin of the reference system is chosen at the level of the dislocation. The foil is divided in three lamellae, as represented in Figure 1.70a. The first and third part, with thicknesses z_1 and z_3, are assumed not to be deformed by the presence of the dislocation. However the two extreme parts of the columns along *z*, i.e. those extending in the lamellae I and III are displaced one relative to the other. The central lamella contains the dislocation situated at z_2' behind the entrance face of this lamella: $z_2' + z_2'' = z_2$. This thickness z_2 is chosen in such a way that $|z_2'|/x >> 1$ and $|z_2''|/x >> 1$, where x is the position of a column, close enough to the dislocation, to make the approximation hold and for which the displacements are still appreciable. Computer simulations suggest that $x \leq 0{,}2\ t_g$ which means that $z_2', z_2'' \geq t_g$. For such a column we can

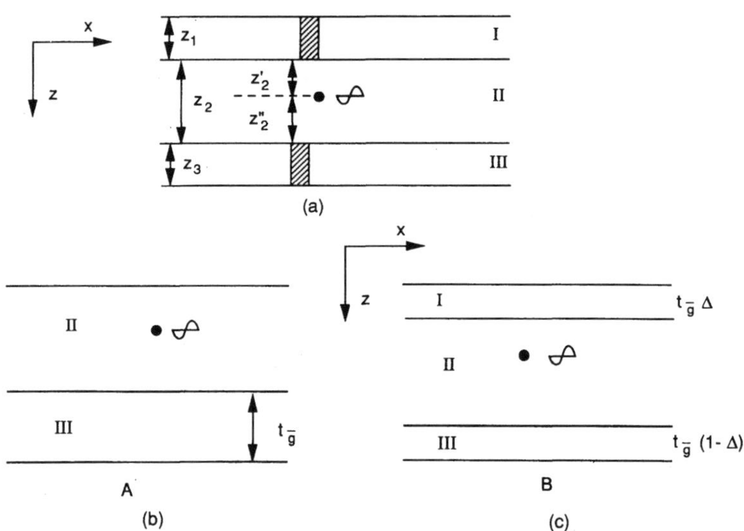

Figure 1.70 Different configurations of a dislocation in a foil consisting of three lamellae: I and III are perfect; II contains the dislocations. The configurations A and B correspond with: A: $z_1 = 0$, $z_3 = t_g$; B: $z_1 = t_g\Delta$; $z_3 = t_g(1\Delta)$.

assume $\theta = -\frac{\pi}{2}$ for the part of the column extending in lamella I and $\theta = +\frac{\pi}{2}$ for the part extending in lamella III. The corresponding displacements are -(1/4)b and +(1/4)b, with respect to the central lamella and the phase shifts $a = -n\frac{\pi}{2}$ and $n\frac{\pi}{2}$ with $n = \mathbf{g.b}$..

Using the matrix formulation we can write the response matrices (§ 1.21) for the first (I) and third (III) part respectively as

$$M_1 = \begin{pmatrix} T_1 & S_1 \exp(-in\pi/2) \\ S_1 \exp(in\pi/2) & T_1 \end{pmatrix} \qquad (1.162)$$

and

$$M_3 = \begin{pmatrix} T_3 & S_3 \exp(in\pi/2) \\ S_1 \exp(-in\pi/2) & T_3 \end{pmatrix} \qquad (1.163)$$

where

$$T_1 = \cos(\pi\sigma z_1) \quad S_1 = i \sin(\pi\sigma z_1) \qquad (1.164a)$$

$$T_3 = \cos(\pi\sigma z_3) \quad S_3 = i \sin(\pi\sigma z_3) \qquad (1.164b)$$

and $s = 1/t_g + (i/\tau_g)$ since $s_g = 0$.
Since the second lamella is not perfect its response matrix depends on the considered column. For a column at position x the response matrix has to be determined by integrating the system of eq. (1.102) where α is now a function of z and x: $\alpha(z,x)$.

The elements T_2 and S_2 in the left column of the response matrix M_2 are obtained by integrating the system ($s_g = 0$!)

$$dT_2/dz = i\pi\sigma S_2 \exp(i\alpha); \quad dS_2/dz = i\pi\sigma T_2 \exp(-i\alpha) \qquad (1.165)$$

with boundary conditions $T_2 = 1$ and $S_2 = 0$ at the entrance face of the second lamellae, i.e. at $z = z'_2$.

The elements of the right column are obtained by integrating the set obtained by interchanging S′ and T in the equations (1.165) i.e. by the substitution $S \to T'$, $T \to S'$ and $\alpha \to -\alpha$ (see § 1.21)

$$dT_2'/dz = i\pi\sigma S_2' \exp(i\alpha);\quad dS_2'/dz = i\pi\sigma T_2' \exp(-i\alpha) \qquad (1.166)$$

which only differs from (1.165) by the boundary conditions which are now $T_2' = 0$ and $S_2' = 1$ at $z = -z_2'$. We note that α depends on z as well as on x.

We introduce the notations $T_2'^{(-)}$ and $S_2'^{(-)}$ for the expressions T_2' and S_2' in which x has been replaced by -x. We note that

$$\alpha(z,-x) = n\,\pi - \alpha(z,x) \qquad (1.167)$$

this follows from purely geometrical considerations (Figure 1.69).

Changing x into -x and, taking into account the relation (1.167) transforms the set of equations (1.166) into the set:

$$\begin{cases} dT_2'^{(-)}/dz = i\pi\sigma S_2'^{(-)} \exp(in\pi)\ \exp(-i\alpha) \\ dS_2'^{(-)}/dz = i\pi\sigma T_2'^{(-)} \exp(-in\pi)\ \exp(i\alpha) \end{cases} \qquad (1.168)$$

which can be rewritten as

$$\begin{cases} \dfrac{d}{dz}\left[T_2'^{(-)} \exp(-in\pi)\right] = i\pi\sigma S_2'^{(-)} \exp(-i\alpha) \\ \dfrac{d}{dz}\ S_2'^{(-)} = i\pi\sigma\left[T_2'^{(-)} \exp(-in\pi)\right]\exp(i\alpha) \end{cases} \qquad (1.169)$$

with still as boundary conditions $T_2'^{(-)} = 0$, $S_2'^{(-)} = 1$ at $z = -z_2'$. We come to the conclusion that the functions T_2' and S_2' satisfy the same set of equations, as well as the same boundary conditions, as the functions $T_2'^{(-)} \exp(-in\pi)$ and $S_2'^{(-)}$. Comparing the two sets of equations (1.166) and (1.169) shows that the following relations must hold

$$S_2' = T_2'^{(-)} \exp(-in\pi);\quad T_2' = S_2'^{(-)}$$

Changing x into -x in both sides of these relations, which preserves these relations, leads to the required expressions for the matrix elements T_2' and S_2'

$$T_2' = S_2^{(-)}\ \exp(in\pi);\ ;\ S_2' = T_2^{(-)} \tag{1.170}$$

The response matrix M_2 is thus finally

$$M_2 = \begin{pmatrix} T_2 & S_2^{(-)}\exp(in\pi) \\ S_2 & T_2^{(-)} \end{pmatrix} \tag{1.171}$$

hereby the dashes were omitted since they were only introduced to keep track of the difference in the boundary conditions.

The response matrix for complete columns over the three parts is now $M = M_3M_2M_1$, which has to operate on $\binom{1}{0}$ in order to obtain the amplitudes of transmitted and scattered beams.

1.30.2. Variation of the image profile with depth position

The matrix formalism developed above can now be used to compare the image profiles of two dislocations differing $t_g\Delta(0<\Delta<1)$ in depth positions in two foils with the same thickness. The two configurations A and B are shown in Figure 1.70b, c.

For A the front part I has zero thickness ($z_1 = 0$) and the part III has a thickness $z_3 = t_g$. In B the front part (I) has a thickness $t_g\Delta$ and the exit part (III) a thickness $z_3 = t_g(1-\Delta)$. We further assume that anomalous absorption can be neglected for foils with a thickness equal or smaller than t_g i.e. in I and III; one then has $\sigma \approx 1/t_g$ (i.e. $\tau_g \to \infty$). Part II containing the dislocation is the same in both combinations A an B. Comparing A and B it is clear that the dislocation is lowered by $t_g\Delta$ in B as compared to A.

The response matrices of the different parts are now in combination B:

$$M_{1,B} = \begin{pmatrix} \cos(\pi\Delta) & i\ \sin(\pi\Delta)\exp(-in\pi/2) \\ i\ \sin(\pi\Delta)\exp(in\pi/2) & \cos\pi\Delta \end{pmatrix} \tag{1.172}$$

and

$$M_{3,B} = \begin{pmatrix} \cos[(1-\Delta)\pi] & i\ \sin[(1-\Delta)\pi]\exp(+in\pi/2) \\ i\ \sin[(1-\Delta)\pi]\exp(-in\pi/2) & \cos[(1-\Delta)\pi] \end{pmatrix} \quad (1.173)$$

or after simplifications:

$$M_{3,B} = \begin{pmatrix} -\cos(\pi\Delta) & i\ \sin(\pi\Delta)\exp(in\pi/2) \\ i\ \sin(\pi\Delta)\exp(-in\pi/2) & -\cos(\pi\Delta) \end{pmatrix} \quad (1.174)$$

For combination $A(\Delta = 0)$ the matrices reduce to

$$M_{1,A} = \begin{pmatrix} 1 & 0 \\ 0 & 1 \end{pmatrix} \text{ and } M_{3,A} = \begin{pmatrix} -1 & 0 \\ 0 & -1 \end{pmatrix} \quad (1.175)$$

For combination $C(\Delta = 1/2)$ one obtains

$$M_{1,C} = \begin{pmatrix} 0 & i\ \exp(-in\pi/2) \\ i\ \exp(in\pi/2) & 0 \end{pmatrix} \quad (1.176)$$

$$M_{3,C} = \begin{pmatrix} 0 & i\ \exp(in\pi/2) \\ i\ \exp(-in\pi/2) & 0 \end{pmatrix} \quad (1.177)$$

which in the case $n = 1$ reduce to

$$M_{1,C} = \begin{pmatrix} 0 & 1 \\ -1 & 0 \end{pmatrix} \text{ and } M_{3,C} = \begin{pmatrix} 0 & -1 \\ 1 & 0 \end{pmatrix} \quad (1.178)$$

Finally, for combination $D(\Delta = 1)$ and $n = 1$ the matrices become

$$M_{1,D} = \begin{pmatrix} -1 & 0 \\ 0 & -1 \end{pmatrix} \text{ and } M_{3,D} = \begin{pmatrix} 1 & 0 \\ 0 & 1 \end{pmatrix} \quad (1.179)$$

We now compare the transmitted and scattered amplitudes for two foils, containing a dislocation at two different levels differing by $t_g(\Delta = 1)$ and by $1/2\ t_g(\Delta = 1/2)$ for the most common case $n = 1$.

For $\Delta = 0$ (combination A) we have

$$\begin{pmatrix} T_A \\ S_A \end{pmatrix} = \begin{pmatrix} -1 & 0 \\ 0 & -1 \end{pmatrix} \begin{pmatrix} T_2 & -S_2^{(-)} \\ S_2 & T_2^{(-)} \end{pmatrix} \begin{pmatrix} 1 & 0 \\ 0 & 1 \end{pmatrix} \begin{pmatrix} 1 \\ 0 \end{pmatrix} \qquad (1.180)$$

and after multiplication of the matrices

$$\begin{pmatrix} T_A \\ T_B \end{pmatrix} = -\begin{pmatrix} T_2 \\ S_2 \end{pmatrix} \qquad (1.181)$$

For $\Delta = 1$ (combination D) one finds similarly

$$\begin{pmatrix} T_D \\ S_D \end{pmatrix} = \begin{pmatrix} 1 & 0 \\ 0 & 1 \end{pmatrix} \begin{pmatrix} T_2 & -S_2^{(-)} \\ S_2 & T_2^{(-)} \end{pmatrix} \begin{pmatrix} -1 & 0 \\ 0 & -1 \end{pmatrix} \begin{pmatrix} 1 \\ 0 \end{pmatrix} = \begin{pmatrix} T_2 \\ S_2 \end{pmatrix} \qquad (1.182)$$

and hence

$$I_{T,A}(x) = I_{T,D}(x) \text{ and } I_{S,A}(x) = I_{S,D}(x) \qquad (1.183)$$

i.e. the image profiles for dislocations differing by one extinction distance in level are the same in the bright field as well as in the dark field image.

For $\Delta = \frac{1}{2}$ (combination C) one finds:

$$\begin{pmatrix} T_c \\ S_c \end{pmatrix} = \begin{pmatrix} 0 & -1 \\ 1 & 0 \end{pmatrix} \begin{pmatrix} T_2 & -S_2^{(-)} \\ S_2 & T_2^{(-)} \end{pmatrix} \begin{pmatrix} 0 & 1 \\ -1 & 0 \end{pmatrix} \begin{pmatrix} 1 \\ 0 \end{pmatrix} = \begin{pmatrix} T_2^{(-)} \\ S_2^{(-)} \end{pmatrix} \qquad (1.184)$$

and hence

$$I_{T,A}(x) = I_{T,C}(-x) \text{ and } I_{S,A}(x) = I_{S,C}(-x) \qquad (1.185)$$

i.e. the image profiles for dislocations situated at levels differing 1/2 t_g are mirror images with respect to $x = 0$ in the bright field image as well as in the dark field image.

1.30.3. Variation of the image profile with thickness

The same formalism can be used to compare the image profiles of dislocations situated at the same distance behind the entrance face in foils differing in thickness. We shall compare the image profiles for configurations E and F. In E the dislocation is at z_2'

behind the entrance face and at z_2'' from the exit face, the total thickness being $z_2 = z_2' + z_2''$. In F the dislocation is still at z_2' behind the entrance face but the total thickness is now $z_2 + 1/2\ t_g$. The response matrix for E is M_2 given by (1.171) and the response matrix of the additional lamella with thickness $1/2\ t_g$ is $M_{3,C}$ (with n = 1). Multiplying these matrices leads to

$$\begin{pmatrix} T_E \\ S_E \end{pmatrix} = \begin{pmatrix} T_2 \\ S_2 \end{pmatrix} \text{ and } \begin{pmatrix} T_F \\ S_F \end{pmatrix} = \begin{pmatrix} -S_2 \\ +T_2 \end{pmatrix} \tag{1.186}$$

from which we can now conclude that

$$I_{T,E}(x) = I_{S,F}(x) \text{ and } I_{S,E}(x) = I_{T,F}(x) \tag{1.187}$$

The bright field image profile of the dislocation in one configuration is thus the same as the dark field image profile in the other configuration, where the foil is $1/2\ t_{\bar{g}}$ thicker, the dislocation being at the same distance from the entrance face.

Using a similar reasoning it was shown [36] that the bright field image profiles of dislocations situated in planes located symmetrically with respect to the foil's central plane are identical, whereas the dark field images of such dislocations are very closely mirror images with respect to x = 0. This is a general property of defect images; it is a direct consequence of anomalous absorption. In particular, stacking fault fringes exhibit this property (see § 1.24.2).

In sufficiently thick foils the bright and dark field images are similar for a dislocation situated close to the entrance face but (pseudo)-complementary (mirror symmetry related) for a dislocation close to the exit face. This can be shown analytically as follows. For a dislocation close to the entrance face we use a model whereby $z_1 = 0$ and z_3 sufficiently large so as to allow the following approximations:

$$T_3 = \cos(\pi\sigma z_3) \approx (½)\exp(-i\pi z_3 / t_g)\exp(\pi z_3 / \tau_g) \tag{1.188}$$

$$S_3 = i\ \sin(\pi\sigma z_3) \approx -(½)\exp(-i\pi z_3 / t_g)\exp(\pi z_3 / \tau_g) \tag{1.189}$$

since $\sigma = 1/t_g + i/\tau_g$.

The matrix M_3 is thus (for n= 1)

$$M_3 = A(z_3)\begin{pmatrix} 1 & -i \\ i & 1 \end{pmatrix} \text{ with } A(z) = (1/2)\exp(-i\pi z/t_g)\exp(\pi z/\tau_g) \quad (1.190)$$

The resulting amplitudes T_L and S_L are then given by

$$\begin{pmatrix} T_L \\ S_L \end{pmatrix} = A(z_3)\begin{pmatrix} 1 & -i \\ i & 1 \end{pmatrix}\begin{pmatrix} T_2 & -S_2^{(-)} \\ S_2 & T_2^{(-)} \end{pmatrix}\begin{pmatrix} 1 \\ 0 \end{pmatrix}$$
$$= A(z_3)\begin{pmatrix} T_2 - iS_2 \\ i(T_2 - iS_2) \end{pmatrix} \quad (1.191)$$

and hence

$$I_{T,L}(x) = I_{S,L}(x) \quad (1.192)$$

i.e. the image profiles are the same in dark field and bright field.

For a dislocation close to the exit face in a sufficiently thick foil one can use as a model: z_1 large and $z_3 = 0$. We then have

$$\begin{pmatrix} T_M \\ S_M \end{pmatrix} = A(z_1)\begin{pmatrix} T_2 & -S_2^{(-)} \\ S_2 & T_2^{(-)} \end{pmatrix}\begin{pmatrix} 1 & i \\ -i & 1 \end{pmatrix}\begin{pmatrix} 1 \\ 0 \end{pmatrix}$$
$$= A(z_1)\begin{pmatrix} T_2 + iS_2^{(-)} \\ i\left(T_2^{(-)} + iS_2\right) \end{pmatrix} \quad (1.193)$$

It is clear that if one changes x to -x in the lower element of the column vector one obtains i times the upper element, i.e. $i\left(T_2 + iS_2^{(-)}\right)$. We conclude that

$$I_{T,M}(x) = I_{S,M}(-x) \quad (1.194)$$

i.e. the image profiles in bright field and dark field are mirror images with respect to n = 0 for a dislocation close to the exit face.

1.31. General Symmetry Properties of Defect Images

1.31.1. Derivation of the bright field symmetry relations [10,G1]

It is well known that the diffraction patterns produced by gratings related by a parallel translation, are the same under the same diffraction conditions. Applied to defect contrast images this means that displacement fields $\mathbf{R}_1(z)$ and $\mathbf{R}_2(z)$ when related by a constant translation $\mathbf{R}_0$ produce the same diffraction images under the same diffraction conditions. This relation can be applied to separate columns in the same foil: columns in which the displacement functions $\mathbf{R}_1(z)$ and $\mathbf{R}_2(z)$ are related by

$$\mathbf{R}_1(z) = \mathbf{R}_2(z) + \mathbf{R}_0 \tag{1.195}$$

produce the same intensity in bright and dark field images.

Less trivial symmetry properties of images can be derived by comparing the transmitted and scattered amplitudes (or intensities) for columns related by symmetry relations of the displacement field. These amplitudes can be obtained by two alternative procedures which both involve the column approximation. The first method, already described in § 1.21, consists in multiplying the response matrices of thin slices within which $s_{eff} = s_g + \mathbf{g} \cdot (d\mathbf{R}/dz)$. On the other hand one can alternatively multiply the response matrices of slices dz of perfect crystal with $s \equiv s_g$, alternating with shift matrices with a phase angle $\alpha(z) = 2\pi\mathbf{g}.\mathbf{R}(z)$ depending on z, which take care of the relative displacements of successive slices according to the displacement field.

The response matrix $M(s_g, dz)$ for a perfect thin slice dz can be obtained from the system of differential equations (1.56) by multiplying both sides with dz, followed by adding T (respectively S) to both the left and the right hand side of these equations. This leads to

$$T + dT = \left(1 - \pi i s_g dz\right)T + \left(\pi i / t_{-g}\right)S\, dz \tag{1.196a}$$

$$S + dS = \left(\pi i / t_g\right)T\, dz + \left(1 + \pi i s_g dz\right)S \tag{1.196b}$$

or in matrix form

$$\begin{pmatrix} T + dT \\ S + dS \end{pmatrix}_{(n+1)dz} = \begin{pmatrix} 1-\pi i s_g dz & (\pi i / t_{-g})dz \\ (\pi i / t_g)dz & 1+\pi i s_g dz \end{pmatrix} \begin{pmatrix} T \\ S \end{pmatrix}_{n\,dz} \tag{1.197}$$

The elements of the thin slice response matrix can of course also be obtained by approximating the elements of the response matrix $M(s_g,z)$ (§ 1.20) retaining only the first order terms in dz of the Taylor expansions.

For a slice which is shifted over $\alpha(n\,dz) \equiv 2\pi \mathbf{g}.\mathbf{R}(n\,dz)$ with respect to the entrance slice at z = 0 the response matrix is constructed by sandwiching the response matrix of the perfect slice between two shift matrices (see § 1.21)

$$\left(\begin{pmatrix} 1 & 0 \\ 0 & \exp[-i\alpha(n\,dz)] \end{pmatrix} \begin{pmatrix} 1-\pi i s_g dz & (\pi i / t_{-g})dz \\ (\pi i / t_g)dz & 1+\pi i s_g dz \end{pmatrix} \begin{pmatrix} 1 & 0 \\ 0 & \exp[i\alpha(n\,dz)] \end{pmatrix} \right) \tag{1.198}$$

Anomalous absorption is further taken into account in the usual phenomenological manner by the substitution $1/t_g \rightarrow 1/t_g + i/\tau_g$. We thus finally obtain for the response matrix of a thin slice at n dz behind the entrance face

$$M_n \equiv \begin{pmatrix} 1 & 0 \\ 0 & '\exp[-i\alpha(n\,dz)] \end{pmatrix} \begin{pmatrix} 1-\pi i s_g dz & \pi i\,(1/t_{-g} + i/\tau_{-g})dz \\ \pi i\,(1/t_g + i/\tau_g)dz & 1+\pi i s_g dz \end{pmatrix}$$
$$\begin{pmatrix} 1 & 0 \\ 0 & \exp[i\alpha(n\,dz)] \end{pmatrix} \tag{1.199}$$

or using a shorter notation:

$$M_n = \$_{-n} M(s_g, dz)\ \$_n \tag{1.200}$$

Alternatively, starting directly from the system of equations (1.104) for a deformed crystal, and following the same reasoning as above, the response matrix for a thin shifted slice can be written as

$$M'_n = \begin{pmatrix} 1-\pi i\left[s_g^- + (d\alpha'/dz)_{n\,dz}\right]dz & \pi i\left[(1/t_{-g}) + (+i/\tau_{-g})\right]dz \\ \pi i\left[(1/t_g)+\right](i/\tau_g)dz & 1+\pi i\left[s_g + (d\alpha'/dz)\right]_{n\,dz} dz \end{pmatrix} \tag{1.201}$$

with

$$\alpha' = \mathbf{g}.\mathbf{R}(z) = \alpha / 2\pi \tag{1.202}$$

The amplitudes emerging from a column can thus be formulated in two different ways

$$\begin{pmatrix} T \\ S \end{pmatrix} = M'_N \ldots M'_2 M'_1 \begin{pmatrix} 1 \\ 0 \end{pmatrix} \tag{1.203}$$

where N is the number of slices or

$$\begin{pmatrix} T \\ S \end{pmatrix} = \$_{-N} M_N \$_N \ldots \$_{-2} M_2 \$_2 \$_{-1} M_1 \$_1 \begin{pmatrix} 1 \\ 0 \end{pmatrix} \tag{1.204}$$

The two approaches can be shown to be equivalent when neglecting terms in $(dz)^2$ and noting that

$$\$\{\alpha[n\ dz]\}\ \$\{-\alpha[(n-1)dz]\} = \$\left\{(d\alpha / dz)_{n\ dz} dz\right\} \tag{1.205}$$

From the expression for $M(s_g, dz)$ [Eq. (1.199)] it is evident that in a centro-symmetric crystal

$$\tilde{M}(s_g, dz) = M(s_g, dz) \tag{1.206}$$

since $t_g = t_{-g}$ and $\tau_g = \tau_{-g}$ The shift matrices have the following properties

$$\tilde{\$}_n = \$_n,\quad \$_n^{-1} = \$_{-n},\quad \$_n^{*} = \$_{-n} \tag{1.207}$$

The shift matrices have further the property that changing **R**(z) into -**R(z)** changes α into -α and hence $n \to -n$ i.e. the following relations are satisfied

$$^{(+)}\$_n = {}^{(-)}\$_{-n} \text{ and } {}^{(+)}\$_{-n} = {}^{(-)}\$ \tag{1.208}$$

where the + and - signs refer to translations **R**(z) and -**R(z)** respectively.

We now consider two deformed crystals of which the structures are related to each other by an inversion through a center. We shall call them (+) and (-) crystals and the matrices referring to these two crystals will be indicated by a left superscript (+) or (-). A translation **R** in the plus crystal becomes through inversion a translation in the opposite sense -**R**, in the minus crystal. Inversion of the structure does not affect the diffraction phenomena according to Friedel's law; moreover if we assume the perfect crystal to be centro-symmetrical the inversion does not change the crystal structure. A point at level z in a column of the plus crystal foil of thickness t is transformed by inversion into a point at level t-z in the corresponding column of the minus crystal foil. The inversion clearly also changes the sense of the displacements in the strain pattern, i.e. **R**(z) in the plus crystal becomes -**R**(z) in the minus crystal. We shall show that two columns along which the displacement fields are respectively $\mathbf{R}_1(z)=\mathbf{R}(z)$ and $\mathbf{R}_2(z)=\mathbf{R}_0-\mathbf{R}(t-z)$, where z is measured from the entrance face in each crystal produce the same transmitted amplitude, i.e. the same bright field image in the two crystals, if the diffraction conditions are the same.

We note that

$$^{(+)}M(s_g,dz) = {}^{(-)}M(s_g,dz) \equiv M \tag{1.209}$$

as a result of Friedel's law. We now consider the properties of the matrices P_n which represent the response of a displaced slice:

$$P_n = \$_{-n} M \,\$_n \tag{1.210}$$

We shall show that

$$^{(+)}\tilde{P}_n = {}^{(-)}P_n \tag{1.211}$$

One has

$$^{(+)}P_n = {}^{(+)}\$_{-n}{}^{(+)}M^{(+)}\$_n \tag{1.212}$$

and hence

$$^{(+)}\tilde{P}_n = {}^{(+)}\tilde{\$}_n{}^{(+)}\tilde{M}^{(+)}\tilde{\$}_{-n} \tag{1.213}$$

since on transposing the order of the multiplications is reversed. We note that the shift matrices do not change on transposing and that $^{(\pm)}\tilde{M} = {}^{(\pm)}M$ and hence taking (1.207) and (1.208) into account:

$$^{(+)}\tilde{P}_n = {}^{(+)}\$_n{}^{(+)}M^{(+)}\$_{-n} = {}^{(-)}\$_{-n}{}^{(-)}M^{(-)}\$_n \equiv {}^{(-)}P_n \tag{1.214}$$

The response matrix of the column in the (+) crystal is given by

$$^{(+)}\wp = {}^{(+)}P_N \ldots {}^{(+)}P_n \ldots {}^{(+)}P_2 {}^{(+)}P_1 \qquad (1.215)$$

and in the (-) crystal by

$$^{(-)}\wp = {}^{(-)}P_1 {}^{(-)}P_2 \ldots {}^{(-)}P_n \ldots {}^{(-)}P_N \qquad (1.216)$$

since the electrons pass through the successive slices in the minus crystal in the reverse order.

We shall show that

$$^{(+)}\tilde{\wp} = {}^{(-)}\wp \qquad (1.217)$$

One has

$$^{(+)}\tilde{\wp} = \left({}^{(+)}P_N \ldots {}^{(+)}P_2 {}^{(+)}P_1\right)^{\sim} \\ = {}^{(+)}\tilde{P}_1 {}^{(+)}\tilde{P}_2 \ldots {}^{(+)}\tilde{P}_N \qquad (1.218)$$

and using (1.214)

$$^{(+)}\tilde{\wp} = {}^{(-)}P_1 {}^{(-)}P_2 \ldots {}^{(-)}P_N = {}^{(-)}\wp \qquad (1.219)$$

The relation (1.219) means that the diagonal elements and in particular $\wp_{11}$ is the same for the two matrices $^{(+)}\wp$ and $^{(-)}\wp$. Since $\wp_{11}$ determines T we can conclude that the transmitted intensity I_T will be the same for both columns. This is generally not the case for I_S because the off-diagonal elements which determine S are not necessarily equal by symmetry in an absorbing crystal.

1.31.2. Applications of the bright field symmetry relations

The use of the condition

$$\mathbf{R}_1(z) = \mathbf{R}_0 - \mathbf{R}_2(t-z) \qquad (1.220)$$

under which the bright field images are the same for $\mathbf{R}_1(z)$ and $\mathbf{R}_2(z)$, will now be illustrated with two simple examples.

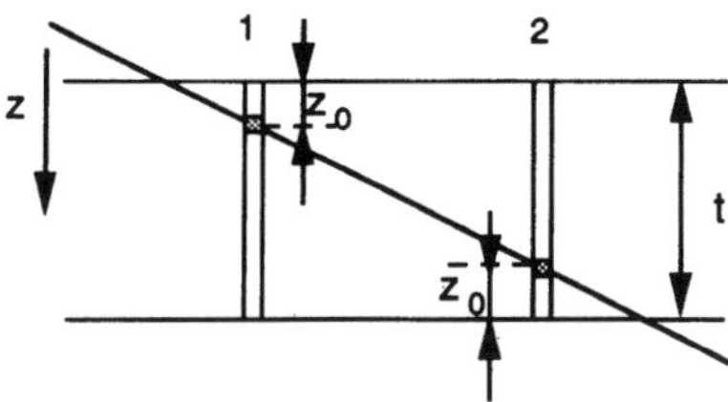

Figure 1.71 Schematic of inclined fault plane intersected by two integration columns 1 and 2; illustrating the derivation of the bright field symmetry properties of the image.

The simplest possible case is that of an inclined translation fault (Figure 1.71) with displacement vector $\mathbf{R}_0$. For column (1), i.e. for $\mathbf{R}_1$ we have $\mathbf{R}_1 = 0$ for $z < z_0$; $\mathbf{R}_1 = \mathbf{R}_0$ for $z \geq z_0$. For column (2), i.e. for $\mathbf{R}_2$ we have $\mathbf{R}_2 = 0$ for $z < t - z_0$; $\mathbf{R}_2 = \mathbf{R}_0$ for $z \geq t - z_0$. We now show that (1.220) is satisfied. If $z < z_0$ then $t - z > t - z_0$ and hence we have $\mathbf{R}_1(z) = 0$ and $\mathbf{R}_2(t - z) = \mathbf{R}_0$ and the relation (1.220) is satisfied. If on the other hand $z > z_0$ then $t - z < t - z_0$ and we have now $\mathbf{R}_1 = \mathbf{R}_0$ and $\mathbf{R}_2(t - z) = 0$ and the relation (1.220) is again satisfied. We can thus conclude that the bright field image of a translation interface will always be symmetrical with respect to the central line of the fringe pattern.

Consider now a screw dislocation with Burgers vector **b** at level z_0 behind the entrance face (Figure 1.72), then

$$\mathbf{R}_1(z) = (\mathbf{b}/2\pi)\ \mathrm{arctg}\left[(z - z_0)/x_0\right] \qquad (1.221)$$

along the column (1) at x_0. We shall now compare this displacement field with that of a screw dislocation at $t - z_0$ behind the entrance face.

$$\mathbf{R}_2(z) = (\mathbf{b}/2\pi)\ \mathrm{arctg}\left\{\left[z - (t - z_0)\right]/x_0\right\} \qquad (1.222)$$

along the columns also at x_0. We further have

$$\begin{aligned} \mathbf{R}_2(t - z) &= (\mathbf{b}/2\pi)\ \mathrm{arctg}\left\{\left[(t - z) - (t - z_0)\right]/x_0\right\} \\ &= -(\mathbf{b}/2\pi)\ \mathrm{arctg}\left[(z - z_0)/x_0\right] = -R_1(z) \end{aligned} \qquad (1.223)$$

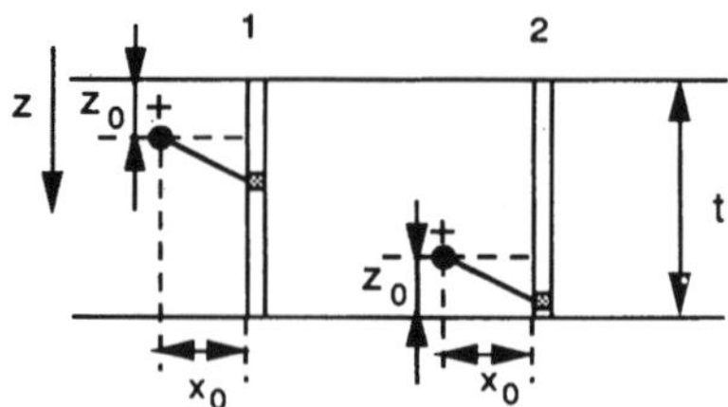

Figure 1.72 Schematic of two screw dislocations of the same sin at two different depths behind the entrance face of the foil, i.e. at z_0 and at t-z_0.

and the condition (1.220) is thus satisfied with $\mathbf{R}_0 = 0$.

We can also compare with a column at $-x_0$, associated with a dislocation of the opposite sign. We then have

$$\mathbf{R}_2(z) = -(\mathbf{b}/2\pi)\ \text{arctg}\ \{[z-(t-z_0)]/(x_0)\} \tag{1.224}$$

and hence

$$\begin{aligned} \mathbf{R}_2(t-z) &= -(\mathbf{b}/2\pi)\ \text{arctg}\ \{[(t-z)-(t-z_0)]/(-x_0)\} \\ &= (\mathbf{b}/2\pi)\ \text{arctg}\ [(z_0-z)/x_0] = -\mathbf{R}_1(z) \end{aligned} \tag{1.225}$$

and relation (1.220) is again satisfied.

The interpretation of these results leads to the statement that the bright field image profile of a screw situated at z_0 behind the entrance face will be the same as that of a similar screw dislocation but situated at a level $t-z_0$. The image of an inclined screw is thus symmetrical with respect to the projection of the central line of the foil. The images of two screws of opposite sign, one situated at z_0 behind the entrance face and the other at a depth $t-z_0$, will be mirror images. Similar conclusions can be drawn for edge dislocations.

1.31.3. Dark field symmetry relations

A similar condition applies to dark field images. It was pointed out by Ball [37] that for $s = 0$ identical dark field images will be obtained for columns in crystals along which the displacement functions $\mathbf{R}_1$ and $\mathbf{R}_2$ are related as follows

$$\mathbf{R}_1(z) = \mathbf{R}_2(t-z) + \mathbf{R}_0 \quad (1.226)$$

(t = foil thickness; $\mathbf{R}_0$ arbitrary translation). A concrete example is provided by the symmetrically oriented domain boundary ($s_1 = -s_2$) represented schematically in Figure 1.74. It is assumed that for the perfect crystal $s = 0$ and that the two structures on the two sides of the boundary have been homogeneously sheared over a small angle. The displacement fields are then given by

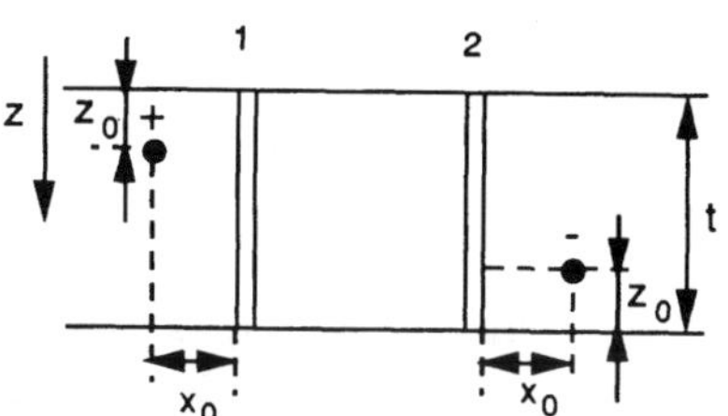

Figure 1.73 Schematic of two screw dislocations of opposite sign at two different depths behind the entrance face of the foil, i.e. the + screw at z_0 and the - screw at t-z_0.

$$\begin{cases} \mathbf{R}_1(z) = k(z - z_0)\mathbf{e}_x & z > z_0 \quad (k = \text{constant}) \\ \mathbf{R}_1(z) = k(z_0 - z)\mathbf{e}_x & z \le z_0 \end{cases} \quad (1.227)$$

and

$$\begin{cases} \mathbf{R}_2(z) = k(z - z_0)\mathbf{e}_x & z > z_0 \\ \mathbf{R}_2(z) = k(z_0 - z)\mathbf{e}_x & z \le z_0 \end{cases} \quad (1.228)$$

If for instance $z < z_0$ we have $t - z > t - z_0$ and we then obtain

$$\mathbf{R}_2(t-z) = k(t - z - z_0)\mathbf{e}_x \quad (1.229)$$

and hence from (1.227)

$$\mathbf{R}_1(z) = \mathbf{R}_2(t-z) + k(2z_0 - t)\mathbf{e}_x \quad (1.230)$$

which is of the form (1.226) with $\mathbf{R}_0 = k(2z_0 - t)\mathbf{e}_x$. The fringe pattern is thus symmetrical in the dark field image. It was shown in § 1.24.3 that this is indeed a property of δ-fringes in the symmetrical orientation.

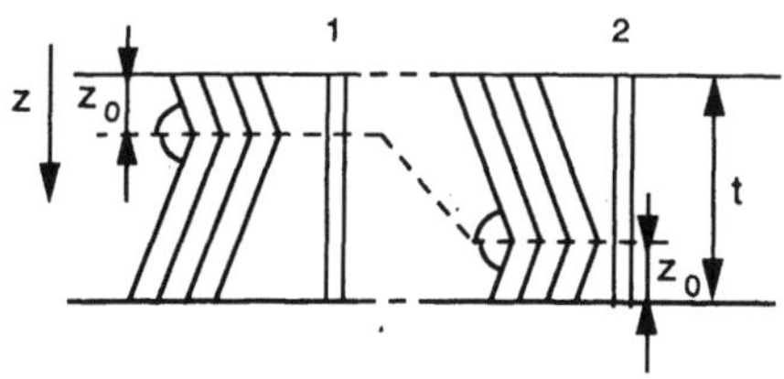

Figure 1.74 Schematic of foil containing an inclined domain boundary in the symmetric orientation (i.e $\mathbf{s}_1 = -\mathbf{s}_2$). The columns 1 and 2 intersect the interface at the levels z_0 and t-z_0 behind the entrance face.

1.31.4. Various applications of the symmetry properties

Dislocation dipoles Dislocation dipoles consist of two parallel dislocations with opposite Burgers vector. If the dislocations are restrained to remain in their respective glide planes they take up a stable configuration which minimizes the elastic energy. For two edge dislocations the regions of expansion in one dislocation tend to overlap the regions of compression in the other one. The configuration is such that the plane formed by the two parallel dislocations encloses an angle ϕ with their slip planes. In the case of two pure edges $\phi = 45°$.

As in the case of dislocation ribbons the image of a dipole is not the superposition of the images of the two separate dislocations. The superposition must be carried out at the level of the strain fields in the framework of the linear elasticity theory. It was shown in [32] that the bright field image of an inclined dipole has a center of symmetry. This symmetry property allows to distinguish the images of a narrow dipole from that of a single dislocation.

Fission tracks [38] [G8] It can be shown quite generally that the bright field image of a fission track modeled by means of equation (1.161) has an inversion center; we hereby make use of the condition (1.220). We shall prove that a column at x_0 for the defect at z_0 produces the same transmitted amplitude as the column at $-x_0$ for the defect at level $t - z_0$.

In the first column at $+x_0$ we have

$$\mathbf{R}_1(z) = \varepsilon\, r_0\left[x\mathbf{e}_x + (z - z_0)\mathbf{e}_z\right] / \left[x^2 + (z - z_0)^2\right] \quad (1.231)$$

For $\mathbf{R}_2(z)$) we similarly have

$$\mathbf{R}_2(z) = \varepsilon\, r_0\left\{x\mathbf{e}_x + \left[z - (t - z_0)\right]\mathbf{e}_z\right\} / \left\{x_2 + \left[z - (t - z_0)\right]^2\right\} \quad (1.232)$$

We now compare a column $\mathbf{R}_1(z)$ at x_0 and a column $\mathbf{R}_2(t - z)$ at $-x_0$. In (1.232) we substitute $z \to t - z$, $x \to -x_0$ and obtain:

$$\begin{aligned} \mathbf{R}_2(t - z) &= \varepsilon\, r_0\left\{-x_0\mathbf{e}_x + \left[(t - z) - (t - z_0)\right]\right\} / \left\{x_0^2 + \left[(t - z) - (t - z_0)\right]^2\right\} \\ &= \varepsilon\, r_0\left\{-x_0\mathbf{e}_x + (-z + z_0)\mathbf{e}_z\right\} / \left[x_0^2 + (z_0 - z)^2\right] = -\mathbf{R}_1(z) \quad (1.233) \end{aligned}$$

The relation (1.220) is thus satisfied with $\mathbf{R}_0 = 0$ and the two columns which were compared produce the same amplitude, which proves the presence of an inversion center.

Finally we compare images made with two different diffraction vectors $\mathbf{g}_1$ and $\mathbf{g}_2$which enclose the same angle with the track line and are mirror images with respect to the plane normal to the track. Two such images are identical because at each level in each column the products $\mathbf{g}_1.\mathbf{R}(z)$ and $\mathbf{g}_2.\mathbf{R}(z)$ are the same since everywhere $\mathbf{R}$ is perpendicular to the track line. The image characteristics of fission tracks or from related line shaped defects allow to make the distinction with images of straight dislocations.

1.32. Various Applications of Diffraction Contrast Images

1.32.1. Overlapping interfaces

The two-beam contrast due to a combination of overlapping planar interfaces situated in parallel planes can be described by the product of the appropriate sequence of matrices associated with the successive slices [G2].

For instance the contrast due to two overlapping stacking faults is simulated by the product of the matrices:

$$\begin{pmatrix} T \\ S \end{pmatrix} = M_3(z_3,s)\ \$(\alpha_2)\ M_2(z_2,s)\ \$(\alpha_1)\ M_1(z_1,s)\begin{pmatrix} 1 \\ 0 \end{pmatrix} \qquad (1.234)$$

of which the meaning is self-evident. If the fault separation is small, i.e. for $z_2 << t_g$, one has to a good approximation

$$\$(\alpha_2)\ M_2(z_2,s)\ \$(\alpha_1) = \$(\alpha_1+\alpha_2)$$

Two overlapping intrinsic faults in closely spaced planes may thus simulate an extrinsic fault.

The two-beam diffraction image due to a cavity is given by

$$\begin{pmatrix} T \\ S \end{pmatrix} = M_3(z_3,s)\ V(z_2,s)\ M_1(z_1,s)\begin{pmatrix} 1 \\ 0 \end{pmatrix} \qquad (1.235)$$

The thickness of the cavity is z_2. The s-value to be used in the vacuum matrix is that characteristic of the preceding part M_1. In the present case normal absorption contrast has to be taken into account since in the area of the cavity the mass-thickness is smaller. The diffraction contrast is usually predominating, especially for small cavities, which can appear either brighter or darker than the perfect foil.

Non-reflecting crystal parts have to be simulated by means of a vacuum matrix as well. For example the contrast due to a microtwin in a face centered cubic crystal is described by the following matrix product

$$\begin{pmatrix} T \\ Se^{i\alpha} \end{pmatrix} = M_3(z_3,s)\ V(z_2,s)\ \$(\alpha)\ M_1(z_1,s)\begin{pmatrix} 1 \\ 0 \end{pmatrix} \qquad (1.236)$$

The vacuum matrix is associated with the non-reflecting twin part with thickness z_2. The shift matrix takes care of the phase shift due to the presence of the microtwin lamella between front and rear lamella with respectively thicknesses z_1 and z_3. The phase angle α may be $\alpha = 0$ or $\alpha = \pm 2\pi/3$ depending on the exact number of atomic layers in the microtwin. The matrices V and \$ commute since both are diagonal; it is physically evident since it is immaterial whether we assume the phase shift to take place before or after the non-reflecting microtwin lamella. If the microtwin is thin it will behave, as far as the contrast is concerned, as a stacking fault.

More examples of the use of response matrices can be found in Ref. [G2].

1.32.2 Determination of the type of stacking fault

Close packed layers of atoms can be stacked in an infinite number of ways which all have nearly the same free energy. Two essentially different types of stacking modes are usually distinguished: the face centered cubic stacking mode with stacking symbol ... ABCABC ... and the hexagonally close-packed mode with stacking symbol ... ABAB ...or ... ACAC ... or ...BCBC ... .

In the face-centered stacking two different types of stacking faults are often distinguished.

If a single atomic layer is extracted and the gap so created is closed by a displacement over a vector $\mathbf{R}_0 = 1/3\,[1,11]$ the resulting sequence is

...BCA↓CABCABC...

This is called an <u>intrinsic stacking fault</u>: it is formed for instance by the precipitation of a layer of vacancies, but it is also generated in the wake of a glissile Shockley partial with Burgers vector 1/6 $[11\bar{2}]$ on a (111) glide plane.

If a single atomic layer is inserted, when for instance a layer of interstitials precipitates, the resulting sequence is

↓
...ABCBABCABC...

This is called an <u>extrinsic stacking fault</u>. The displacement vector is now $\mathbf{R}_0$ = -1/3 [111], i.e. the opposite of the previous one. A single glide dislocation cannot generate such a fault. In both faults two triplets in the hexagonal configuration occur, in a different configuration however.

For the detailed interpretation of partial dislocation - fault configurations in face centered cubic metals, the distinction between intrinsic and extrinsic stacking faults is important. It was shown that this information can be obtained from the nature of the edge fringes in stacking fault images [39]. In particular in Ref. [40] it was demonstrated how this information can be obtained from a single dark field image made in a well defined reflection.

One can distinguish three classes of reflections in the face centered cubic structure, depending on whether or not h+k+l = threefold, threefold + 1 or threefold - 1. The reflections for which h +

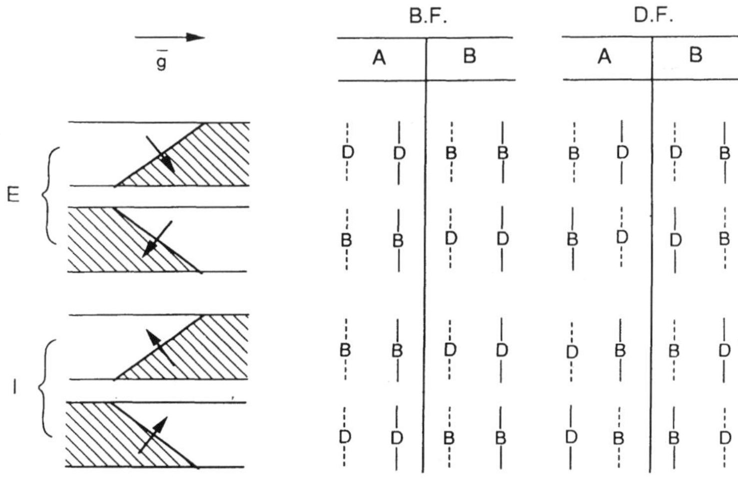

Table II Determination of the type of stacking fault in the FCC structure.

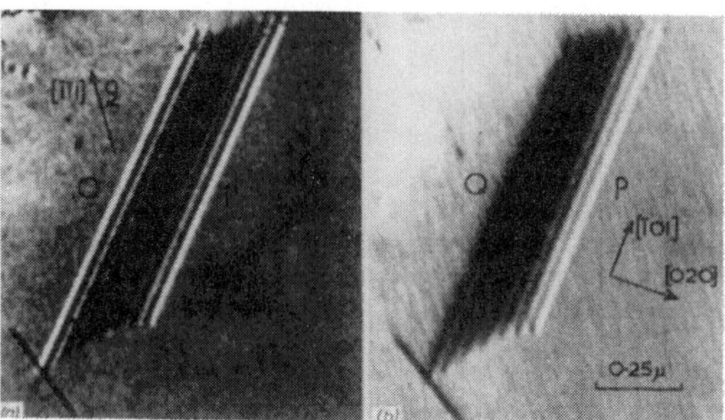

Figure 1.75 Bright and dark field image pair of a stacking fault in a small stacking fault energy alloy. From Table II it can be concluded that the fault is intrinsic.

k+l = threefold lead to $\alpha = k.2\pi$ and therefore do not give rise to a fringe pattern. Reflections such as {200}, {2$\bar{2}$0} and {440} for which h + k+ l = threefold - 1, will be called type A, whereas reflections such as {1$\bar{1}$1}, {220} and {400} for which h + k + l = threefold + 1 will be called type B reflections.

The edge fringes in bright and dark fields for all possible combinations of the type of active reflections (A or B), the sense of inclination of the fault planes and the nature of the fault (E or I) are represented schematically in Table II where the diffraction vector is assumed to point to the right. One notes that for a given type of fault and a given type of vector the nature of the edge fringes in the dark field image is independent of the sense of inclination of the fault plane. A simple rule can thus be formulated: if in the dark field image the *g*-vector, its origin being put in the center of the fringe pattern, points towards a bright fringe and the operating reflection **g** is of type A the fault is intrinsic. If one of the parameters, either the nature of the edge fringes or the class of the operating reflection, changes, also the conclusion changes. The nature of the edge fringes in the bright field image further allows to determine the sense of inclination of the fault plane. In applying the present method one must be aware of the fact that the nature of the edge fringes is only well defined in sufficiently thick foils, where anomalous absorption is important.

In Figure 1.75 a bright and dark field image pair is shown which allows to conclude that the fault which is being imaged is intrinsic.

1.33. Moiré Patterns

1.33.1. Intuitive considerations

Electron microscopic specimens consisting of two similar thin films superposed with a small orientation difference produce interference patterns consisting of parallel fringes when a **g**-vector parallel to the film plane is active in the two components of the sandwich. In the bright field image this fringe pattern results from the interference between the doubly transmitted and the doubly scattered beam, which enclose a small angle. This angle is usually revealed in the diffraction pattern by a doubling of the spots. By doubly transmitted beam is meant the beam transmitted through the first thin film and subsequently through the second; the doubly scattered beam is formed in a similar way.

The geometrical features of such fringes provide useful information in a number of cases. A geometrical analogue, consisting of the superposition of two line patterns, the lines representing lattice planes, one of them containing a dislocation, is shown in Figure 1.76. In (b) the two patterns have the same line spacing, but the directions of the lines enclose a small angle. In (a) the directions of the lines are the same, but their spacing is slightly different. The moire′ pattern or the superposition pattern shows a magnified representation of a dislocation [41]. Moiré patterns can thus provide "geometrical" magnification, which was especially

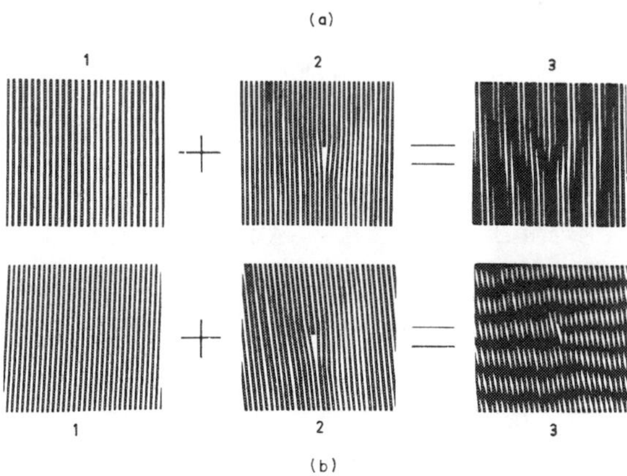

Figure 1.76 Geometrical analogue illustrating the formation of moiré patterns. (a) parallel moiré b) rotation moiré
One of the two superposed foils contains a dislocation.

useful at a time when atomic resolution was not possible. With the development of atomic resolution microscopy, moiré imaging lost some of its importance: the geometrical features are still useful however [42].

1.33.2. Theoretical considerations

Consider a composite crystal consisting of two plan parallel slabs I and II. Let part II of a column be derived from part I by the displacement field $\mathbf{u}(\mathbf{r})$. The phase shift between the waves diffracted locally by the two parts of the crystal is then $2\pi(\mathbf{g}+\mathbf{s}).\mathbf{u} \cong 2\pi\mathbf{g}.\mathbf{u}$. This expression is of the same form as the phase shift introduced by a stacking fault, the main difference being that $\mathbf{u}$ is not a constant vector $\mathbf{R}$ but now depends on $\mathbf{r}$ and hence that α is also a function of $\mathbf{r}$. The transmitted and scattered amplitudes are then given by the set of equations (1.100 and 1.103), in which α enters through the periodic factor exp $i\alpha$. Without solving the system of equations it is clear that the loci of the points of equal intensity (i.e. the fringes) are given by exp $i\alpha$ = constant, i.e. by α = constant + $k2\pi$ (k = integer).

Assuming $\mathbf{r}$ to be a lattice vector $\mathbf{g}.\mathbf{r}$ = integer and hence for small difference vectors:

$$\Delta\mathbf{g}.\mathbf{r} + \mathbf{g}.\Delta\mathbf{r} = 0 \quad \text{with} \quad \Delta\mathbf{r} = \mathbf{u}(\mathbf{r}) \tag{1.237}$$

and thus

$$\alpha \equiv 2\pi\mathbf{g}.\mathbf{u} = -2\pi\Delta\mathbf{g}.\mathbf{r} \tag{1.238}$$

Provided $\mathbf{u}(\mathbf{r})$ is such that $\Delta\mathbf{g}$ does not depend on $\mathbf{r}$ which is true for moiré patterns, the lines of equal intensity are given by $\Delta\mathbf{g}.\mathbf{r}$= constant + k (k = integer). This equation represents a set of parallel straight lines perpendicular to $\mathbf{K} = -\Delta\mathbf{g}$ where $\mathbf{K}$ can be considered as the wavevector of the fringe system, with wavelength $\Lambda = 1/K$.

In the case of a rotation moiré $K= 2g \sin(\theta/2) \approx g\theta$ (for small θ); or expressed in terms of the interplanar spacing d_g of the active reflection

$$\Lambda_r = d_g / \theta \tag{1.239}$$

The fringes are parallel to $\mathbf{g}$ for small θ.

For parallel moiré patterns $\Delta\mathbf{g} = \mathbf{g}_2 - \mathbf{g}_1$ with $\mathbf{g}_2$ parallel to $\mathbf{g}_1$ or in terms of interplanar spacings:

$$\Delta \mathbf{g} = (1/d_2) - (1/d_1) = (d_1 - d_2)/d_1 d_2 \tag{1.240}$$

and

$$\Lambda_p = d_1 d_2 / (d_1 - d_2) \tag{1.241}$$

The fringes are again perpendicular to $\Delta\mathbf{g}$, i.e. also perpendicular to $\mathbf{g}_1$ and $\mathbf{g}_2$.

If an orientation difference as well as a spacing difference are present mixed moiré patterns are found. One can always decompose $\Delta\mathbf{g}$ in a component perpendicular and a component parallel to $\mathbf{g}$:

$$\Delta \mathbf{g} = \Delta \mathbf{g}_{\parallel} + \Delta \mathbf{g}_{\perp} \tag{1.242}$$

The fringes being still perpendicular to $\Delta\mathbf{g}$ they enclose an angle β with the direction of $\mathbf{g}$ given by $\mathrm{tg}\ \beta = \Delta \mathbf{g}_{\perp} / \Delta \mathbf{g}_{\parallel}$.

One further has

$$|\Delta \overline{\mathbf{g}}|^2 = |\Delta \overline{\mathbf{g}}_{\perp}|^2 + |\Delta \overline{\mathbf{g}}_{\parallel}|^2 \tag{1.243}$$

and hence:

$$1/\Lambda^2 = 1/\Lambda_{\parallel}^2 + 1/\Lambda_{\perp}^2 \tag{1.244}$$

The intensity variation of the fringe system can be found in a similar way as for stacking faults. For a quantitative theory of the intensity profiles we refer to Ref. 43.

A semi-quantitative theory, which is sufficient for most purposes will now be formulated within the framework of the kinematical theory. The scattered amplitude is due to the interference between two beams

(i) the beam scattered by the first component of the sandwich and transmitted through the second;
(ii) the beam transmitted through the first lamella and scattered by the second.

The resulting scattered amplitude is then:

$$S = S_1 T_2^{(-)} + T_1 S_1 \exp i\alpha \tag{1.245}$$

This expression is formally equivalent to (1.137) and has a similar meaning. The main difference is that in the case of a planar translation interface α is a constant which is no longer the case here since $\alpha = -2\pi\Delta\mathbf{g}.\mathbf{r} = 2\pi\mathbf{K}.\mathbf{r}$.

Depending on the expressions used for S_1, S_2, T_1, and T_2 either a kinematical or a dynamical theory is obtained. Since applications of the quantitative features of moiré patterns seem to have lost some of their interest, we shall limit ourselves to a description in terms of the kinematical theory which contains all the essential features. In this approximation the symbols have the following meaning:

$$T_1 = T_2^{(-)} = 1 \tag{1.246a}$$

$$S_1 = \exp(\pi i s z_1)\sin(\pi s z_1)/\pi s \tag{1.246b}$$

$$S_2 = \exp(\pi i s z_2)\sin(\pi s z_2)/\pi s \tag{1.246c}$$

where we have moreover assumed $s_1 = s_2 = s$ which is a reasonable approximation for the usual geometry used for the observation of moiré patterns i.e. with the incident beam perpendicular to the sandwich.

This leads to

$$I_s = SS^* = I_S^{(1)} + I_S^{(2)} + 2\sqrt{I_S^{(1)}\, I_S^{(2)}}\cos[\pi s(z_1 + z_2) + \alpha] \tag{1.247}$$

where $I_S^{(1)}$ and $I_S^{(2)}$ are the intensities scattered by the separate lamellae.

The explicit expression, which is also obtained as a limiting case of the dynamical theory, can be written as

$$\begin{aligned} I_s = \left[4/(\pi s)^2\right]\sin^2[1/2\ \pi s(z_1 + z_2)]\cos^2[1/2\ \pi s(z_1 - z_2)] \\ -4\sin(\pi s z_1)\sin(\pi s z)\sin^2[\pi(x - x_0/\Lambda)] \end{aligned} \tag{1.248}$$

where $x_0/\Lambda = (z_1 + z_2)(\tfrac{1}{2}s)$

From these expressions we can conclude that the positions of the moiré fringes, as defined by x_0, depend on the total thickness of the sandwich and hence the fringe positions are influenced by surface steps. Furthermore x_0 depends on s, which means that the fringe positions depend on the specimen tilt. The fringe contrast can be deduced from (1.247). Taking as a convenient measure the difference between minimum and maximum the contrast is

$$C = 4\left[I_S^{(1)} I_S^{(2)}\right]^{1/2} \qquad (1.249)$$

The dynamical theory [43] without anomalous absorption leads to an analogous expression

$$C = 4\left[I_S^{(1)} I_S^{(2)} I_T^{(1)} I_T^{(2)}\right]^{1/2} \qquad (1.250)$$

which shows that the contrast is optimum if the four intensities in this formula are all equal, i.e. for

$$I_S^{(1)} = I_S^{(2)} = I_T^{(1)} = I_T^{(2)} = 1/2$$

(reminding that $I_S^{(1)} + I_T^{(1)} = 1$ and $I_T^{(1)} + I_T^{(2)} = 1$.

As we have seen for coherent domain boundaries $\Delta\mathbf{g}$ is perpendicular to the interface. When the interface is perpendicular to the incident beam, which is the usual geometry for moiré patterns, the projection of $\Delta\mathbf{g}$ on the interface thus vanishes and no moiré fringes are formed. The fringe pattern imaging this type of interface has therefore a different origin. The image for an inclined interface consists of the δ-fringes discussed above and which are perpendicular to the projection of $\Delta\mathbf{g}$ on the foil plane (§ 1.24.3) i.e. to the intersection lines of the interface with the foil surfaces.

If $\Delta\mathbf{g}$ has an arbitrary orientation with respect to the contact plane between the two crystal parts, $\Delta\mathbf{g}$ has a perpendicular component as well as a parallel component with respect to the interface and the image can be a complicated mixture of both types of images. The parallel component gives rise to moiré type fringes; the perpendicular component to δ-type fringes.

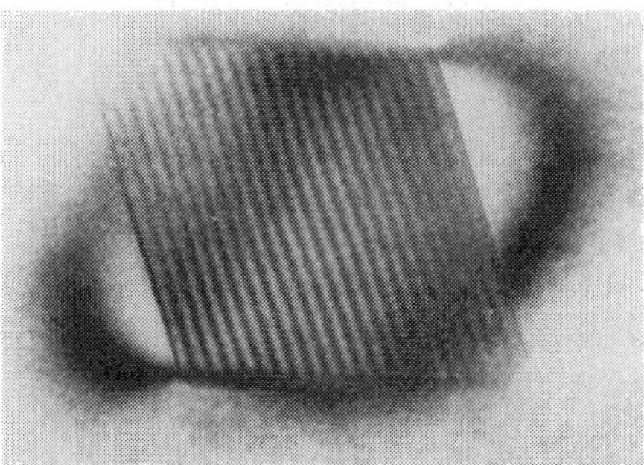

Figure 1.77 Moiré pattern formed at the interface between voidite and the diamond matrix. Note the extinction contours revealing the strain field.

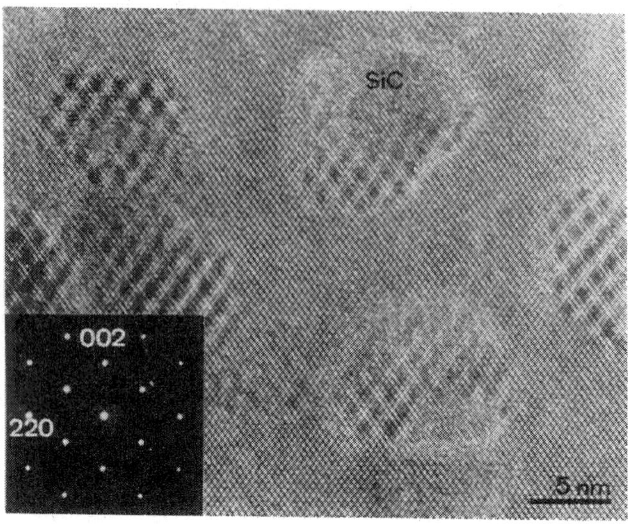

Figure 1.78 Moiré pattern at the interface between silicon and a silicon carbide precipitate.

1.33.3. Applications

An important application of parallel moiré fringes consists in the determination of the lattice parameter of one of the two components in a sandwich, the lattice parameter of the other one being known. This can be of interest for the identification of plate-like coherent precipitates in a matrix with a known lattice parameter. Moire′ fringes formed at the interface between "voidite" and diamond are shown in Figure 1.77 whereas Figure 1.78 shows the moiré fringes at the interface between silicon and SiC precipitate particles.

Moiré fringes have also been used as a tool in the study of dislocations. Ending moiré fringes reveal the emergence points of dislocations in one of the two components of the sandwich. The number N of supplementary half-fringes depends on the reflection used to produce the dislocation image; it is given by N = **g.b.** This number is independent of the character of the dislocation. Supplementary half fringes can therefore not be interpreted as meaning necessarily that the corresponding dislocation has edge character. Partial dislocations bordering stacking faults are revealed by a "fractional" number of supplementary half fringes, i.e. along the trace of the stacking fault the moiré fringes are shifted discontinuously over a fraction **g.b** such as 1/3, 2/3 of the interfringe distance.

The moiré fringes are also shifted by a surface step in one of the components. The fringe shift is not only a function of the step height but also of the deviation parameter and hence of the specimen orientation.

If two or more diffraction vectors are active in both components of the sandwich a crossed grid of moiré fringes is formed, which has the rotation symmetry of the two films.

High Resolution Imaging

1.34. Transfer in the Microscope

1.34.1. Image formation

The main property of a lens is to focus a parallel beam into a point of the back focal plane of the lens. If a lens is placed behind a diffracting object, each parallel diffracted beam is focused into another point of the back focal plane, whose position is given by the reciprocal vector **g** characterizing the diffracted beam. The wavefunction $\psi(\mathbf{R})$ at the exit face of the object can be considered as a planar source of spherical waves (Huyghens principle) (**R** is taken in the plane of the exit face). The amplitude of the diffracted wave in the direction given by the reciprocal vector **g** (or spatial frequency) is given by the Fourier transform of the object function, i.e.

$$\psi(\mathbf{g}) = \underset{\mathbf{g}}{\mathfrak{F}}\ \psi(\mathbf{R}) \tag{1.252}$$

The intensity distribution in the diffraction pattern is given by $|\psi(\mathbf{g})|^2$. The back focal plane visualizes the square of the Fourier transform (i.e. the diffraction pattern) of the object. If the object is periodic, the diffraction pattern will consist of sharp spots. A continuous object will give rise to a continuous diffraction pattern. In the second stage of the imaging process, the back focal plane acts, in its turn, as a set of Huyghens sources of spherical waves which interfere, through a system of lenses, in the image plane. This stage in the imaging process is described by an inverse Fourier transform which reconstructs the object function $\psi(\mathbf{R})$ (usually enlarged) in the image plane (Figure 1.79). The intensity in the image plane is then given by $|\psi(\mathbf{R})|^2$.

In practice, not all the diffracted beams can be allowed to take part in the imaging process. Indeed, the object sees the objective lens under a maximal angle α. In electron microscopy, the outermost beams are strongly influenced by spherical and chromatic aberration and have to be eliminated using an objective aperture. Usually, the aperture is very small (some tens of μm) and limits the diffracted beams to within a very small solid angle (typically 1°).

During the second step in the image formation, which is described by the inverse Fourier transform, the electron beam **g** undergoes a phase shift $\chi(\mathbf{g})$ with respect to the central beam caused by spherical aberration and defocus and damped by incoherent damping functions $D(\alpha,\Delta,g)$ so that the wavefunction $\Phi(\mathbf{R})$ at the image plane is finally given by (for a derivation: see the appendix)

$$\Phi(\mathbf{R}) = \mathcal{F}^{-1}_{R}\, T(\mathbf{g})\, \psi(\mathbf{g})$$

with

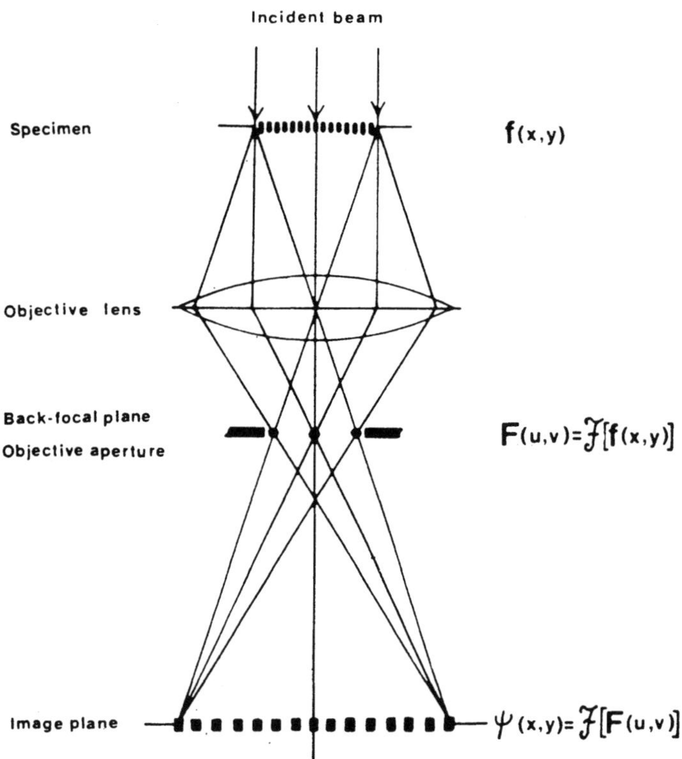

Figure 1.79 Schematic representation of the image formation by the objective lens in a transmission electron microscope. The corresponding mathematical operations are indicated (see text).

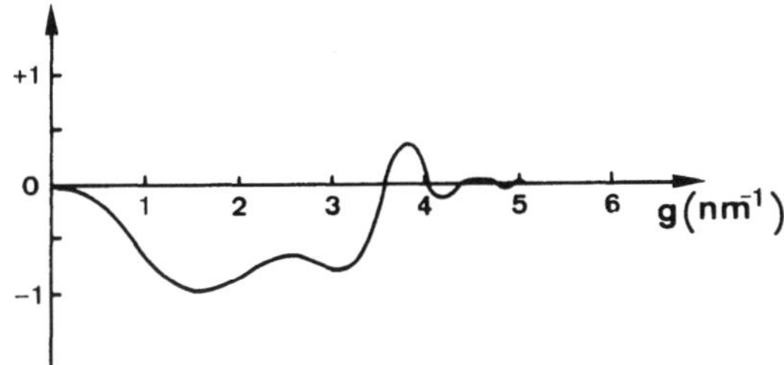

Figure 1.80 Typical transfer function (for 100 KeV microscope) including damping envelope at optimum defocus.

$$T(\mathbf{g}) = A(\mathbf{g}) \exp[-i\chi(\mathbf{g})] D(\alpha, \Delta, \mathbf{g}) \qquad (1.253)$$

the transfer function. The image intensity is then

$$I(\mathbf{R}) = |\phi(\mathbf{R})|^2 \qquad (1.254)$$

(1.253) is called the coherent approximation; it is valid for thin objects. For thicker objects one uses the concept of transmission cross coefficient (TCC). Here, the Fourier components of the image intensity are given by

$$I(\mathbf{g}) = \mathfrak{F}_g(I(\mathbf{R})) = \int \psi(\mathbf{g}+\mathbf{g}') \, \tau(\mathbf{g}+\mathbf{g}',\mathbf{g}') \, \psi^*(\mathbf{g}')d\mathbf{g}' \qquad (1.255)$$

with τ the transmission cross coefficient which describes how the beams $\mathbf{g}'$ and $\mathbf{g}+\mathbf{g}'$ are coupled in the Fourier component $I(\mathbf{g})$. For a derivation of (1.255) see the appendix.

1.34.2. Impulse response function

Calling $t(\mathbf{R})$ the Fourier transform of the transfer function, (1.253) can be rewritten as a convolution product

$$\Phi(\mathbf{R}) = \psi(\mathbf{R}) * t(\mathbf{R}) \qquad (1.256)$$

For an hypothetical ideal pointlike object, $\psi(\mathbf{R})$ would be a delta function so that $\Phi(\mathbf{R}) = t(\mathbf{R})$, i.e. the microscope would reveal $t(\mathbf{R})$ which therefore is called impulse response function. If the transfer function would be constant (i.e. perfectly flat) up to $g = \infty$, the impulse response function would be a delta function so that $\Phi(\mathbf{R}) = \psi(\mathbf{R})$, i.e. the wavefunction in the image plane represents exactly the wavefunction of the object. In a sense the image is perfect. However,

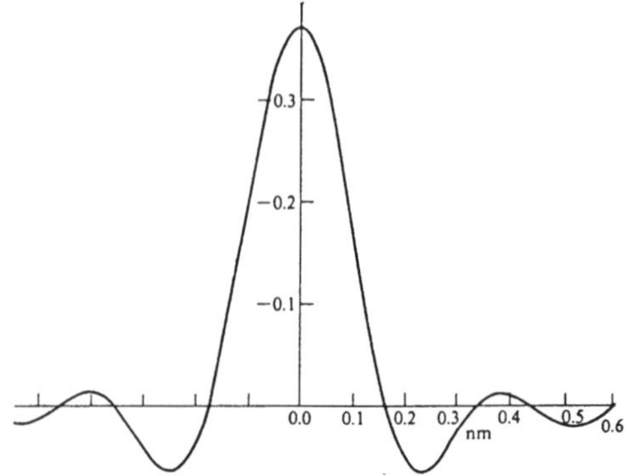

Figure 1.81 Impulse response function.

in practice the transfer function cannot be made arbitrarily flat as is shown in Figure 1.80. The impulse response function is still peaked as shown in Figure 1.81.

Hence, as follows from (1.256), the object wavefunction $\Phi(\mathbf{R})$ is then smeared out (blurred) over the width of the peak. This width can then be considered as a measure for the resolution in the sense as originally defined by Rayleigh. The width of this peak is the inverse of the width of the constant plateau of the transfer function in Figure 1.80. In fact the constant phase of the spatial frequencies g assures that this information is transferred forward, i.e. retains a local relation to the structure.

All information beyond this plateau is still contributing to the image but with a wrong phase. It is scattered outside the peak of the impulse response function and it is thus redistributed over a larger area in the image plane.

1.34.3. Optimum focus

The optimal imaging can be achieved by making the transfer function as constant as possible. From (A2) it is clear that oscillations occur due to spherical aberration and defocus. However, the effect of spherical aberration which, in a sense, makes the objective lens too strong for the most inclined beams, can be compensated somewhat by slightly underfocussing the lens. The optimum defocus value (also called Scherzer defocus) for which the plateau width is maximal, is given by

$$\varepsilon = -1.2(\lambda C_s)^{1/2} = -1.2 \text{ Sch} \tag{1.257}$$

with 1 Sch $= (\lambda C_s)^{1/2}$ the Scherzer unit.

The transfer function for this situation is depicted in Figure 1.80. The phase shift $\chi(g)$ is nearly equal to $-\pi/2$ for a large range of spatial coordinates g. The Scherzer plateau extends nearly to the first zero, given by

$$g \approx 1.5 C_s^{-1/4} \lambda^{-3/4} \tag{1.258}$$

This result was first obtained by Otto Scherzer [44].

1.34.4 Imaging at optimum focus: phase contrast microscopy

In an ideal microscope, the image would exactly represent the object function and the image intensity for a pure phase object function would be

$$|\Phi(\mathbf{R})|^2 = |\psi(\mathbf{R})|^2 = |\exp[i\varphi(\mathbf{R})]|^2 = 1 \tag{1.259}$$

i.e. the image would show no contrast. This can be compared with imaging a glass plate with variable thickness in an ideal optical microscope. Also thin material objects in the transmission electron microscope behave as phase objects (1.285).

Assuming a weak phase object (WPO) one has

$$\varphi(\mathbf{R}) \leq 1$$

so that

$$\psi(\mathbf{R}) \approx 1 + i\varphi(\mathbf{R}) \tag{1.260}$$

The constant term 1 contributes to the central beam (zeroth Fourier component) whereas the term $i\varphi$ mainly contributes to the diffracted beams. If the phases of the diffracted beams can be shifted over $\frac{\pi}{2}$ with respect to the central beam, the amplitudes of the diffracted beams are multiplied with $\exp\left(\frac{i\pi}{2}\right) = i$. Hence the image term $i\varphi(\mathbf{R})$ becomes $-\varphi(\mathbf{R})$. It is as if the object function has the form

$$\Phi(\mathbf{R}) = 1 - \varphi(\mathbf{R}) \approx \exp[-\varphi(\mathbf{R})]$$

i.e. the phase object now acts as an amplitude object. The image intensity is then

$$|\Phi(\mathbf{R})|^2 \approx 1 - 2\varphi(\mathbf{R}) \tag{1.261}$$

which is a direct representation of the phase of the object. In optical microscopy, this has been achieved by F. Zernike using a quarter wavelength plate. In electron microscopy however, the phase shift can be made approximately $-\frac{\pi}{2}$ for a range of beams if one operates at optimum focus where phase contrast is realized by a fortunate balance between spherical aberration and defocus (Figure 1.80). Furthermore, for a thin object the phase is proportional to the projected potential of the object (1.285) so that the image contrast can be interpreted directly in terms of the projected structure of the object.

1.34.5. Resolution

General considerations In principle the characteristics of an electron microscope can be completely defined by its transfer function, i.e. by the parameters C_s, Δf, Δ and α.. A clear definition of resolution is not easily given for an electron microscope. For instance, for thick specimens, there is not necessarily a one-to-one correspondence between the projected structure of the object and the wavefunction at the exit face of the object so that the image does not show a simple relationship.

If one wants to determine a "resolution" number, this can only be meaningful for thin objects. Furthermore one has to distinguish between structural resolution as the finest detail that can be interpreted in terms of the structure, and the information resolution or information limit which is the finest detail that can be resolved by the instrument, irrespective of a possible interpretation. The information resolution may be better than the structural resolution. With the present electron microscopes, individual atoms cannot yet be resolved within the structural resolution.

Structural resolution (point resolution) As shown in § 1.34.2 the electron microscope in the phase contrast mode at optimum focus directly reveals the projected potential, i.e. the structure, of the object provided the object is very thin. All spatial frequencies g with a nearly constant phase shift are transferred forward from object to image. Hence the resolution can be obtained from the first zero of the transfer function as:

$$\rho_s = \frac{1}{g} \approx 0.65\ C_s^{1/4}\ \lambda^{3/4} = 0.65\ \mathrm{Gl} \qquad (1.262)$$

with $\mathrm{Gl} = C_s^{1/4}\lambda^{3/4}$ the Glaser unit. This value is generally accepted as the standard definition of the structural resolution of an electron microscope. It is often also called the point resolution. It is also equal to the width of the impulse response function. The information beyond the intersection ρ_s is transferred with a non-constant phase and, as a consequence, is redistributed over a larger image area.

Information resolution The information resolution can be defined as the finest detail that can be resolved by the instrument. It corresponds to the maximal diffracted beam angle that is still transmitted with appreciable intensity, i.e. the transfer function of the microscope (A1) is a spatial band filter which cuts all information beyond the information resolution. For a thin specimen, this limit is mainly determined by the envelope of chromatic aberration (temporal incoherence) and beam convergence (spatial incoherence). In principle beam convergence can be reduced using a smaller illuminating aperture and a larger exposure time. If chromatic aberration is predominant, the damping envelope function is given

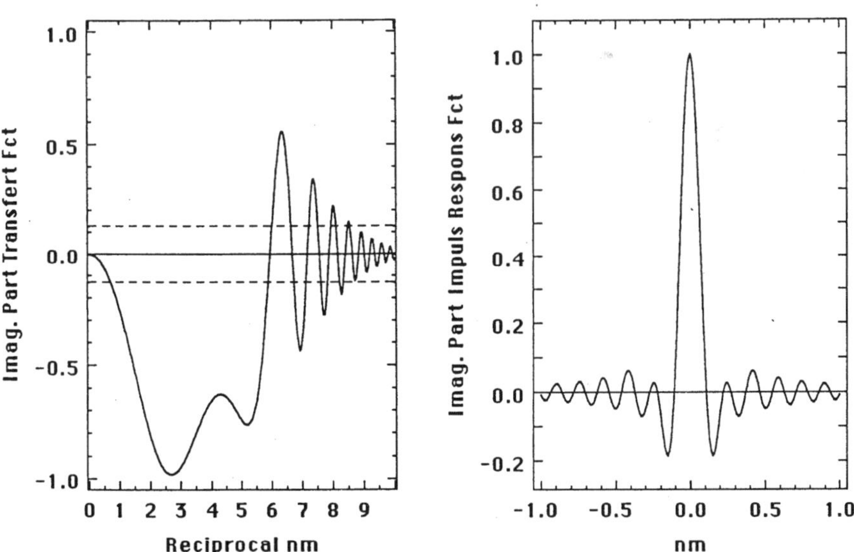

Figure 1.82 Phase transfer function left and corresponding impulse response function for a 300 KeV instrument (C_s = 0.7 nm, C_c = 1.3 nm, ΔE = 0.8 eV).

by (A5) (see appendix) from which the resolution can be estimated as

$$\rho_l = \frac{1}{g} = \left(\frac{\pi\lambda\Delta}{2}\right)^{1/2} \tag{1.263}$$

For a typical 100 KeV instrument, $\Delta = 5$ nm, $\lambda = 3.7$ pm, one obtains

$$\rho = 0.17 \text{ nm}$$

which is much smaller than the structural resolution for such an instrument.

Ultimate resolution The information between ρ_s and ρ_l is present in the image, albeit with the wrong phase. Hence this information is redistributed over the image. However, it can be restored by means of holographic methods (see & 1.38). In that case ρ_l is the ultimate instrumental resolution. Using a field emission gun (FEG) the spatial as well as the temporal incoherence can be reduced so as to push the information resolution towards 0.1 nm. Figure 1.82 shows the phase transfer function and impulse response function (IRS) of a 300 KeV instrument with FEG. Here the information limit extends to 0.1 nm but a large amount of information with the wrong phase is present between ρ_s and ρ_l, i.e. in the tails of the IRS, and has to be restored by holographic methods combined with image processes. However, the ultimate resolution will be limited by the object itself. Indeed, since the electron interacts with the atoms of the object through the electrostatic potential, the atom itself is the ultimate probe, the width of which is also of the order of 0.05 to 0.1 nm.

A new definition of resolution based on information theory According to the theory of Shannon [45] the maximal information rate of a communication channel is given by

$$C = B \log_2(1 + S/N) \text{ bits/sec} \tag{1.265}$$

where B is the bandwidth of the channel, S is the average signal power and N the noise power at the output of the channel.

Eq. (1.265) can be applied to microscopy in the following way: The bandwidth of a channel is defined as the number of independent degrees of freedom that the channel can transmit per unit time. In microscopy the bandwidth can be defined as the number of independent degrees of freedom that the microscope can transmit per unit area. Consider for example a square unit cell of size $a \times a$

The diffraction pattern consists of reflections at the positions of the reciprocal lattice with mesh $\frac{1}{a}\times\frac{1}{a}$.

Each reciprocal node contains one independent degree of freedom. (In practice, each node contains two (real and imaginary) numbers which, for a real object, are symmetry related by Friedel's law). The total number of degrees of freedom, transmitted within the information resolution ρ_1 of the microscope is then equal to

$$\frac{\pi\left(\frac{1}{\rho_l}\right)^2}{\left(\frac{1}{a}\right)^2} = \frac{\pi a^2}{(\rho_l)^2} \tag{1.266}$$

The number of degrees of freedom per unit area is then

$$B = \frac{\pi}{(\rho_l)^2} \tag{1.267}$$

which is independent of the choice of a.

Eq. (1.265) can now be interpreted as follows. The microscope transmits about three independent degrees of freedom per unit $(\rho_l)^2$. If we consider the noise level N as the smallest significant piece of information, the signal + noise can be expressed in units N as (S+N)/N, which in binary code equals $\log_2(1+S/N)$ bits. Hence, each degree of freedom carries, on the average, $\log_2(1+S/N)$ bits of information.

Gabor [46] stated that Shannon's communication theory is of little use for electron microscopy, since it was developed for situations where two persons communicate by means of a vocabulary of commonly agreed messages, whereas in electron microscopy, the sender is an object which is usually unknown.

As a consequence, electron microscopy can only resolve structures for which sufficient information is obtained from other techniques (e.g. X-ray diffraction) so that the structure model only contains a small number of unknown parameters, which is much less than the information capacity of the electron microscope. Examples are the characterization of defects (type, Burgers vector), the structure of building block structures (e.g. mixed layer compounds) which are determined uniquely by their stacking sequence, the structure of binary alloys ordering on a known lattice, etc. Most of these structures can be determined unambiguously even at low resolution. On the other hand, it is amazing that most high resolution images are only interpreted by visual comparison with computer simulations. Indeed, this is a very poor technique which allows only to discriminate between a limited number of plausible structure models and which therefore requires considerable prior information.

However, suppose HREM would be able to resolve individual atoms. Since all possible atom types are known, a structure can then be characterized completely by the positions of its constituent atoms. The atoms can thus be considered as the messages of Shannon and the argument of Gabor does not hold. In this way a structure could be completely resolved by HREM without prior knowledge. However, the requirement is that the number of unknowns (e.g. atom coordinates) is less than the capacity of the microscope, i.e. 3 per unit $(p_l)^2$.

In this way resolution gets a completely new meaning. If the structure (in projection) contains less than about 1.5 atoms per $(p_l)^2$, the position of each atom can in principle be determined with an average precision of $\log_2(1+S/N)$ bits. This opens quite new perspectives, comparable to X-ray crystallography where, using comparable information (diffracted beams) the atom positions can be determined with high precision. If on the other hand the resolution is insufficient to determine the individual atoms, i.e. the number of atoms exceeds 1.5 per $(p_1)^2$ the required information exceeds the capacity of the microscope channel. In a sense the channel is then blocked and no information can be obtained without much a priori knowledge.

In a real object the first electron "sees" the projected structure of the object. Hence it is important to notice that the requirement of less than 1.5 atoms per unit $(p_l)^2$ has to be fulfilled for the projected object. This requirement can most easily be met when studying a crystal along a simple zone axis in which the atoms are aligned along columns parallel to the beam direction. However for more complicated zone axes, the number of atoms in projection increases and the channel may be blocked. Also in amorphous objects the number of different atoms in projection increases with depth, so that, except for very thin amorphous objects, the information channel is blocked and the images only reveal information about the imaging characteristics of the microscope rather than about the object [47].

Concluding we propose to define the resolving capacity of the electron microscope as the number of independent degrees of freedom (parameters) that can be determined per unit area (per **Å**^{2} or nm^2). In order to determine a structure completely without prior knowledge it is essential that the number of atom coordinates does not exceed the resolving capacity and that this information can be extracted unambiguously from the images. For this purpose holographic methods are most promising (see further on).

1.35. Transfer in the Object

1.35.1. Dynamical electron diffraction

As is clear from (1.253) the calculation of the image wavefunction $\Phi(\mathbf{R})$ requires the knowledge of $\psi(\mathbf{R})$, i.e. the wavefunction at the exit face of the object. This can be obtained by numerically solving the Schrödinger equation in the object. For convenience we will now follow a simplified more intuitive approach, which leads to the correct results. For a more rigorous treatment we refer to Volume 1, Chapter 2.

If we assume that the fast electron, in the direction of propagation (z-axis) behaves as a classical particle with velocity

$v = \frac{hk}{m}$ we can consider the z-axis as a time axis with

$$t = \frac{mz}{hk} \tag{1.268}$$

Hence we can start from the time-dependent Schrödinger equation

$$-\frac{\hbar}{i}\frac{\partial\psi}{\partial t}(\mathbf{R},t) = H\psi(\mathbf{R},t) \tag{1.269}$$

with

$$H = -\frac{\hbar^2}{2m}\Delta_{\mathbf{R}} - eU(\mathbf{R},t) \tag{1.270}$$

with $U(\mathbf{R},t)$ the electrostatic crystal potential, m and k the relativistic electron mass and wavelength and $\Delta_{\mathbf{R}}$ the Laplacian operator acting in the plane ($\mathbf{R}$) perpendicular to z.

Using (1.268) we then have

$$\frac{\partial\psi(\mathbf{R},z)}{\partial z} = \frac{i}{4\pi k}\left(\Delta_{\mathbf{R}} + V(\mathbf{R},z)\right)\psi(\mathbf{R},z) \tag{1.271}$$

with

$$V(\mathbf{R},z) = \frac{2me}{\hbar^2}U(\mathbf{R},z) \tag{1.272}$$

This is the well-known high energy equation in real space which can also be derived from the stationary Schrödinger equation in the forward scattering approximation (1.263) (see Volume 1, Chapter 2).

In high resolution electron microscopy of crystalline objects, the object is usually oriented along a zone axis, so that the electrons are travelling parallel to the atom columns. If the periodicity along the column direction is not too large (less than 1-2 nm), the fast electron does not feel this variation. In fact it sees the potential as constant along z. In other words, the effect of higher order Laue zones or upper layer lines is negligible. This is the projection approximation, which is usually valid for most high resolution conditions. Now (1.271) becomes

$$\frac{\partial\psi(\mathbf{R},z)}{\partial z} = \frac{i}{4\pi k}(\Delta_{\mathbf{R}} + V(\mathbf{R}))\,\psi(\mathbf{R},z) \tag{1.273}$$

with

$$V(\mathbf{R}) = \frac{2me}{\hbar}\,\frac{1}{z}\int_0^z U(\mathbf{R},z)dz \tag{1.274}$$

the potential, averaged (projected) along z. In the time-dependent Schrödinger picture (1.269) the electron walks as a function of time in a two-dimensional potential of projected atom columns.

Eq. (1.273) can also be transformed to reciprocal space.

Assuming $V(\mathbf{R})$ to be periodic in two dimensions, we can expand it in Fourier series

$$V(\mathbf{R}) = \sum_{\mathbf{g}} V_{\mathbf{g}} \exp 2\pi i\mathbf{g}.\mathbf{R} \tag{1.275}$$

with **g** in the zone plane. V_g are known as structure factors. Similarly we have

$$\psi(\mathbf{R}) = \sum_{\mathbf{g}} \psi_{\mathbf{g}}(z) \exp 2\pi i\mathbf{g}.\mathbf{R} \tag{1.276}$$

$\psi_g(z)$ represents the amplitude of the beam **g** at a depth z. Substitution in (1.273) then yields

$$\frac{d\psi_g}{dz} = i\pi\left[2s_g\psi_g(z) + \sum_{g'} V_{g-g'}\,\psi_{g'}(z)\right] \tag{1.277}$$

with

$$s_g = g^2 / 2k$$

the excitation error, which is approximately equal to the distance between the reciprocal node **g** and the Ewald sphere, measured along z. This system of coupled first order differential equations has been derived in the early sixties (Tournarie, Howie-Whelan, Sturkey [48]). Most of the image simulation programs are based on a numerical solution of the dynamical equation in real space (1.273), or reciprocal space (1.277) or a combination of both.

1.35.2. Solution of the dynamical equation

The dynamical equation (1.271) or (1.273) is a mixture of two equations, each representing a different physical process

$$\frac{\partial\psi(\mathbf{R},z)}{\partial z} = \frac{i}{4\pi k}\,\Delta_R\psi(\mathbf{R},z) \tag{1.279}$$

is a complex diffusion-type of equation, which represent the free electron propagation and whose solution can be represented formally as

$$\psi(\mathbf{R},z) = \exp\left(\frac{i\Delta_R z}{4\pi k}\right)\psi(\mathbf{R},0) \tag{1.280}$$

from (1.277) or in reciprocal space

$$\psi(\mathbf{R},z) = \exp\left(\frac{i\pi g^2 z}{k}\right)\psi_g(0)$$

$$= \exp\left[2\pi i s_g z\right]\psi_g(0) \tag{1.281}$$

each beam amplitude is multiplied with a phase shift, which increases with increasing excitation error. Beams with **g** on the Ewald sphere have zero phase shift. A direct product in reciprocal space is equal to a convolution product in real space so (1.281) can be rewritten in real space as

$$\psi(\mathbf{R},z) = \exp\left(\frac{i\pi k R^2}{z}\right) * \psi(\mathbf{R},0) \tag{1.282}$$

which is the explicit form of (1.280). The propagation is thus described by a convolution product in real space and a direct product in reciprocal space.

The other part of (1.273) is a differential equation

$$\frac{\partial\psi(\mathbf{R},z)}{\partial z} = \frac{i}{4\pi k} V(\mathbf{R},z)\,\psi(\mathbf{R},z) \tag{1.283}$$

which represents the scattering of the electron by the crystal potential. It can be readily integrated in real space, yielding

$$\psi(\mathbf{R},z) = \exp\left(\frac{i}{4\pi k} V(\mathbf{R})z\right)\psi(\mathbf{R},0) \tag{1.284}$$

with $V(\mathbf{R})$ the projected potential as defined in (1.274).

The wavefunction in real space is multiplied with a phase factor which is proportional to the electrostatic potential of the object projected along z and which is called the phase object function. In the case of an incident plane wave $(\psi(\mathbf{R},0) = 1)$ and a very thin object, (1.284) becomes

$$\psi(\mathbf{R},z) = 1 + \frac{i}{4\pi k} V(\mathbf{R})z \tag{1.285}$$

which is the weak phase object approximation. The direct product in real space (1.284) now yields a convolution product in reciprocal space

$$\psi_g(z) = \mathfrak{F}_g\left(\exp\left[\frac{i}{4\pi k} V(\mathbf{R})z\right]\right) * \psi_g(0) \tag{1.286}$$

The solution of the complete dynamical equation (1.273) can be written formally as

$$\psi(\mathbf{R},z) = \exp\left(\frac{i}{4\pi k}[\Delta_R + V(\mathbf{R})]z\right)\psi(\mathbf{R},0) \tag{1.287}$$

For the explicit calculation, slice methods are the most appropriate. Here the crystal is cut into thin slices with thickness ε perpendicular to the incident beam.

If the slice thickness is sufficiently small, the solution within one slice is approximated by

$$\psi(\mathbf{R},z+\varepsilon) = \exp\left(\frac{i}{4\pi k}\Delta_R\varepsilon\right)\exp\left(\frac{i}{4\pi k}V(\mathbf{R})\varepsilon\right)\psi(\mathbf{R},z)$$

or explicitly

$$\psi(\mathbf{R},z+\varepsilon) = \exp\left(\frac{i\pi kR^2}{\varepsilon}\right) * \left\{\exp\left(\frac{i}{4\pi k}V(\mathbf{R})\varepsilon\right)\psi(\mathbf{R},0)\right\} \qquad (1.288)$$

In practice the wavefunction is sampled in a network of closely spaced points. In each point the wavefunction is multiplied with the phase object function. Then the wavefunction is propagated to the next slice and so on. Calling N the number of sampling points, the phase object requires a calculation time proportional to N^2. In reciprocal space, direct and convolution products are interchanged yielding

$$\psi_g(z+\varepsilon) = \exp\left(\frac{i\pi g^2\varepsilon}{k}\right)\left[\mathfrak{F}_g\left(\exp\left[\frac{i}{4\pi k}V(\mathbf{R})\varepsilon\right]\right) * \psi_g(z)\right] \qquad (1.289)$$

Now the calculation time of the propagation is proportional to N the number of beams whereas the scattering in the phase object gives a calculation time proportional to N^2.

An elegant way to speed up the calculation has been proposed in [49]. Here the phase object is calculated in real space, and the propagator in reciprocal space. Between each a Fast Fourier Transform is performed, the calculation time of which is only proportional to $N \log_2 N$. In the standard slice programs the object is assumed to be a perfect crystal. Defects are treated by the periodic continuation method in which the defect is artificially repeated so as to create an artificial supercrystal. For details we refer to Volume 1, Chapter 4.

In the real space method, proposed in [50], the whole calculation is performed in real space but due to the forward scattering of the electrons, the propagation effect is limited to a local area so that the calculation time remains proportional to N This is particularly interesting for treating extended or aperiodic structures. Furthermore, this approach allows to incorporate higher order Laue zone effects. Recently [51] a new approach has been proposed based on Wigner functions in which the calculations are done in a six-dimensional space, combining the advantages of real and reciprocal space. However, although conceptually beautiful, the merits of this technique are not yet clear.

1.36. A Simple Intuitive Theory: Electron Channelling

1.36.1. Principle

Although the slice methods are valuable for numerical purposes, they do not provide much physical insight in the diffraction process. There is need for a simple intuitive theory that is valid for larger crystal thicknesses. In our view, a channelling theory fulfills this need. Indeed, it is well known that, when a crystal is viewed along a zone axis, i.e. parallel to the atom columns, the high resolution images often show a one-to-one correspondence with the configuration of columns provided the distance between the columns is large enough and the resolution of the instrument is sufficient. This is for instance the case in ordered alloys with a column structure [52][53]. From this, it can be suggested that, for a crystal viewed along a zone axis with sufficient separation between the columns, the wave function at the exit face mainly depends on the projected structure, i.e. on the type of atom columns. Hence, the classical picture of electrons traversing the crystal as plane-like waves in the directions of the Bragg beams which stems from the X-ray diffraction picture and upon which most of the simulation programs are based, is in fact misleading. The physical reason for this "local" dynamical

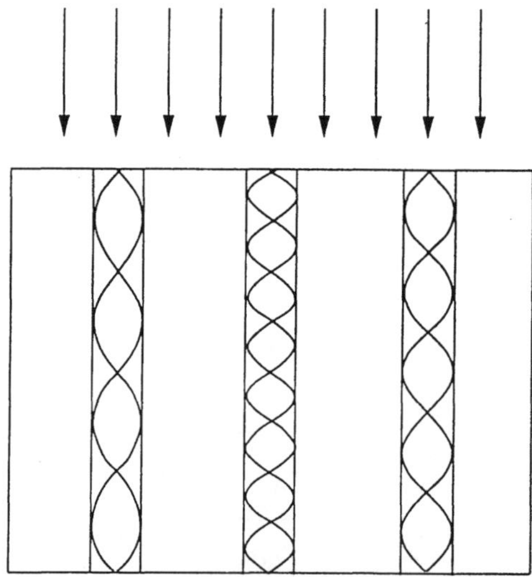

Figure 1.83 Schematical representation of electron channelling.

diffraction is the channelling of the electrons along the atom columns parallel to the beam direction. Due to the positive electrostatic potential of the atoms, a column acts as a guide or channel for the electron [54] within which the electron can scatter dynamically without leaving the column (Figure 1.83). It has been proposed [55] to exploit this so-called atom column approximation to speed up the dynamical diffraction calculations by assembling the wavefunction at the exit face using parts that have been calculated for each atom column separately.

The importance of channelling for interpreting high resolution images has often been ignored or underestimated, probably because of the fact that for historical reasons, dynamical electron diffraction is often described in reciprocal space. However, since most of the high resolution images of crystals are taken in a zone axis orientation, in which the projected structure is the simplest, but in which the number of diffracted beams are the largest, we believe that a simple real-space channelling theory yields a much more useful and intuitive, albeit approximate, description of the dynamical diffraction, which allows to provide an intuitive interpretation of high resolution images, even for thicker objects.

1.36.2. Perfect crystal

In a perfect zone axis orientation, neglecting higher order Laue zones, if we consider the depth proportional to the time, the dynamical equation (1.273) represents the walk of an electron in a two-dimensional assembly of potential wells, corresponding with the different projected columns.

The solution of (1.271) can be expanded in eigenfunctions of the Hamiltonian

$$\psi(\mathbf{R},z) = \sum_n C_n \phi_n(\mathbf{R}) \exp\left(-i\pi \frac{E_n}{E} \frac{z}{\lambda}\right) \qquad (1.290)$$

where

$$H\phi_n(\mathbf{R}) = E_n \, \phi_n(\mathbf{R}) \qquad (1.291)$$

with

$$H = -\frac{\hbar^2}{2m}\Delta_R - e\,U(\mathbf{R}) \qquad (1.292)$$

and

$$E = \frac{\hbar^2 k^2}{2m} \qquad (1.293)$$

the incident electron energy. λ is the electron wavelength. For $E_n < 0$ the states are bound to the columns. We now rewrite (1.290) as

$$\psi(\mathbf{R},z) = \sum_n C_n \phi_n(\mathbf{R})\left[1 - i\pi\frac{E_n}{E}\frac{z}{\lambda}\right]$$
$$+ \sum_n C_n \phi_n(\mathbf{R})\left[\exp\left(-i\pi\frac{E_n}{E}\frac{z}{\lambda}\right) - 1 + i\pi\frac{E_n}{E}\frac{z}{\lambda}\right] \qquad (1.294)$$

The coefficients C_n are determined from the boundary condition

$$\sum_n C_n \phi_n(\mathbf{R}) = \psi(\mathbf{R},0) \qquad (1.295)$$

In case of plane wave incidence one thus has

$$\sum_n C_n \phi_n(\mathbf{R}) = 1 \qquad (1.296)$$

and from (1.291) and (1.292)

$$\sum_n C_n \phi_n(\mathbf{R})E_n = H\psi(\mathbf{R},0) = H.1 = -e\,U(\mathbf{R}) \qquad (1.297)$$

Now (1.294) becomes

$$\psi(\mathbf{R},z) = 1 + i\pi\frac{e\,U(\mathbf{R})}{E}\frac{z}{\lambda}$$
$$+ \sum_n C_n \phi_n(\mathbf{R})\left[\exp\left(-i\pi\frac{E_n}{E}\frac{z}{\lambda}\right) - 1 + i\pi\frac{E_n}{E}\frac{z}{\lambda}\right] \qquad (1.298)$$

The first two terms yield the well-known weak phase object approximation (1.285).

In the third term only these states will appear in the summation for which

$$|E_n| \geq \frac{E\lambda}{z} \qquad (1.299)$$

In case the object is very thin, so that no state obeys (1.299), the weak phase object approximation is valid. For a thicker object, only bound states will appear with very deep energy levels, which are localized near the column cores. Furthermore, a two-dimensional projected column potential has only very few deep states, and when the overlap between adjacent columns is small only the radial symmetric states will be excited. In practice, for most types of atom columns, only one state appears, which can be compared with the 1 s state of an atom.

In the case of an isolated column of type i, taking the origin in the center of the column, we then have

$$\psi_i(\mathbf{R},z) = 1 + i\pi e \frac{U_i(\mathbf{R})}{E}\frac{z}{\lambda} + C_i\,\phi_i(\mathbf{R})\left[\exp\left(-i\pi\frac{E_i}{E}\frac{z}{\lambda}\right) - 1 + i\pi\frac{E_i}{E}\frac{z}{\lambda}\right] \tag{1.300}$$

A very interesting consequence of this description is that, since the states ϕ_i are very localized at the atom cores, the wavefunction for the total crystal can be expressed as a superposition of the individual column functions

$$\psi(\mathbf{R},z) = 1 + i\pi e \frac{U(\mathbf{R})}{E}\frac{z}{\lambda} + \sum_i C_i\,\phi_i(\mathbf{R}-\mathbf{R}_i)\left[\exp\left(-i\pi\frac{E_i}{E}\frac{z}{\lambda}\right) - 1 + i\pi\frac{E_i}{E}\frac{z}{\lambda}\right] \tag{1.301}$$

with

$$U(\mathbf{R}) = \sum_i U_i(\mathbf{R}-\mathbf{R}_i) \tag{1.302}$$

If all the states other than the ϕ_i have very small energies i.e.

$$|E_n| << \frac{E\lambda}{z} \tag{1.303}$$

then (1.294) can be simplified as

$$\psi(\mathbf{R},z) = \sum_n C_n\,\phi_n(\mathbf{R}) + \sum_n C_n\,\phi_n(\mathbf{R})\left[\exp\left(-i\pi\frac{E_n}{E}\frac{z}{\lambda}\right) - 1\right] \tag{1.304}$$

so that (1.301) in that case becomes

$$\psi(\mathbf{R},z) = 1 + \sum_i C_i\phi_i(\mathbf{R}-\mathbf{R}_i)\left[\exp\left(-i\pi\frac{E_i}{E}\frac{z}{\lambda}\right)-1\right] \tag{1.305}$$

Expressions (1.301) and (1.305) are the basic result of this channelling theory.

The interpretation of (1.305) is simple. Each column i acts as a channel in which the wavefunction oscillates periodically with depth. The periodicity is related to the "weight" of the column, i.e. proportional to the atomic number of the atoms in the column and inversely proportional to their distance along the column. The importance of these results lies in the fact that they describe the dynamical diffraction for larger thicknesses than the usual phase grating approximation and that they require only the knowledge of one function ϕ_i per column (which can be tabelized similar to atom scattering factors or potentials). Furthermore, even in the presence of dynamical scattering, the wavefunction at the exit face still retains a one-to-one relation with the configuration of columns. Hence this description is very useful for interpreting high resolution images and to provide a possible answer to the direct retrieval problem (see § 1.37). Eq. (1.305) applies to light columns, such as Si(111) or Cu(100) with an accelerating voltage up to about 200 KeV. When the atom columns are "heavier" and the accelerating voltage higher (which due to the relativistic correction also increases the effective strength of the potential, then (1.301) has to be used. This is for example the case for Au(100).

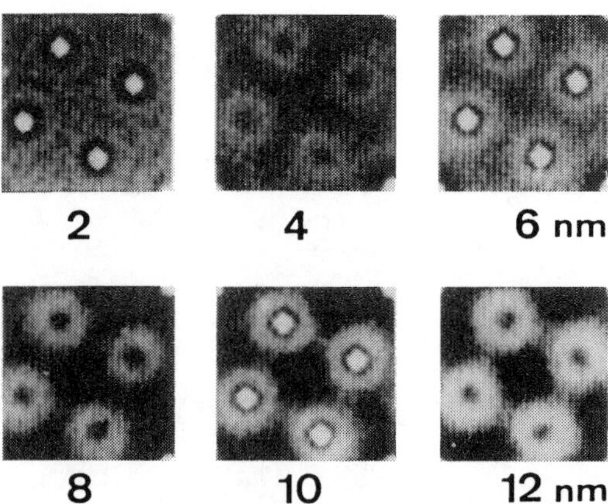

Figure 1.84 Electron density as a function of depth in Au_4Mn (see text).

Figure 1.84 shows the electron density $|\psi(\mathbf{R},t)|^2$ as a function of depth in a Au_4Mn alloy crystal for 200 KeV incident electrons. The corners represent the projection of the Mn column. The square in the center represents the four Au columns. The distance between adjacent columns is 0.2nm. The periodicity along the direction of the column is 0.4nm. From these results it is clear that the electron density in each column fluctuates nearly periodically with depth. For Au this periodicity is about 4nm and for Mn 13 nm. These periodicities are nearly the same as for isolated columns so that the influence of neighboring columns in this case is still small. The energies of the respective s states are respectively about 250 eV and 80 eV.

When the atoms are heavy and the accelerating voltage very high (0.5 to 1 MeV), a larger number of s states come in and the result becomes more complicated. When the crystal is viewed along a higher index zone axis, the distance between adjacent columns decreases whereas the weight of the columns also decreases.Hence the bound states broaden and overlap between adjacent columns starts to occur. This can be incorporated in the theory using perturbation theory. When the overlap between columns is too large one has to consider them as a kind of molecules [55]. The localization can also be improved by using higher voltages. It has to be stressed that the derived results are only valid in a perfect zone axis orientation. A slight tilt can destroy the symmetry and excite other, non-symmetric states, so that the results become much more complicated. It is interesting to note that channelling has usually been described in terms of Bloch waves [56,57]. However as follows from the foregoing, channelling is not a mere consequence of the periodicity of the crystal but occurs even in an isolated column parallel to the beam direction. In fact, even for an isolated column, the problem can be treated mathematically by making the column artificially periodical so as to generate a basis of functions (Bloch functions) to expand the wavefunction. In this view, the Bloch character is only of mathematical importance. Even in a crystal in which the distance between the adjacent columns is sufficient (e.g. 0.2 nm), this is the case. Bloch wave calculations then yield the same 1s states as found in our simplified treatment. Only when the overlap between columns increases or when the beam is inclined, the other Bloch states become physically important.

1.36.3. Channelling and defects

The channelling effect still occurs in the presence of defects such as translation interfaces, twin interfaces, dislocations, provided the columns parallel to the incident beam are not disrupted. To

Figure 1.85 Deviation from the ideal channelling at a translation interface in Au_4Mn (see text). Left model, center electron density at the exit face, right difference image.

demonstrate this we take again the Au_4Mn alloy of Figure 1.84 in which we now introduce translation interfaces (antiphase boundaries) (Figure 1.85 left). The electron density, calculated with the periodic combination method is shown in the center (thickness 8 nm). If the channelling along the individual columns would not be affected by the interface, the electron density at both sides of the interface would be identical to that of the perfect crystal shifted over the displacement vector of the interface. In order to reveal the deviation from this ideal situation, we subtract this "ideal" electron density at both sides and display the difference with increased contrast (Figure 1.85 right). From this it is clear that the deviation from the ideal channelling condition is very small and occurs only very close to the interface.

1.36.4. Diffraction pattern

Fourier transforming the wavefunction (1.305) at the exit face of the object yields the wavefunction in the diffraction plane, which can be written as

$$\psi(\mathbf{g},z) = \delta(\mathbf{g}) + \sum_i \exp(-2\pi i\mathbf{g}.\mathbf{R}_i)\, F_i(\mathbf{g},t) \qquad (1.306)$$

(In the case of heavy columns we will have to use (1.301) instead). In a sense the simple kinematical expression for the diffraction amplitude holds, provided the scattering factor for the atoms is replaced by a dynamical scattering factor for the columns, in a sense as obtained in [58] and which is defined by

$$F_i(\mathbf{g},z) = \left[\exp\left(\frac{-i\pi E_i}{E}\frac{z}{\lambda}\right) - 1\right] C_i\, f_i(\mathbf{g}) \qquad (1.307)$$

with $f_i(\mathbf{g})$ the Fourier transform of $\phi_i(\mathbf{R})$. It is clear that the dynamical scattering factor varies periodically with depth. This periodicity may be different for different columns.

In case of a monoatomic crystal, all F_i are identical. Hence $\psi(\mathbf{g},z)$ varies perfectly periodically with depth. In a sense the electrons are periodically transferred from the central beam to the diffracted beams and back. The periodicity of this dynamical oscillation (which can be compared with the Pendelösung effect) is called the dynamical extinction distance. It has for instance been observed in Si(111). An important consequence of (1.306) is the fact that the diffraction pattern can still be described by a kinematical type of expression so that existing results and techniques that have been based on the kinematical theory remain valid to some extent for thicker crystals in zone orientation. Examples are

- diffraction at periodical stacking of translation interfaces, twin interfaces and mixed layer compounds (see Chapter 4)
- diffuse scattering from substitutionally ordering alloys with a column structure (see Chapter 2)
- diffraction contrast at defects. In particular the extinction rule based on the **g.R** criterion (§ 1.24) remains valid if the defect is parallel to the incident beam.

1.36.5. High resolution images

The wavefunction in the image plane can be written as the convolution product of the wavefunction at the exit face of the crystal with the impulse response function $t(\mathbf{R})$ of the electron microscope (§ 1)

$$\psi(\mathbf{R}) = 1 + \sum_i \left[\exp\left(\frac{-iE_i}{E}\frac{z}{\lambda}\right) - 1\right] C_i \phi_i(\mathbf{R} - \mathbf{R}) * t(\mathbf{R}) \qquad (1.308)$$

If the microscope is operated close to optimum focus and in axial mode, the impulse response function is sharply peaked (Figures 1.81, 1.82).

If the distance between the columns is larger than the width of the impulse response function $t(\mathbf{R})$, the overlap between convolution products $\phi_i * t(\mathbf{R})$ of adjacent sites can be assumed to be small so that the image intensity is

$$I(\mathbf{R}) = |\psi(\mathbf{R})|^2 = \sum_1 4\, C_i^2 \sin^2\left(\frac{E_i z}{2E\lambda}\right) |\phi_i(\mathbf{R} - \mathbf{R}) * t(\mathbf{R})|^2 \qquad (1.309)$$

Each column is thus imaged separately. The contrast of a particular column varies periodically with thickness. The periodicity can be different for different types of columns. It is interesting to note that the functions ϕ_i as well as t(**R**) are symmetrical around the origin, provided the objective aperture is centered around the optical axis. Hence, the image of a column is rotationally symmetric around the position $\mathbf{R}_l$ of the columns. The intensity at $\mathbf{R}_l$ is a maximum or a minimum. The positions of the columns can thus be determined from the positions of the intensity extrema.

In case the resolution of the microscope is insufficient to discriminate the individual columns, or the focus is not close to optimum, the overlap between the convolution products of adjacent columns cannot be avoided and the interpretation of the contrast is not straightforward. In that case image simulation is required.

1.37. Interpretation of High Resolution Images

1.37.1. Weak phase objects

For a weak phase object ((1.285),(1.301)) the electron microscope operating in the optimum focus mode (1.261) reveals the phase contrast, which is given by

$$I(\mathbf{R}) = 1 - 2\pi i e \frac{U(\mathbf{R})z}{E\lambda} \tag{1.310}$$

The contrast directly reveals the projected potential of the object.

Atoms, or groups of atoms, are imaged with black contrast and holes with white contrast. Not always the highest resolution is required to allow a direct interpretation. This is very much dependent on the structure itself.

Figure 1.86 shows an image of a tunnel structure, which even at moderate resolution (0.4 nm) can be interpreted in terms of the projected structure (inset).

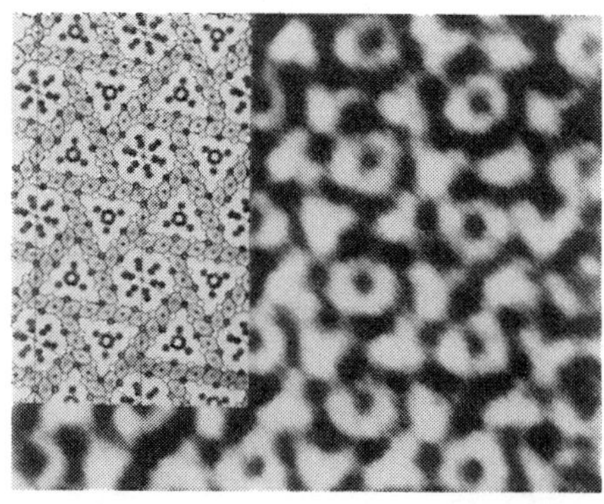

Figure 1.86 Moderate resolution image of the tunnel structure $Ba_{1-p}Cr_2Se_{4-p}$ [59].

1.37.2. Small defocus images

We assume a phase object, imaged with a small defocus ε. If we are interested in details much larger than the point resolution, the spherical aberration is not important and the phase transfer function mainly describes a simple propagation over ε in the sense as discussed in (1.280).

Hence the wavefunction in the image plane is (compare with 1.288)

$$\psi(\mathbf{R}) = \exp\left(\frac{i\Delta_{\mathbf{R}}\varepsilon}{4\pi k}\right) \exp\left[\frac{i}{4\pi k} V(\mathbf{R})z\right] \qquad (1.311)$$

For small defocus values

$$\psi(\mathbf{R}) = \left(1 + \frac{i\Delta_{\mathbf{R}}\varepsilon}{4\pi k}\right) \exp\left[\frac{i}{4\pi k} V(\mathbf{R})_z\right]$$

$$= \left\{1 - \frac{\varepsilon}{(4\pi k)^2} \Delta_{\mathbf{R}}(V(\mathbf{R})z) - \frac{i\varepsilon}{(4\pi k)^3}\left[\nabla_{\mathbf{R}}(V(\mathbf{R})z)\right]^2\right\} \exp\left(\frac{i}{4\pi k} V(\mathbf{R})z\right) \qquad (1.312)$$

so that the image intensity is given up to first order in ε as

$$I(\mathbf{R}) = 1 - \frac{2\varepsilon}{(4\pi k)^2} \Delta_{\mathbf{R}}(V(\mathbf{R})_z) \qquad (1.313)$$

and since the electrostatic potential obeys Poisson's law

$$\Delta_{\mathbf{R}}\,(V(\mathbf{R})) = -\,4\pi\rho(\mathbf{R}) \qquad (1.314)$$

with $\rho(\mathbf{R})$ the charge density (also averaged along z) we finally obtain

$$I(\mathbf{R}) = 1 - \frac{\varepsilon\,\rho(\mathbf{R})z}{2\pi k^2} \qquad (1.315)$$

Close to zero focus the image contrast is proportional to the projected charge densities. At underfocus atoms, or groups of atoms, are imaged with black contrast. The contrast disappears in the Gaussian image plane ($\varepsilon = 0$) and reverses with reversing the focus.

1.37.3. Substitutional alloys with a column structure

A particular situation can occur in the case of a substitutional binary alloy with a column structure. In a substitutional binary alloy, the two types of atoms occupy positions on a regular lattice, usually FCC. Since the lattice, as well as the types of the atoms and the average composition is known, the problem of structure determination is then reduced to a binary problem of determining which atom is located at which lattice site.

Particularly interesting are the alloys in which columns are found parallel to a given direction and which consist of atoms of the same type. Examples are the gold-manganese system and other FCC alloys [60-62]. If viewed along the column direction, which is usually $[001]_{FCC}$, the high resolution images contain sufficient information to determine unambiguously the type and position of the individual columns. Even if the resolution of the microscope is insufficient to resolve the individual lattice positions, which have a separation of about 0.2 nm, it is possible to reveal the minority columns only, which is sufficient to resolve the complete structure. This can be done using dark field superlattice imaging, which proceeds as follows.

Let us assume a binary system consisting of columns of A-atoms, which are the majority atoms, and columns of B-atoms. We call the fraction of columns respectively m_A and m_B; $m_A + m_B = 1$ with $m_A > m_B$. The wavefunction at the exit face of the object (1.305) can be written in a condensed notation as $\phi = \sum_i \phi_i$, where $\phi_i = \phi_A$ if the ith column is of the type A and $\phi_i = \phi_B$ if the ith column is of the type B.

This can be formulated making use of the Flinn occupation parameter σ_i which can take two values: $\sigma_i = m_A$ when the site i is occupied by a B-column and $\sigma_i = -m_B$ when the site i is occupied by an A-atom column. Introducing the average wavefunction

$$<\phi> = m_A\phi_A + m_B\phi_B, \qquad (1.316)$$

we have

$$\phi_A = <\phi> - m_B\Delta\phi, \quad \phi_B = <\phi> + m_A\Delta\phi,$$

with

$$\Delta\phi = \phi_B - \phi_A.$$

We can then rewrite ϕ as:

$$\phi = \sum_i <\phi> + \sum_i \sigma_i \Delta\phi \tag{1.317}$$

The meaning of this expression is obvious; the wavefunction associated with the site i is equal to the average value increased by $m_A\Delta\phi$ if the site i is occupied by a B-atom and decreased by $m_B\Delta\phi$ if the site i is occupied by an A-atom. The first term

$$\sum_i <\phi\ (\mathbf{R}-\mathbf{R}_i, t)>,$$

which contains the information about the "average" FCC lattice, contributes to the FCC reflections only. On the other hand, the second term,

$$\sum_i \sigma_i\ \Delta\phi(\mathbf{R}-\mathbf{R}_i, t), \tag{1.318}$$

which contains the information about the occupation of the lattice sites, contributes only to the superlattice reflections, if the system is ordered, or to the diffuse intensity if the system is not ordered. When

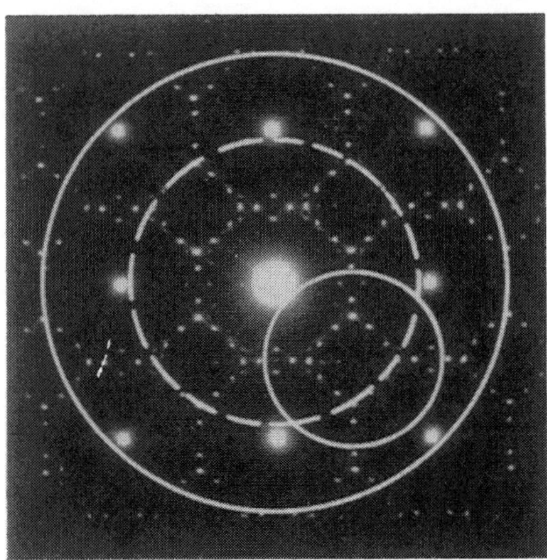

Figure 1.87 Schematic view of the position and size of the aperture (small circle) with respect to the diffraction pattern of an FCC based alloy is Au_4Mn [100].

the structure is imaged without FCC reflections (Figure 1.87) the wavefunction in the image plane is now (1.318)

$$\psi(\mathbf{R}) = \sum_i \sigma_i \, \Delta\phi(\mathbf{R}-\mathbf{R}_i,t) * t(\mathbf{R}), \tag{1.319}$$

the wavefunction has the same shape at each lattice site multiplied by a factor σ_i which is m_B for an A-column and $-m_A$ for a B-column.

If it is assumed that the overlap between adjacent functions $\Delta\phi * t$ can be neglected, the image intensity is given by

$$|\psi(\mathbf{R})|^2 \approx \sum_i \sigma_i^2 \, |\Delta\phi(\mathbf{R}-\mathbf{R}_i) * t(\mathbf{R})|^2 \tag{1.320}$$

i.e. the atom columns are imaged as bright peaks, the shape of the peaks being given by $|\Delta\phi(\mathbf{R}) * t(\mathbf{R})|^2$and the height of the peaks being proportional to m_A^2 for the B-atom columns and m_B^2 for the A-atom columns (Figures 1.35 and 1.88).

As a consequence the minority atoms B are imaged as brighter dots than the majority atoms A, the ratio of brightness being equal to $(m_A / m_B)^2$. When this ratio is large (e.g. >10), the minority atoms will be visible as bright dots on a dark background. It has to be noted that, in case the objective aperture is not centered around the optical axis of the microscope, the positions of the dots may be systematically shifted.

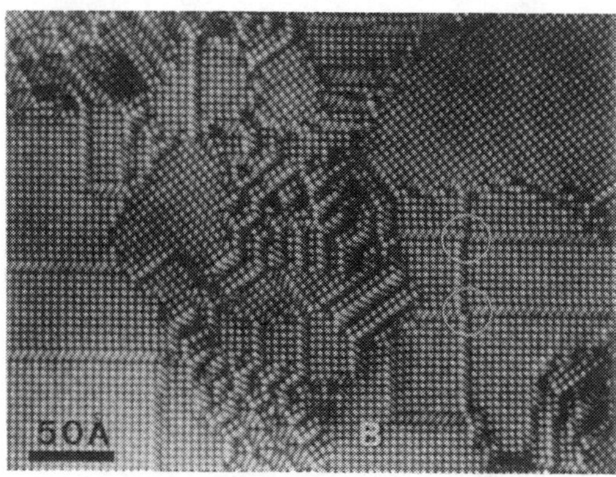

Figure 1.88 Dark field superlattice image of Au_4Mn. Orientation and translation variants are revealed [60].

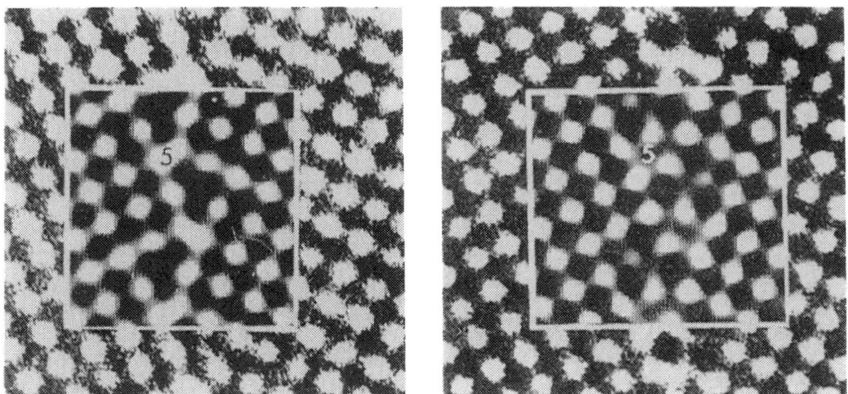

Figure 1.89 (310) $\Sigma = 5$ grain boundary in Germanium viewed along [001]. (Courtesy J.L. Rouviére)

1.37.4. Defects

Most experimental high resolution work on crystals with planar defects is done with the crystal in zone orientation so as to minimize the complexity of the projected structure. We will now discuss two different situations whether the planar interface is parallel or inclined to the zone axis.

Planar interfaces parallel to the zone axis Figure 1.88 shows an experimental high resolution micrograph of Au_4Mn in the dark field superlattice imaging mode using a 200 KeV electron microscope with 0.25 nm point resolution. The bright dots represent manganese columns; their configuration can be compared directly with the structure model of Figure 1.87. Two orientation variants are present, as well as a number of anti-phase boundaries.

The imaging characteristics in which the manganese atoms are revealed as bright dots are preserved even close to the interfaces. This allows to deduce the displacement vectors and the orientations directly from the images.

For non-coherent interfaces the situation is much more complicated. Although the atom columns can still be imaged by bright dots, the exact positions of the dots do not necessarily correspond to the exact positions of the columns so that comparison with simulated images is necessary. An example is given in Figure 1.89 showing a (310) $\Sigma = 5$ grain boundary in Germanium viewed along [001] which contains many dislocations. As in agreement with (1.305) the intensity at the column positions varies periodically with depth. The left hand side shows an image revealing black columns.

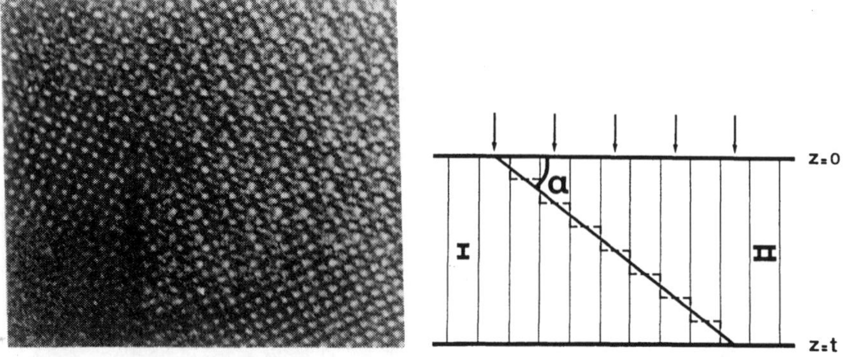

Figure 1.90 Au_4Mn viewed along (0012) showing the overlap of two orientation variants [63]. Left: high resolution image, right: schematical cross section view.

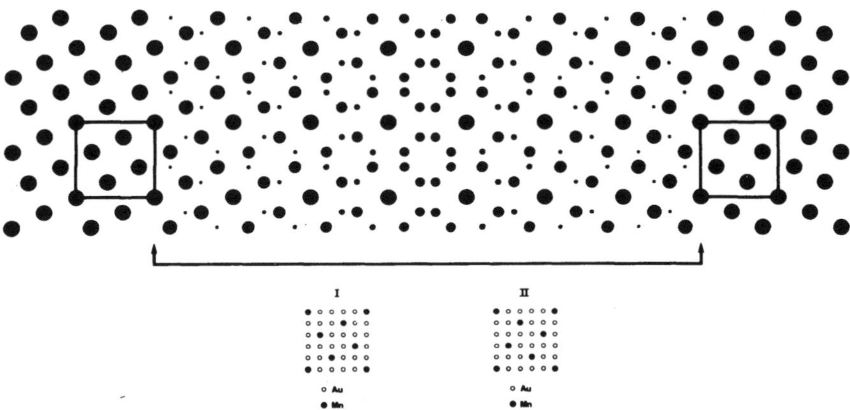

Figure 1.91 Projected model for the structure in Figure 1.90 [63].

The right hand side shows white columns. If possible it is preferable to search for a thickness for which the columns are imaged as bright dots. It should be noted that, although the position of the columns can be determined with an accuracy of 0.01 nm, they do not correspond to the exact column projection, probably because of the fact that the intercolumn distance is of the same order of magnitude as the point resolution, which in this case is 0.17 nm (400 KeV).

Interfaces inclined to the incident beam Figure 1.90 shows an example of a high resolution image of the Au_4Mn system viewed along [001] in which an inclined planar interface separates two

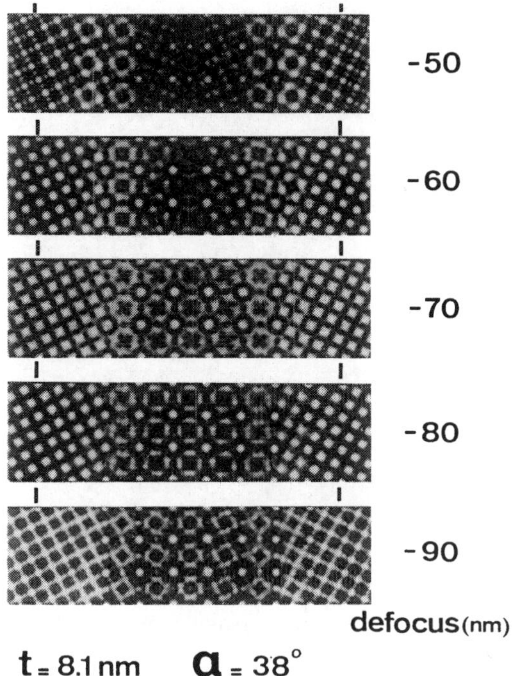

Figure 1.92 Simulated images using the model in Figure 1.91. Thickness and wedge angle are indicated [63].

orientation variants. Since Au_4Mn is an alloy the underlying lattice of which is continuous along the interface, the respective columns of both variants overlap. The projected structure is shown in Figure 1.91. It can be expected that in the superlattice dark field mode the overlapping minority columns (Mn) will still be revealed as bright dots, situated on a coincidence site lattice. This is confirmed by extensive image simulations, some of which are shown in Figure 1.92.

Another very special case is the stacking fault tetrahedron in Silicon which is limited by (111) stacking fault planes intersecting along stair rod dislocations and which have been observed in low stacking fault energy metals and alloys.

In ion implanted and annealed Silicon, the sizes of the SFT are suitable to be studied with high resolution electron microscopy. When the images are taken with the incident beam along the [110] zone axis, the images show a V shaped discontinuity in the rows of bright dots. Within the V zone, the dots are displaced due to the fact that the atom columns are intersected by two stacking faults.

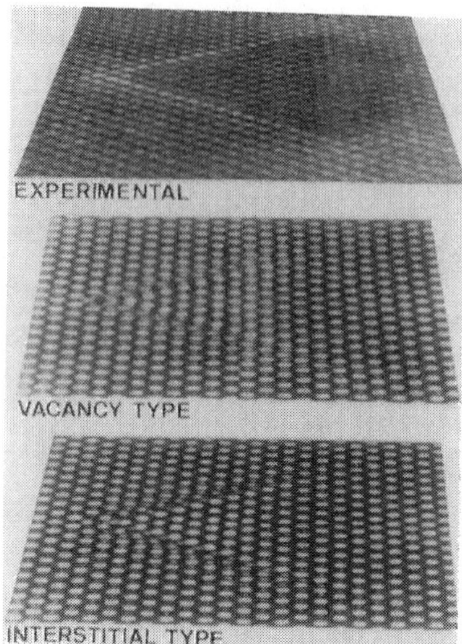

Figure 1.93 Stacking fault tetrahedron (SFT) in Silicon, viewed under glancing angle. Top: experimental image. Middle: simulated image for a vacancy of SFT. Bottom: simulated image for an interstitial type of SFT. From this it can be concluded that the SFT is of the vacancy type. (Courtesy W. Cowne, H. Bender) [64]

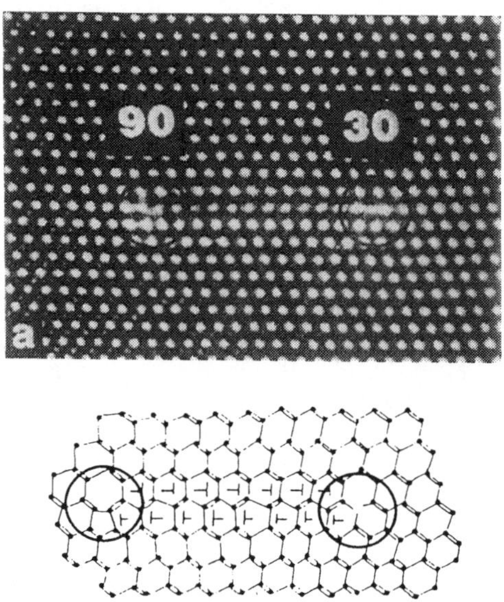

Figure 1.94 60° dissociated dislocation in Si. (Courtesy J. Thibault-Desseaux)

The displacement of the dots is related to the length of the displaced column, i.e. the displacement is largest near the point of the V and decreases gradually with increasing distance from this point. Image calculations were performed with the real space method for tetrahedra of different sizes using atom positions derived from models in the literature [63], for the vacancy type of tetrahedron as well as for the interstitial type. Approximately 10^5 atoms are involved in the calculations. By carefully looking along the rows at a glancing incidence (Figure 1.93), it is clear that the displacement of the rows of bright dots for the interstitial type tetrahedron is directed towards the point, whereas for the vacancy type, the displacement is in the opposite sense, as in agreement with experiment. Hence it can be concluded that the stacking fault tetrahedra in Si are of the vacancy type.

Dislocations When the dislocations are parallel to the zone axis of observation and when the resolution of the microscope is sufficient to discriminate the individual atoms the dislocation structure can be unravelled. Figure 1.94 represents the high resolution images of a 60° dissociated dislocation in Si. The dislocation is dissociated in two Shockley partials, one 90° and one 30°, enclosing an intrinsic

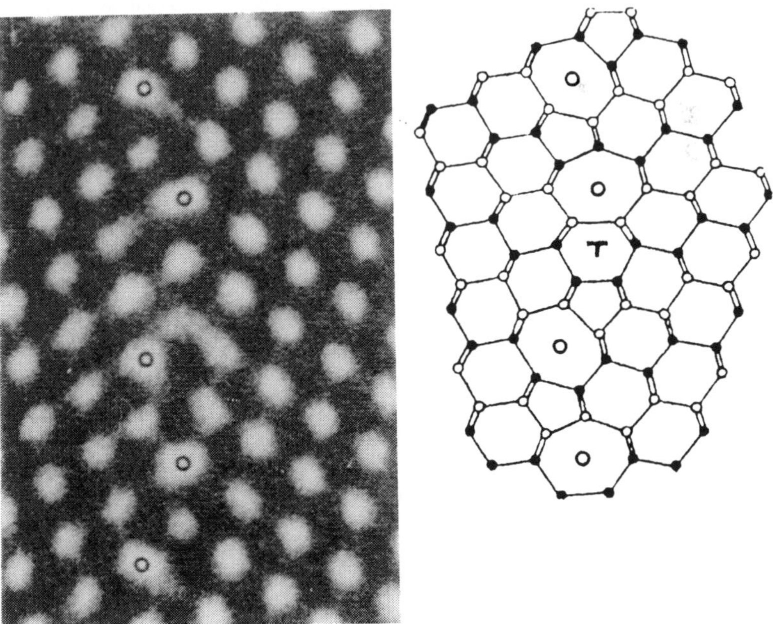

Figure 1.95 Σ9 grain boundary in Si. (Courtesy J. Thibault-Desseaux)

stacking fault. The main feature is that the partials consist of well defined structural units that also occur in the dislocation cores. Figure 1.95 shows the dislocation structures occurring at a Σ9 grain boundary in Si. Here also the same structural units can be observed.

1.37.5 Building block structure

It often happens that a family of crystal structures exists, of which all members consist of a stacking of simpler building blocks but with a different stacking sequence. This is for instance the case in mixed layer compounds, polytypes, periodical twins, but also periodic interfaces such as antiphase boundaries and crystallographic shear planes can be considered as mixed layer systems. Even substitutional binary alloys with a column structure can be considered as a special case of a building block system.

In order to characterize the stacking sequence of unknown members of the family, one can search for a member with a known stacking sequence. Often, this will be a simple member, the structure of which has been determined by X-ray diffraction. From this system, high resolution micrographs are taken. The image characteristics of each type of building block in these images are then called the imaging code: Since the high energy electrons are scattered forward, it can be expected that the image of a building block will not be severely affected by its surroundings, provided the crystal is not too thick and the focus is close to the optimum focus. Here the thickness for which this imaging code remains valid is estimated to be of the order of 20 nm. However, for larger building blocks, this thickness may even be larger. Now the building blocks of the unknown members are imaged with the same code, provided the experimental conditions (thickness, focus) are the same, so that the stacking sequence can be read directly from the high resolution micrographs.

An example is shown in Figure 1.96 for a polytype of the series YSeF [56]. This system is built on two types of blocks, respectively labelled S and T, which can be deduced from each other by a glide mirror plane. The width of the blocks is 0.31 nm. As can be deduced from micrographs of known sequences, the block S is imaged as a white dot and T as a black dot. This difference stems from multiple diffraction effects. Due to the relatively poor resolution of the instrument (≈0.6 nm) three successive S blocks are not resolved from each other. Now the sequence of the unknown polytype 12M can be deduced directly from the micrograph in Figure 1.96 and identified as SSSTSTSTSTST.

In general the technique of the imaging code is applicable up to larger crystal thicknesses (a few 10 nm) and also to other imaging modes.

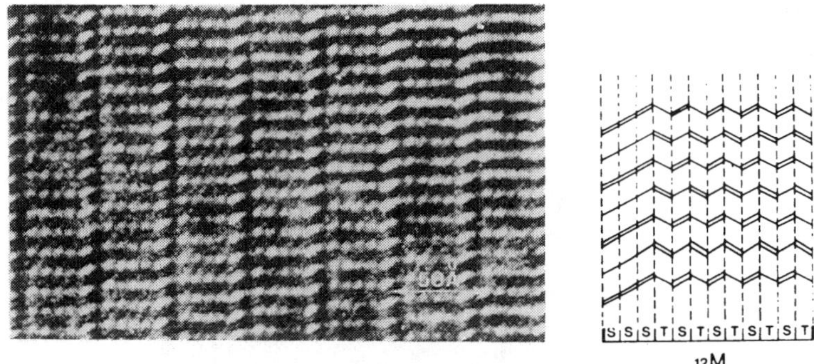

Figure 1.96 High resolution micrograph of the polytype YSeF 12M. A schematical drawing is shown on the right. Even though this image has been taken with a resolution of about 0.5 nm the one-to-one correspondence with the building blocks of the structure is maintained [65].

1.37.6. Interpretation using image simulation

When no obvious image code is available or when quantitative structural information at high resolution is required, the only reliable way to interpret the images is by comparing them with computer simulations carried out for all plausible structure models. For details and discussion about image simulation we refer to Volume 1, Chapter 4. In practice however, a number of parameters such as object thickness, accurate focus, etc. ... are only roughly known. Furthermore, the comparison is often made visually. The selectivity of the procedure can be enhanced by comparing complete focus and/or thickness series. Nevertheless, the technique is so tedious and so insensitive that, thus far, it has been only successful if the number of plausible structure models is very limited. In our view it is astonishing that such an expensive technique as HREM is so dependent on the availability of prior information obtained from other techniques.

1.38. Future Prospects: Direct Methods

HREM would be much more powerful if a direct method existed to extract the structural information directly and quantitatively from the electron micrographs. A direct method should consist of three stages. First the wavefunction has to be reconstructed in the image plane (phase retrieval). This phase problem can be solved in two ways, by using holography [66] or using the focus as an externally controllable parameter [67]. In the next step, the wavefunction has to be calculated at the exit face of the object, which is rather straightforward. The final step consists in retrieving the

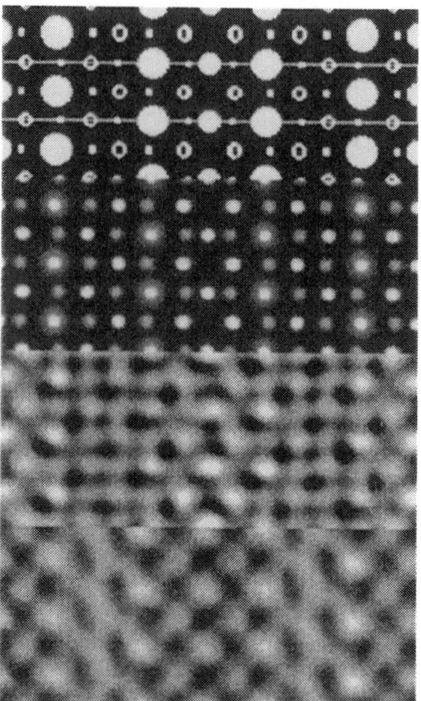

Figure 1.97 Experimentally retrieved wavefunction a the exit face of the high Tc superconductor $Y_1Ba_2Cu_3O_7$ from top to bottom. Structure model, phase of the wavefunction (simulated), experimental phase, high resolution image at optimum focus. (Courtesy M. Op de Beeck)

projected structure of the object. If the object is a phase object, the phase of the wavefunction directly yields the electrostatic potential. If the object is thicker but still consists of columns along the beam direction, the wavefunction at the exit face still shows a one-to-one correspondence with the column configuration so that the channelling theory of § 1.35.4 may be used to solve this problem [68].

Recently significant progress is being made in direct methods so that, for thin objects, direct structure retrieval at the 0.1 nm level seems within reach. The first results are shown in Figure 1.97.

Appendix: Image formation

The wavefunction in the image plane is given by

$$\phi(\mathbf{R}) = \mathcal{F}^{-1}_{\mathbf{R}} A(\mathbf{g}) \exp[-i\chi(\mathbf{g})] \mathcal{F}_{\mathbf{g}} \psi(\mathbf{R}) \tag{A1}$$

$A(\mathbf{g})$ represents the physical aperture with radius g_A selecting the imaging beams: thus

$$A(\mathbf{g}) = \begin{matrix} 1 \text{ for } |g| \leq g_A \\ 0 \text{ for } |g| > g_A \end{matrix}$$

The total phase shift due to spherical aberration and defocus is

$$\chi(\mathbf{g}) = \frac{1}{2}\pi C_s \lambda^3 g^4 + \pi\varepsilon\lambda g^2 \tag{A2}$$

with C_s: the spherical aberration coefficient
ε: the defocus
λ: the wavelength

The imaging process is also influenced by spatial and temporal incoherence effects. Spatial incoherence is caused by the fact that the illuminating beam is not parallel but can be considered as a cone of incoherent plane waves (beam convergence). The image then results from a superposition of the respective image intensities. Temporal incoherence results from fluctuations in the energy of the thermally emitted electrons, in the fluctuation of the lens currents and of the accelerating voltage. All these effects cause the focus ε to fluctuate. The final image is then the superposition (integration) of the images corresponding with the different incident beam directions **K** and focus values ε, i.e.

$$I(\mathbf{R}) = \int_{\varepsilon} \int_{\mathbf{K}} |\phi(\mathbf{R},\mathbf{K},\varepsilon)|^2 f_s(\mathbf{K}) f_T(\varepsilon) \, d\mathbf{K} \, d\varepsilon \tag{A3}$$

where $\phi(\mathbf{R},\mathbf{K},\varepsilon)$ denotes that the wavefunction in the image plane also depends on the incident wavevector **K** and on the defocus ε. $f_s(K)$ and $f_T(\varepsilon)$ are the probability distribution functions of **K**, respectively ε. Expressions (A1), (A2) and (A3) are the basic expressions describing the whole real imaging process. They are also used for the computer simulation of high resolution images. However the computation of (A3) requires the computation of $\psi(\mathbf{R})$ for a large number of defocus values and beam directions, which in practice is a horrible task. For this reason (A3) has often been approximated. In order to study the effect of chromatic aberration and beam convergence (on a more intuitive basis) we will use a well-known approximation for (A3) introduced by Anstis and O'Keefe [69]. We assume a disk-like effective source function

$$f_s(\mathbf{K}) = \begin{cases} 1 \text{ for } |\mathbf{K}| < \dfrac{\alpha}{\lambda} \\ 0 \text{ for } |\mathbf{K}| > \dfrac{\alpha}{\lambda} \end{cases}$$

with α the apex angle of the illumination cone. We assume further that the integrations over defocus and beam convergence can be performed coherently, i.e. over the amplitudes rather than the intensities. This latter assumption is justified when the intensity of the central beam is much larger than the intensities of the diffracted beams so that cross products between diffracted beam amplitudes can be neglected. We assume that the defocus spread $f_T(\varepsilon)$ is a Gaussian centered on ε with a half width Δ. Assuming the object function $\psi(\mathbf{R})$ to be independent of the inclination $\mathbf{K}$, which is only valid for thin objects, one then finally finds that the effect of the chromatic aberration, combined with beam convergence, can be incorporated by multiplying the transfer function with an effective aperture function

$$D(\alpha,\Delta,\mathbf{g}) = B(\Delta,\mathbf{g})\, C(\alpha,\Delta,\mathbf{g}) \tag{A4}$$

where

$$B(\Delta,\mathbf{g}) = \exp(-1/2\ \pi^2\lambda^2\Delta^2\mathbf{g}^4) \tag{A5}$$

representing the effect of the defocus spread, and

$$C(\alpha,\Delta,\mathbf{g}) = 2J_1(|\mathbf{q}|)/|\mathbf{q}| \tag{A6}$$

with J_1 the Bessel function and $|\mathbf{q}| = (\mathbf{q}.\mathbf{q})^{1/2}$ which may be a complex function for a complex $\mathbf{q}$

$$\mathbf{q} = 2\pi\alpha\mathbf{g}\left[\varepsilon + \lambda g^2(\lambda C_s - i\pi\Delta^2)\right] \tag{A7}$$

$C(\alpha,\Delta,\mathbf{g})$ represents the combined effect of beam convergence and defocus spread. Some corrections have to be made when the convergence disk of a diffracted beam cuts the physical aperture [69].

The total image transfer can now be described as

$$\phi(\mathbf{R}) = \mathfrak{F}^{-1}_{\mathbf{R}}\, A(\mathbf{g}) \exp[-i\chi(\mathbf{g})]\ \mathfrak{F}_{\mathbf{g}}\ \psi(\mathbf{R}) \tag{A8}$$

i.e. the effective aperture yields a damping envelope function for the phase transfer function. Other approximations for including the effects of beam convergence and chromatic aberrations [70] using a Gaussian effective source lead to a similar damping envelope function. Experimentally obtained transfer functions confirm this behavior.

In (A8) the incoherent effects are approximated by a coherent envelope function. Hence it is called the coherent approximation. It is usually valid for thin objects. A full treatment of incoherent effects requires the calculation of the double integral in (A3). Another approximation which is valid for thicker objects is based on the concept of the transmission cross coefficient (TCC) [71]. Here it is assumed that beam convergence and defocus spread do not influence the diffraction in the object. Hence in (1.253) they do not appear in the object wavefunction but only in the phase transfer function. Now the wavefunction in the image plane (1.253) can be written as

$$\phi(\mathbf{R},\mathbf{K},\varepsilon) = \mathfrak{F}_R^{-1}\, T(\mathbf{g},\mathbf{K},\varepsilon)\, \psi(\mathbf{g}) \tag{A9}$$

with

$$T(\mathbf{g},\mathbf{K},\varepsilon) = A(\mathbf{g}) \exp(-i\chi(\mathbf{g},\mathbf{K},\varepsilon)) \tag{A10}$$

Substituting (1.255) into (A3) then yields after Fourier transforming

$$I(\mathbf{g}) = \mathfrak{F}_g(I(\mathbf{R})) = \int \psi(\mathbf{g}+\mathbf{g}')\, \tau(\mathbf{g}+\mathbf{g}',\mathbf{g}')\, \psi_*(\mathbf{g}')d\mathbf{g}' \tag{A11}$$

with

$$\tau(\mathbf{g}+\mathbf{g}',\mathbf{g}') = \int_\varepsilon \int T^*(\mathbf{g}+\mathbf{g}',\mathbf{K},\varepsilon)\, T(\mathbf{g}',\mathbf{K},\varepsilon)\, d\mathbf{K}\, d\varepsilon \tag{A12}$$

τ is the transmission cross coefficient, it describes how the beams $\mathbf{g}'$ and $\mathbf{g}+\mathbf{g}'$ are coupled to yield the Fourier component $\mathbf{g}$ of the image intensity.

Applications and Case Studies

1.39. Introduction

It is obvious that the number of applications of electron microscopy has become so large that an exhaustive review of all applications, even when restricted to materials science, is an almost impossible task. We shall therefore limit this review to a few typical examples of those applications for which electron microscopy and electron diffraction have made an essential contribution, leading to conclusions which could not have been obtained by other techniques.

For many problems high resolution images, diffraction contrast images and diffraction patterns provide complementary information. A separation on the basis of these different aspects would be artificial. In this survey we therefore focussed on the problem, rather than on the technique.

The interpretation of the images is only meaningful in terms of the underlying solid state phenomena. We shall therefore sketch for each application as briefly as possible the framework in which the images acquire their significance and are interpreted. The choice of the examples which is admittedly subjective is mainly motivated by the availability of suitable and aesthetic photographs.

1.40. The fine structure of dislocations

1.40.1 Measuring the stacking fault energy

In most materials the dislocations are not simple line defects but consist in fact of two or more partial dislocations connected by strips of stacking fault or of out-of-phase boundary. The simplest situation arises when glide takes place between two close-packed layers of identical "spherical" atoms in an elemental face centered cubic crystal. The glide motion along the (111) plane in the $[\bar{1}10]$ direction then follows the valleys, i.e. it takes place in two steps, each performed by the motion of a partial, the first with Burgers vector $\mathbf{b}_1$ = 1/6 $[\bar{2}11]$, the second with Burgers vector $\mathbf{b}_2$ = 1/6 $[\bar{1}2\bar{1}]$, enclosing an angle of 60° and leading to a symmetry translation 1/2 $[\bar{1}10]$ along the (111) glide plane. Between the two partials a stacking fault ribbon with a displacement vector, equal to one of the Burgers vector of the partials, is present [72].

The two partial dislocations repel one another since their Burgers vectors enclose an acute angle. In an infinite solid this repulsion is proportional with 1/d (d = partial separation) and its

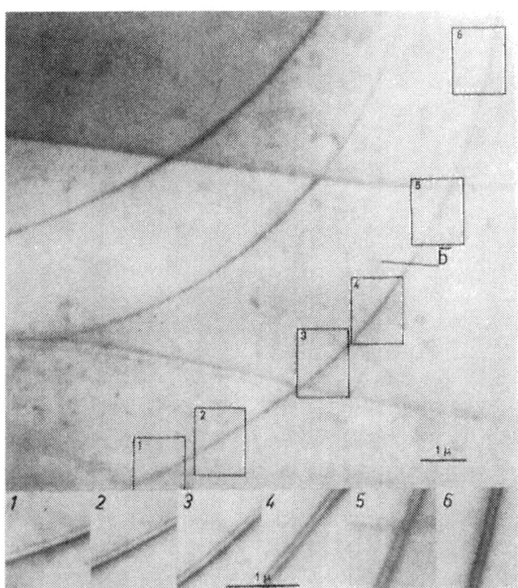

Figure 1.98 Curved dislocation in the (0001) plane of graphite. Several segments are reproduced as magnified insets. The direction **b** of the total Burgers vector, as determined by extinction experiments, is indicated. Note the systematic change in width orientation [73].

magnitude is a function of the orientation of the partials, i.e. of their character (screw or edge). The presence of the stacking fault ribbon causes an effective attractive force per unit length between the two dislocations, which is independent of their separation and numerically equal to the stacking fault energy γ. An equilibrium separation is thus established. Assuming the repulsive force law to be known it is then possible to deduce the stacking fault energy from the measured equilibrium separation of the partials. Dislocation ribbons are thus sensitive probes for measurements of the stacking fault energy, a quantity which is difficultly accessible in any other direct way. The following relations apply in an infinite isotropic solid:

$$d = d_0\left\{1 - \left[2\nu/(2-\nu)\right] \cos 2\phi\right\}$$

with

$$d_0 = \mu b^2(2-\nu)/8\pi\gamma(1-\nu)$$

where ϕ is the angle between the total Burgers vector and the ribbon direction; μ is the shear modulus and ν Poisson's ratio.

The orientation dependence of the ribbon width could be verified on an image such as Figure 1.98, which represents a curved dislocation in a graphite foil. The Burgers vectors were determined

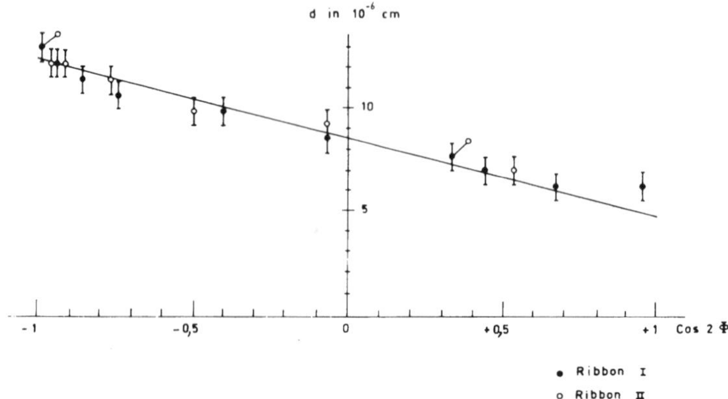

Figure 1.99 Plot of ribbon width d as a function of cos 2ϕ [73].

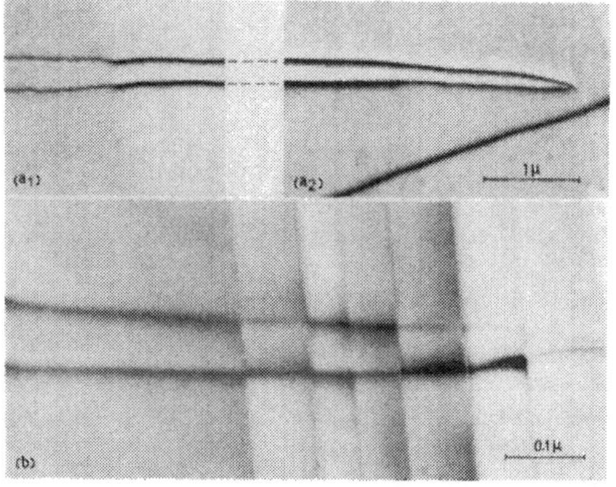

Figure 1.100 A wide ribbon in SnS_2 gradually approaching the surface. As the ribbon crosses surface steps it becomes discontinuously narrower. The ribbon closes where it emerges in the surface [74].

using the method described in § 1.29.4. Plotting d as a function of cos 2ϕ the slope of the straight line so obtained gives the effective value of the Poisson ratio as well as the intercept d_0 to be used in the second relation, which then yields a value for the stacking fault energy (Figure 1.99).

Using this method it is implicitely assumed that the repulsive force between dislocations is proportional to 1/d which is only the case in an infinite solid. In a thin foil the repulsive force between dislocations parallel to the foil surfaces decreases with decreasing distance to the specimen surfaces. This behaviour can be observed, as in Figure 1.100, where a ribbon gradually approaching the

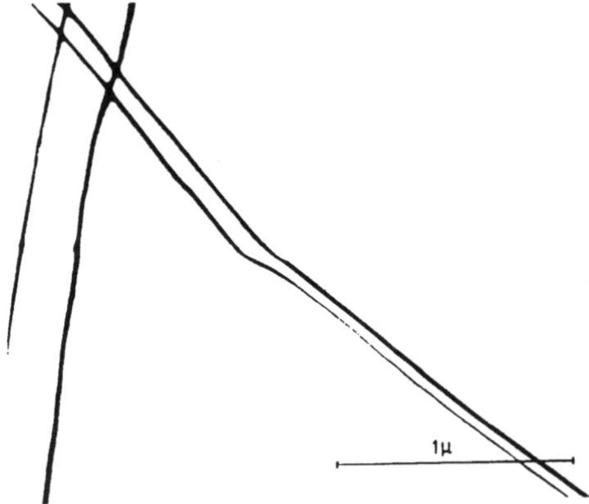

Figure 1.101 Dislocation ribbon in SnS_2 . Refraction, accompanied by a width change occurs on passing underneath a surface step [74].

Figure 1.102 High resolution images of dislocation ribbons in silicon as viewed along the close-packed direction parallel with the partial dislocation lines. (Courtesy H. Bender)

surface in a wedge shaped lamella of tin disulphide closes as it emerges at the surface.

The energy of a dislocation ribbon depends on its distance to the surface. As a result the shape of minimum energy of a dislocation ribbon crossing a surface step is not a straight line; "refraction" of the ribbon as well as a change in width occur on passing underneath the surface step (Figure 1.101). The *index of refraction* is the ratio of the total energies of the ribbon in the two parts of the foil on either side of the surface step.

These images prove that such surface effects are not negligible. When measuring stacking fault energies care should thus be taken to use foils of maximum thickness compatible with the visibility of the dislocation and moreover take the widest ribbon as the most representative one. The width of narrow stacking fault ribbons can best be determined by imaging in the weak-beam mode (§ 1.28.4).

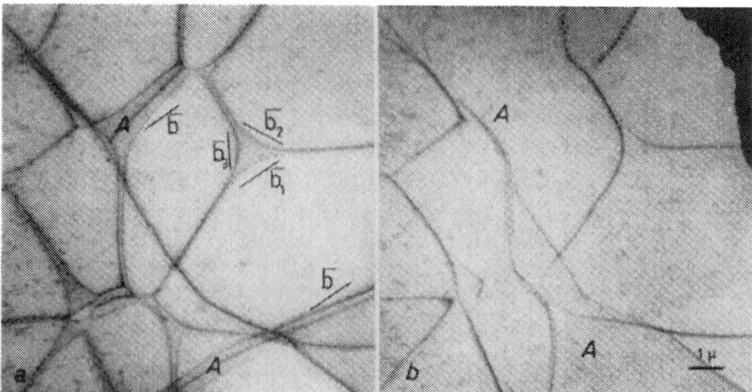

Figure 1.103 (a) widely extended dislocation node of partial dislocations in graphite. In A a triple ribbon is present; the three partials have the same Burgers vector as follows from the contrast experiment in (b) where the three partials are simultaneously out of contrast. Nevertheless the contrast at the three partials is different in (a) [73].

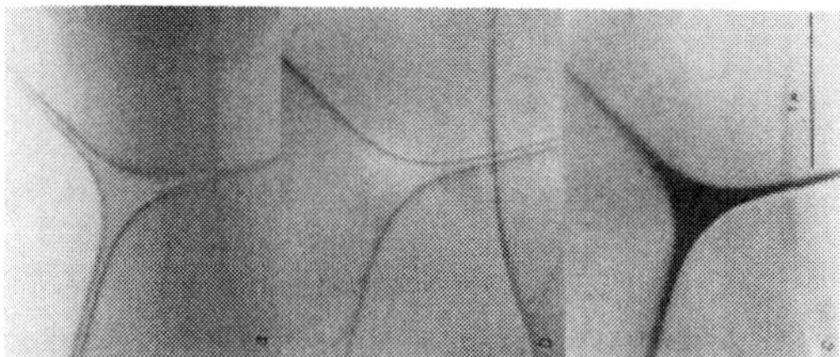

Figure 1.104 Isolated extended node in the (0001) plane of AlN [75].

An alternative method of measuring ribbon widths consists in the use of high resolution images with the viewing direction along the partial dislocations. Such images are shown in Figure 1.102 for ribbons in Silicon; they reveal the stacking fault as well as the structure of the partial dislocations. The ribbon width is in this geometry not so sensitive to the foil thickness as in the parallel configuration discussed above.

Other geometrical configurations involving stacking faults can be used, such as the separation of partials in triple ribbons in graphite (Figure 1.103) and in close packed structures, or the radius of curvature of single partials in a network of extended-contracted nodes (Figure 1.104). In the latter case one has approximately $\gamma = 1/2\mu \mathbf{b}^2/R$ (R = radius of curvature; **b** = Burgers vector of the partial). Isolated extended nodes such as the one reproduced in Figure 1.104 (observed in AlN) are particularly suitable. More accurate relations are discussed in Ref. [74,G3].

1.40.2. Multiribbons

Ordering in alloys based on close packed structures leads to long symmetry translations along the glide directions in the close-packed glide planes. As a result ribbons consisting of several partials, separated either by stacking faults or by out-of-phase boundaries result. The equilibrium separation of superdislocations (i.e. perfect dislocations with respect to the basic lattice, but partial dislocations with respect to the ordered structure) can be used to derive values of the anti-phase boundary energy in the same way as described above for stacking faults. In Ni_4Mo as many as ten partial dislocations are connected by faults and anti-phase boundaries [G3].

The dislocations involved in glide between the close-packed layers of anions (X) in layered ionic sandwich crystals of the CdI_2 type AX_2 (XAXXAX ...) or AX_3 are of particular interest. The glide motion takes place between the two weakly Van der Waals bonded close-packed anion layers. Dislocations can thus dissociate into two or more Shockley partials. Although in the close-packed layers, between which glide takes place, all X atoms are equivalent, the A cations in the adjacent central layers of the sandwiches may form configurations which impose a large unit mesh in the glide plane, either because not all octahedral cation sites are occupied (e.g. in $CrCl_3$, $CrBr_3$) or because the cations form metal-metal bonded clusters leading to a slight deformation of the close-packed layers (e.g. in $NbTe_2$, $TaTe_2$) and the creation of a superperiod.

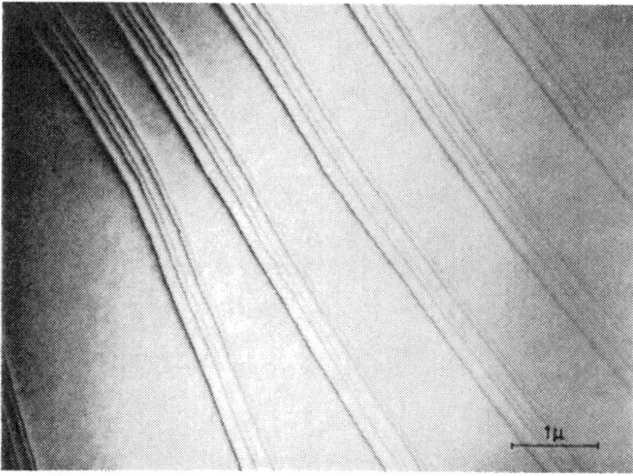

Figure 1.105 Sixfold ribbons of partials in the (0001) plane of $CrCl_3$. The Burgers vectors of the partials form a zigzag glide path [76].

In the chromium trihalides [76] multiribbons containing either four or six partials are observed. Assuming glide along the close-packed anion layers to take place by the propagation of Shockley partials, two types of stacking fault ribbons can be distinguished:

(i) faults violating only the chromium stacking, i.e. involving only third neighbours

(ii) faults violating the stacking of the chromium ions as well as that of the anions, i.e. involving next nearest neighbours.

Intuitively it is clear that the type (ii) faults will have a larger energy than those of type (i). The sixfold ribbons correspond with a "straight" zigzag glide path along the close-packed directions in the (0001) glide plane of the anion sublattice; they contain the two types of faults in an alternating fashion, the outer ribbons corresponding with high energy faults. Diffraction contrast images of such ribbons are reproduced in Figure 1.105. The outer ribbons are clearly the narrowest ones. The structure of the fourfold ribbons can similarly be related to the structure.

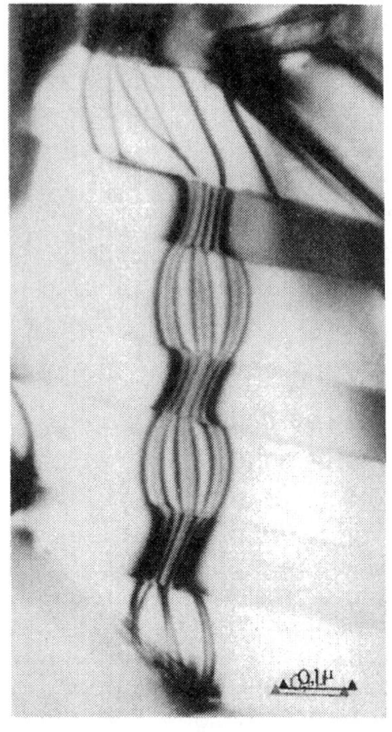

Figure 1.106 Sixfold ribbon of partials in $NbTe_2$ intersecting domain boundaries along which the underlying structure changes 60° in orientation. In half of the domains the sixfold ribbons separate into three twofold ribbons which form bulges as a result of repulsive forces [77].

Also in $NbTe_2$ [77], which has a deformed CdI_2 structure, sixfold ribbons occur. In this structure the Nb ions form clusters of three parallel close-packed niobium rows, having a somewhat smaller separation than in the ideal hexagonal structure which probably occurs only in the temperature range where the crystal is grown. The resulting structure then becomes monoclinic on cooling. The unit mesh in the glide plane is now a centered rectangle, which can adopt three different but equally probable orientations differing by 60°. As a consequence the room temperature structure is fragmented in domains corresponding with the three possible orientations of the clustered niobium rows. The monoclinic symmetry causes the glide paths along the three close-packed directions within the same domain to become non-equivalent.

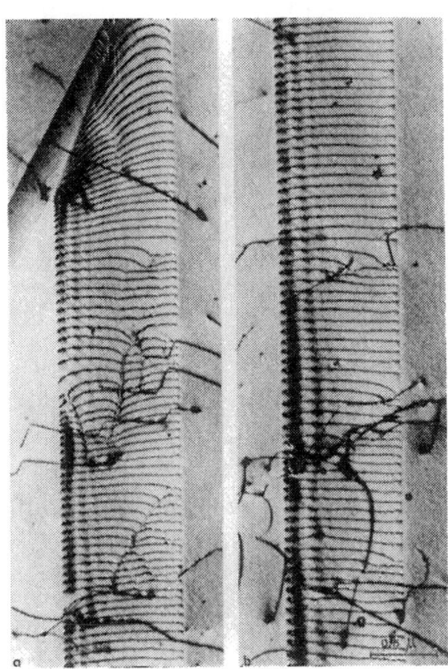

Figure 1.107 Procession of dislocations confined to a glide plane in stainless steel. Note the wavy contrast close to the surfaces and its absence in the central part of the foil. (High voltage electron micrograph).

The zigzag glide paths in the direction enclosing an angle of 30° with the long side of the rectangular mesh consist of six partials whereas the glide path along the other close-packed direction (i.e. along the short side of the rectangle) repeats after two partials. The Burgers vector is conserved all along dislocation lines. Hence when a sixfold ribbon passes through a domain wall the glide path changes its orientation relative to the underlying structure. A sixfold ribbon in one domain is thus transformed into three separate twofold ribbons in the adjacent domain. Whereas in the sixfold ribbon the six partials are held together by stacking faults, this is no longer the case with the three twofold ribbons which repel one another and hence develop "bulges". The image of Figure 1.106 illustrates the described behaviour of a sixfold ribbon intersecting a set of parallel domain boundaries, in $NbTe_2$ [77].

1.40.3 Plastic deformation: glide dislocations

Plastic deformation is one of the subjects that has been intensely studied at an early stage by means of diffraction contrast (see e.g. EMTC and EFP). High voltage electron microscopy (≈1000 kV) has been of considerable interest in this respect because it is possible to study thicker foils, which are more representative of the bulk material than the thin foils required at 100 kV. Figure 1.107 shows a procession of glide dislocations in FCC stainless steel, confined to their (111) glide plane, as observed in high voltage electron microscopy. The strictly planar arrangement implies that the dislocations are dissociated and that for this reason the cross-glide is a difficult process. The dissociation is too small to be directly observable at this resolution, but it has been found from other images that the stacking fault energy is rather small in stainless steel. Note the periodic contrast of the dislocations in the vicinity of their emergence points in the foil surfaces and the absence of such contrast in the central part of the foil. In Figure 1.108 which refers to a low stacking fault energy alloy (Cu-Ga) the dissociation is clearly visible; stacking fault fringes can be observed between the partials.

Figure 1.109 shows a network of intersecting glide dislocations confined to the (111) glide plane in a FCC copper alloy (Cu-Ga) with low stacking fault energy. One set of dislocation nodes is dissociated and gives rise to the dark triangular areas; the other set is contracted. Such nodes allow to deduce the stacking fault energy from the curvature of the partial dislocations forming the extended nodes.

The image of Figure 1.110 shows glide dislocations in the

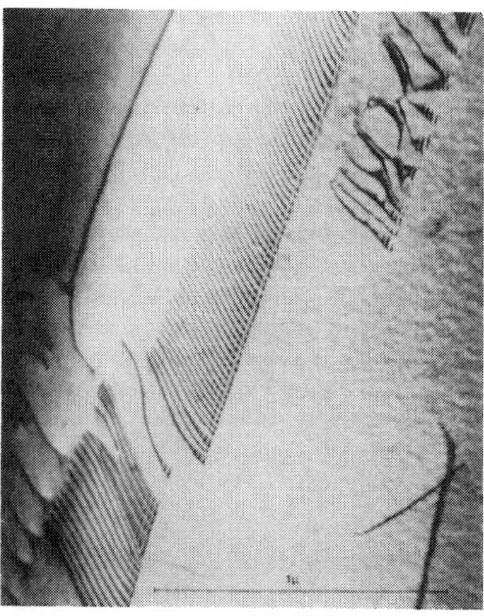

Figure 1.108 Ribbons dissociated in Shockley partials observed in Cu-Ga [78].

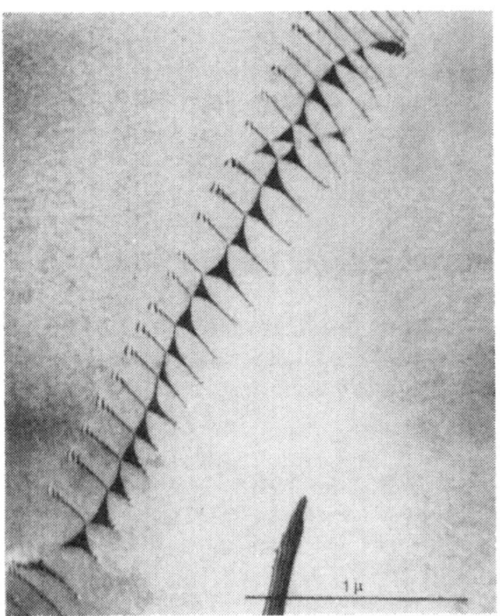

Figure 1.109 Network of dissociated dislocations in a Cu-Ga alloy with stacking fault energy [78].

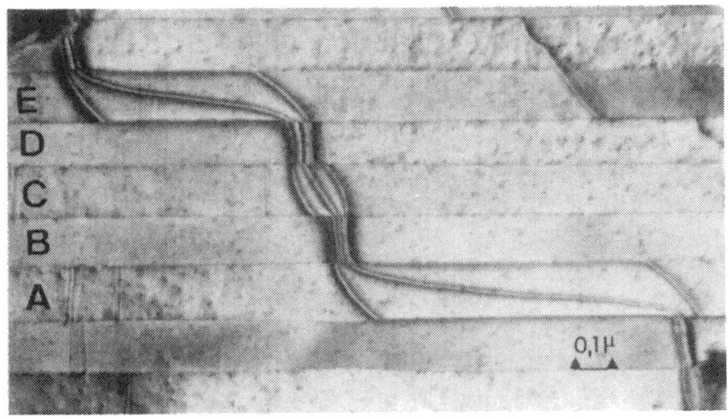

Figure 1.110 Glide dislocations in the layer plane $NbTe_2$. Note the interaction between dislocations and twin domain walls [77].

layer plane (001) of $NbTe_2$, which is parallel to the foil plane, the specimen having been obtained by cleavage. In every other domain the dislocation multiribbons consist of six partials; in the remaining domains the dislocations are simple ribbons, as described above in § 1.40.2.

The image illustrates the strong interactions of the glide dislocations with the domain walls. On entering a domain in which the sixfold ribbons would have to be formed, the single ribbons line up with the domain wall, minimizing in this way the generation of stacking faults. This leads to an effective interaction between dislocation ribbons and domain walls.

1.40.4. The structure of subgrain boundaries

Small angle grain boundaries can be described in terms of arrays of dislocation lines. Diffraction contrast electron microscopy has contributed significantly to firmly establishing dislocation models for such boundaries.

A general subgrain boundary is characterized by five parameters describing its geometry: the rotation axis, the rotation angle and the normal to the contact plane. By the combined use of the spot diffraction pattern and of the Kikuchi line pattern these parameters can be determined. The diffraction contrast image then allows to visualize the geometry of the dislocation lines, and using the extinction criterion to determine their Burgers vectors. If the rotation axis is parallel to the contact plane the boundary is a tilt boundary and the dislocation configuration consists of parallel lines. If on the other hand the rotation axis is perpendicular to the contact

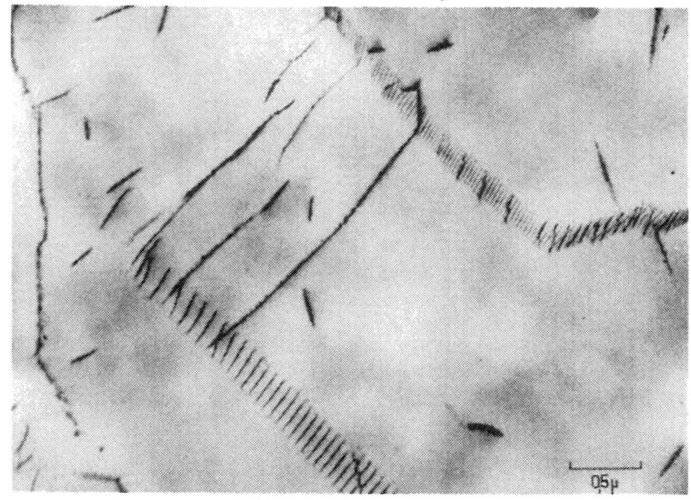

Figure 1.111 Tilt boundary consisting of sets of parallel dislocations in Niobium. Some of the dislocations are decorated by small particles. (Courtesy A. Fourdeux and A. Berghezan).

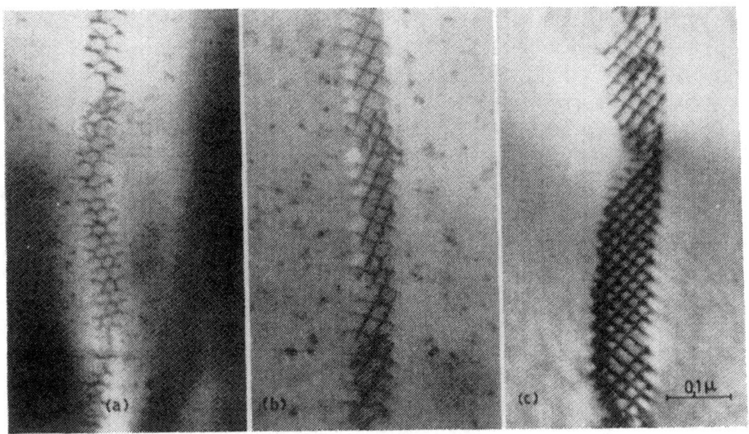

Figure 1.112 Twist boundaries in platinum (a) hexagonal network (b) and (c) square networks [79]

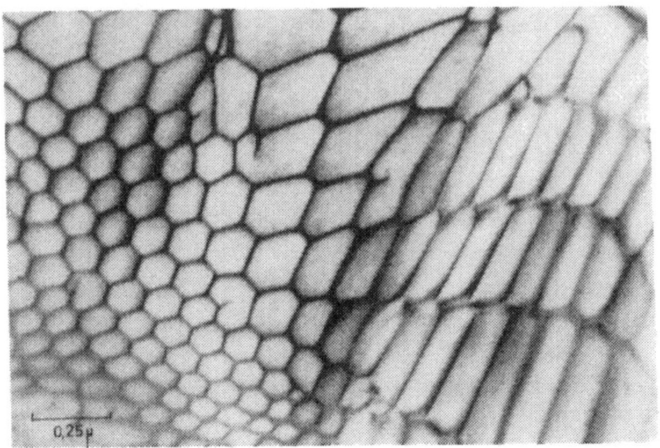

Figure 1.113 Hexagonal network of undissociated dislocations (0001) plane of zinc [80].

plane the boundary consists of a network of interacting dislocations. Depending on the symmetry of the contact plane this network may ideally consist of square meshes or of hexagonal meshes.

Figure 1.111 shows two tilt boundaries, in body centered niobium, consisting of parallel dislocation lines. Some of the dislocations are decorated by small precipitate partials.

Images of twist boundaries in platinum are reproduced in Figure 1.112; (a) is a hexagonal network containing three intersecting families of dislocations with Burgers vectors enclosing angles of 120°; (b) and (c) represent square networks, consisting of dislocations with mutually perpendicular Burgers vectors.

Figure 1.113 shows a well-developed hexagonal network of undissociated dislocations in the (0001) plane of hexagonal zinc. All dislocations are mobile in the plane of the boundary. The right part of the boundary moved along the (0001) glide plane during the exposure, leading to blurring of the image. In the left part of the image, some of the dislocations, leaving the network and terminating in the foil surfaces, have become sessile; they hereby pinned the network in that part.

Figure 1.114 shows a hexagonal network of widely extended dislocations in the basal plane of graphite. The network is in fact a glissile twist boundary. From the curvature of the partials in the extended nodes one can deduce the stacking fault energy.

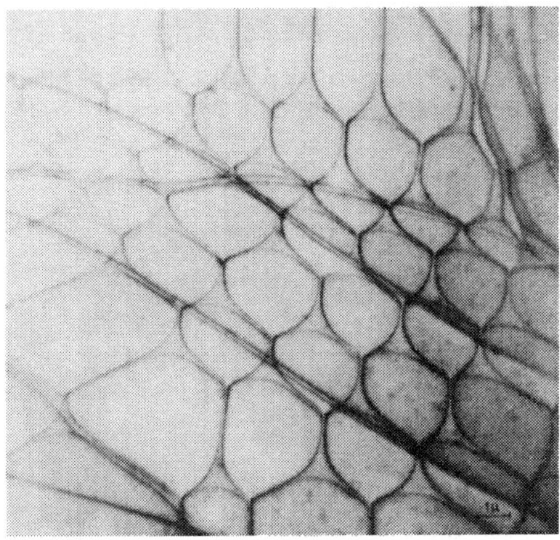

Figure 1.114 Network of dissociated dislocations in the (0001) plane of graphite [73].

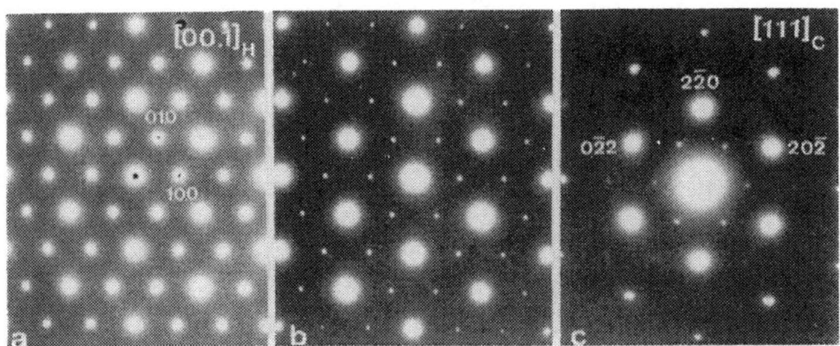

Figure 1.115 Evolution under electron irradiation of the diffraction pattern of Fullerite (C^{60}) from (a) hexagonal to cubic (c) [81].

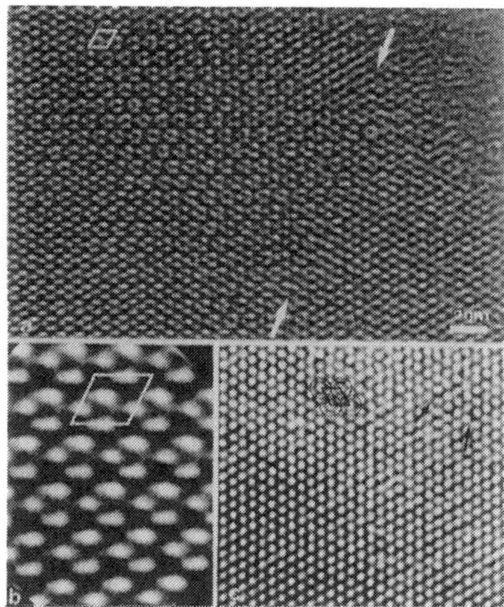

Figure 1.116 High resolution images of a Fullerite microcrystal consisting mainly of C_{60} molecules [81].

1.41. Small particles

Certain materials can only be obtained as small crystals because only a limited quantity of substance is available. Other particles have to be very small because this is essential for their use, such as catalysts, silver halides in photographic emulsions, or magnetic particles for recording. The crystallographic study of such particles can only be performed by means of electron microscopy.

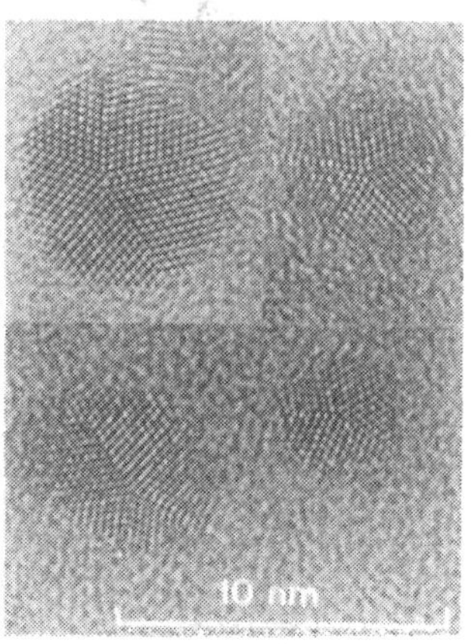

Figure 1.117 High resolution image of multiply twinned silver microcrystals obtained by evaporation in vacuum on an amorphous substrate. (Courtesy C. Coessens)

An interesting example is Fullerite, a material consisting of large C_{60} molecules. These molecules form spherical shells of carbon atoms interconnected as hexagons and pentagons adopting the topology of a soccer ball. In the as grown microcrystals, from organic solvents, the layers of hollow spheres are stacked in the hexagonal fashion. It could be shown that under electron irradiation in the microscope vacuum the stacking transforms "in-situ" into a cubic one, i.e. the microcrystal undergoes a phase transformation. Figure 1.115 shows the evolution of the diffraction pattern whereas Figure 1.116 shows the high resolution image of a Fullerite crystallite along the [111] zone [81].

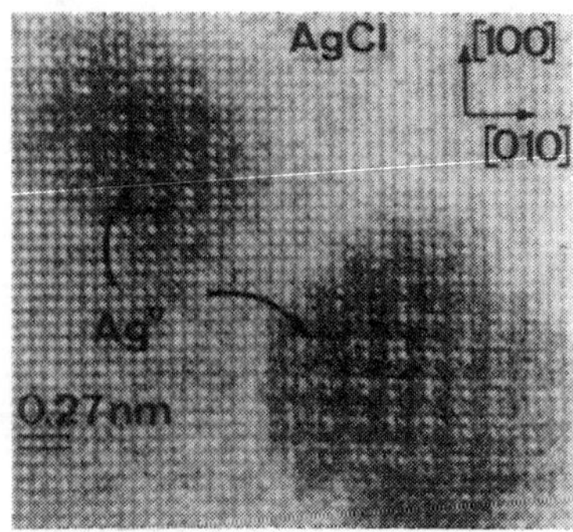

Figure 1.118 Silver chloride microcrystal exposed to electron irradiation irradiation. An expitaxial layer of metallic silver has been formed on the surface and gives rise to a coincidence pattern [82].

Very small silver particles obtained by evaporation in vacuum on an amorphous carbon substrate exhibit pentagonal symmetry due to multiple twinning on {111} planes. Since the angle between two {111} planes in the FCC lattice is 70° 1/4 five repeat twins on {111} around the same axis leave an angular gap of 8° 3/4, i.e. a grain boundary should be present. Most particles seem to contain five equivalent coherent twin interfaces (Figure 1.117), this suggests that in such cases the fivefold axis is in fact a discontinuity line, accommodating the angular misfit.

Although the silver halides are very radiation sensitive it is nevertheless possible to observe sufficiently small microcrystals under high resolution conditions, during some short period of time.

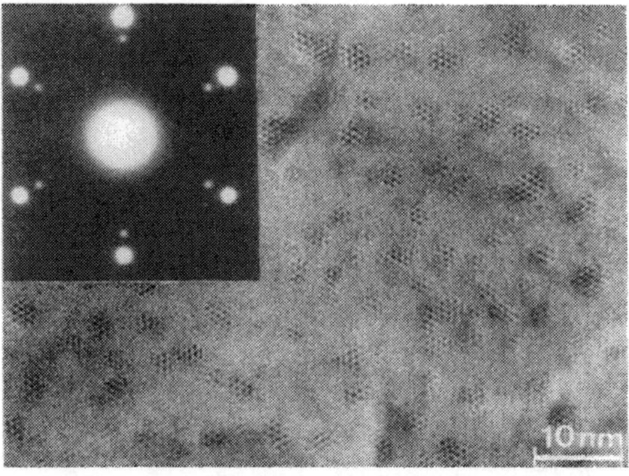

Figure 1.119 Small particles of lead embedded in an Aluminium matrix. Inset is the corresponding diffraction pattern [83].

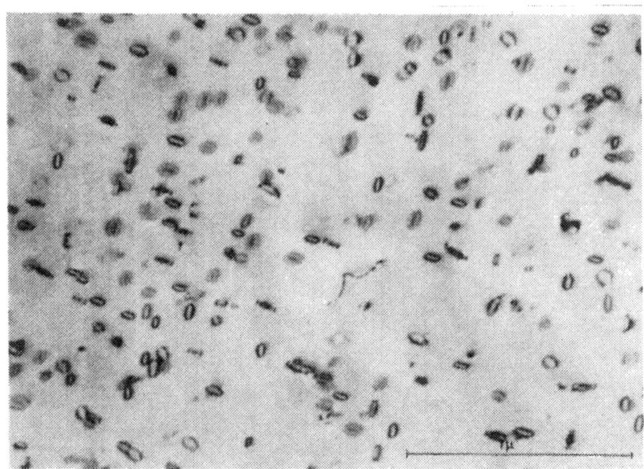

Figure 1.120 Unfaulted dislocation loops in quenched aluminium [84].

The "print-out" process could in this way be followed "in-situ". Surface layers of AgCl are "peeled off" and simultaneously epitaxial layers of metallic silver are formed. Figure 1.118 shows the superposition pattern of photolytic silver on a cube plane of AgCl.

Small particles of lead, embedded in an aluminium matrix formed by ion implantation, are imaged in Figure 1.119. The diffraction pattern reveals the orientation relationship between matrix and precipitate as well as the difference in lattice parameter between the two. The high resolution image reveals a two-dimensional moiré pattern; under high resolution conditions this dot pattern can also be interpreted as a coincidence pattern of atomic columns [83].

1.42. Point defect clusters

Vacancies in quenched metals form disc shaped agglomerates in $(111)_{FCC}$ or $(0001)_{HCP}$ layers, limited by Frank type dislocation loops. If the stacking fault energy is large enough the loop is "unfaulted" since energy is gained by nucleating a Shockley partial and sweeping the loop, transforming the sessile Frank loop into a perfect glissile loop. Such unfaulted loops in quenched aluminum are shown in Figure 1.120.

If the stacking fault energy is small enough, which is for instance true in gold and in Ni-Co alloys, the Frank loop is transformed into a stacking fault tetrahedron consisting of four intersecting triangular stacking faults in {111} planes, limited along their intersection lines by edge type stair rod dislocations with a

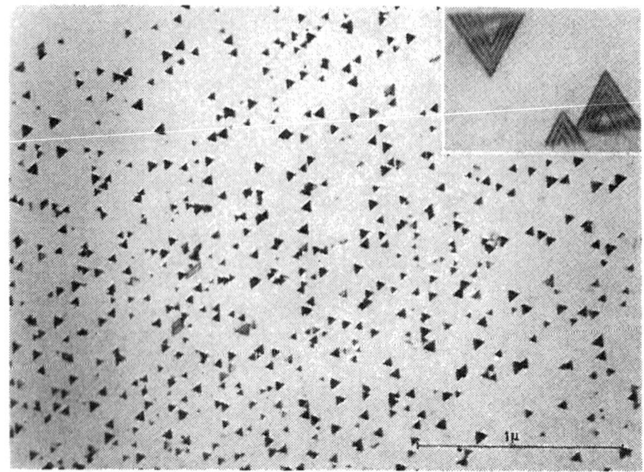

Figure 1.121 Diffraction contrast image of stacking fault tetrahedra in quenched gold. The inset shows a magnified image [85].

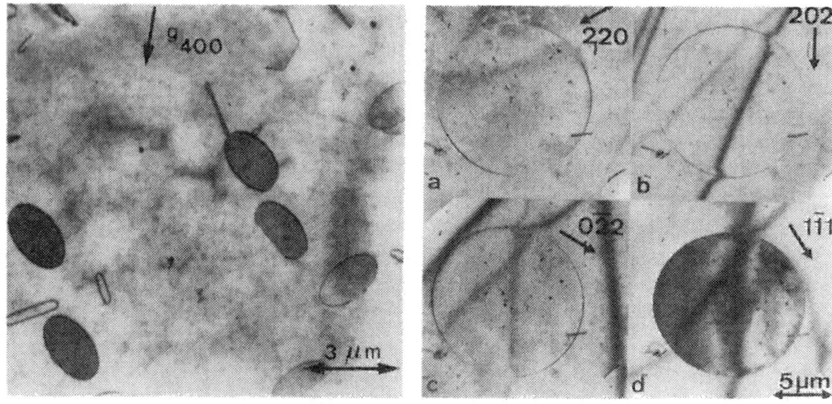

Figure 1.122 Faulted Frank loops in Silicon due to interstitials (courtesy H. Bender) (i) Loops exhibiting stacking fault contrast (ii) contrast experiment on faulted loop in Silicon

Burgers vector of the type 1/6 [81]. For intermediate values of the stacking fault energy the Frank loops may remain faulted. In Figure 1.121 stacking fault tetrahedra in gold are imaged in diffraction contrast.

Faulted Frank loops in silicon are imaged in diffraction contrast in Figure 1.122(i). The presence of the stacking fault causes contrast inside the loop. Figure 1.122(ii) shows a contrast experiment on an extrinsic Frank type dislocation loop in silicon. Note the presence of a line of no contrast perpendicular to the active g-vector. Note also the deformation of the extinction contours where they cross the dislocation loop. For $\mathbf{g} = [1\bar{1}1]$ the loop exhibits stacking fault contrast as do the loops in Figure 1.122(i).

Figure 1.123 Diffraction pattern of a quasi-crystal of Al-Mn exhibiting fivefold symmetry. (Courtesy Van Tendeloo)

1.43. Quasi-crystals

This new class of materials, characterized by the presence of "non-crystallographic" symmetry elements, such as 5, 8, 10 and 12-fold rotation axis and the simultaneous absence of translation symmetry, was first observed by electron diffraction and electron microscopy [86]. Figure 1.123 shows the diffraction pattern of a quenched Al-Mn alloy along the fivefold zone axis and the corresponding high resolution image is reproduced in Figure 1.124: note the absence of translation symmetry.

1.44. Mixed layer compounds

High resolution electron microscopy is particularly suitable for the study of homologous series of mixed layer compounds such as $As_2Te_3(GeTe)_n$ These compounds consist of a variable number n of GeTe layers, which have a slightly deformed sodium chloride structure, alternating with single five layered lamellae of As_2Te_3.

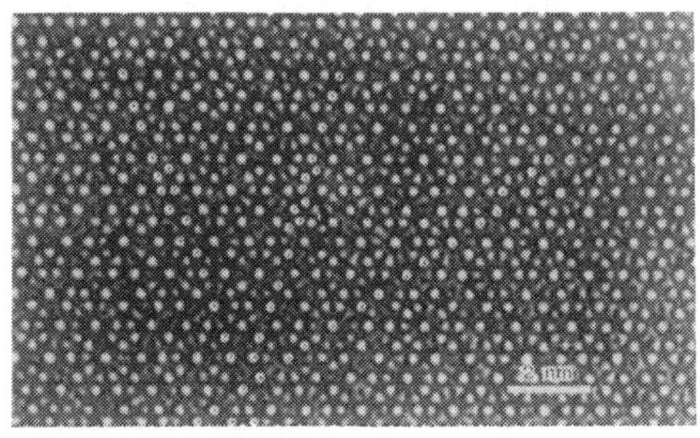

Figure 1.124 High resolution image of an icosahedral quasi-crystal of Al-Mn; note the absence of transition symmetry and the presence of pentagonal dot arrangements. (Courtesy Van Tendeloo)

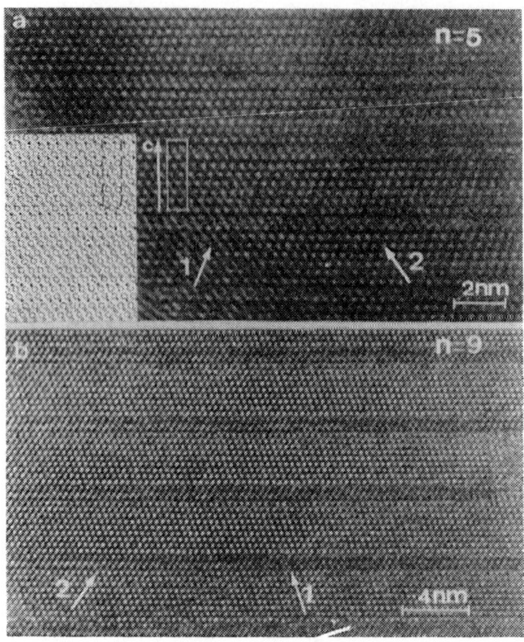

Figure 1.125 HREM images of $(As_2Te_3)(GeTe)_n$ with n = 5 and n = 9 as viewed along the close-packed atom rows [87].

The (111) layers of GeTe fit perfectly on to the close-packed layers of the As_2Te_3 lamellae giving rise to hexagonal or rhombohedral layer structures depending on the n-value. High resolution images made along close-packed rows of the layers, reveal directly the stacking. In particular the As_2Te_3 lamellae can clearly be distinguished from the blocks of GeTe layers and the number n of GeTe layers can be observed directly. In Figure 1.125 the n values are for instance n = 5 and n = 9 [87].

In the Y-Ba-Cu-O system of superconducting compounds a number of layered structures with various compositions were identified and imaged [88]. In orthorhombic (quasi-tetragonal) $YBa_2Cu_3O_{7-\delta}$ the succession of layers is

$$... CuO - BaO - CuO_2 - Y - CuO_2 - BaO ...$$

The CuO layer consists of Cu-O-Cu-O chains parallel with the b_0 direction. This single chain layer can be replaced by a layer of double chains $(CuO)_2$ along b_0. If each CuO layer is replaced by a

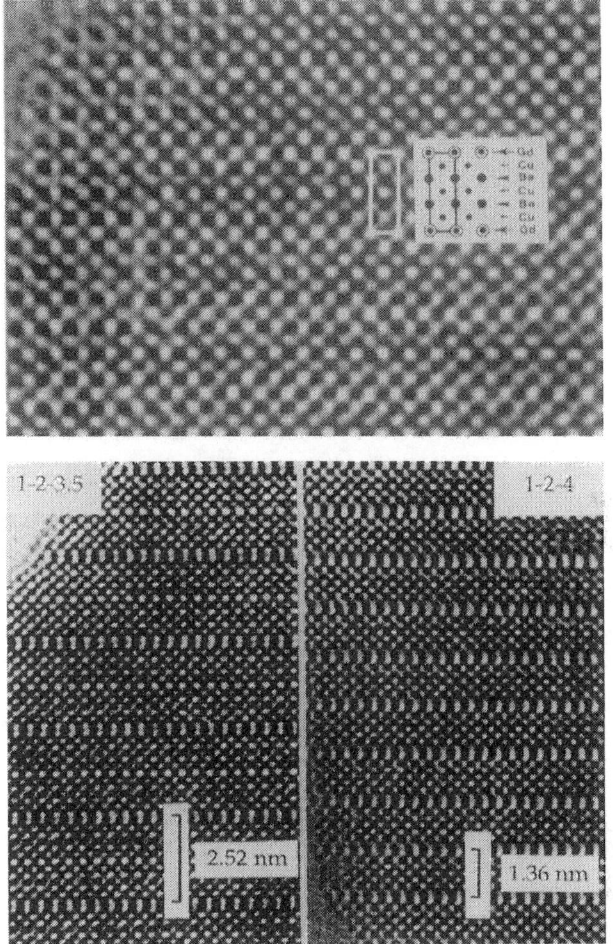

Figure 1.126 Comparison of the structures of the 1-2-3, 1-2-4 and 1-2-31/2 compounds in the Y-Ba-Cu-O system (a)$YBa_2Cu_3O_{7-\delta}$ (1 – 2 – 3); (b) $YBa_2Cu_4O_8$ (1 – 2 – 4); (c) $Y_2Ba_4Cu_7O_{15}$(1 – 2 – 3.5); (CourtesyT. Krekels) [88]

double layer $(CuO)_2$ the composition is $YBa_2Cu_4O_8$. A view along the b_0 direction shows the open channels in the double chain layers as rows of elongated bright dots, whereas along the a_0 direction the double chain layer is revealed as a staggered double row of bright dots. In $Y_2Ba_4Cu_7O_{15}$ only every other CuO layer is replaced by a double chain layer. The structure images of these different phases are compared in Figure 1.126.

Figure 1.127 High resolution image of SnNbS3 and the corresponding simulated image as an inset [89].

1.45. The MTS_3 compounds

The MTS_3 compounds (M = Rb, Sn, ...; T = Nb, Ta, ...) form a remarkable series of mixed layer compounds. They consist of a regular alternation of TS_2 layers having a trigonal prismatic NbS_2 type of structure and of two-layer lamellae MS having a slightly deformed sodium chloride structure. The striking feature is that successive layers have different rotation symmetries. The TS_2 layers have hexagonal symmetry whereas the MS layers have fourfold symmetry. The distance between successive atom rows is the same for the close-packed rows in the TS_2 layers as for the <100> rows in the sodium chloride layer. One family of atom rows of one layer fits in the "grooves" formed by atom rows in the adjacent layer and hence the lattice parameter along the direction normal to these rows is common for both types of lamellae. On the other hand the a_0 parameter along the rows is very different for the two types of lamellae. Along this direction the two types of layers "modulate" one another. The diffraction pattern reveals this mutual modulation; along the zone normal to the layers it exhibits the superposition of a hexagonal grid of strong spots due to the TS_2 layers and a square grid of strong spots due to the MS layers. Moreover each spot of one grid is surrounded by an array of weaker satellite spots having the geometry of the other grid. Such a diffraction pattern is reproduced in Figure 1.127 together with the corresponding high resolution image (see also § 13.7.3).

The high resolution image exhibits a quasi-periodic variation of the brightness of the dots which image the atom columns. The brightest dots form a quasi-periodic pattern based on a centered elongated rectangular unit mesh. Strictly speaking the pattern is not periodic since there is no simple relation between the a_0 parameters of the two types of lamellae.

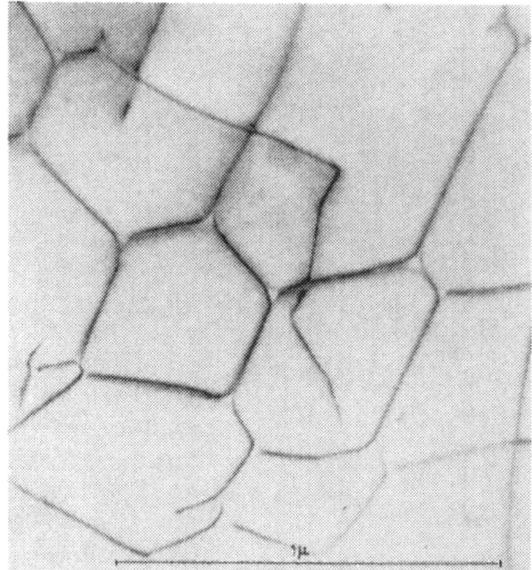

Figure 1.128 Network of extended dislocations in silicon: all nodes are dissociated [90].

Figure 1.129 Stacking fault in 2H wurtzite in HREM [91].

1.46. Planar interfaces

It is well known that two simple types of stacking faults can occur in the face-centered cubic structure. The intrinsic fault, formed either by the extraction of a layer, or by glide of a Shockley partial, is represented by the stacking symbol abcabcacabc ... . The extrinsic fault, formed for instance by the precipitation of interstitials in a Frank loop corresponds with the stacking symbol abcabacabc ... . The two types of faults have comparable energies in certain materials. In a

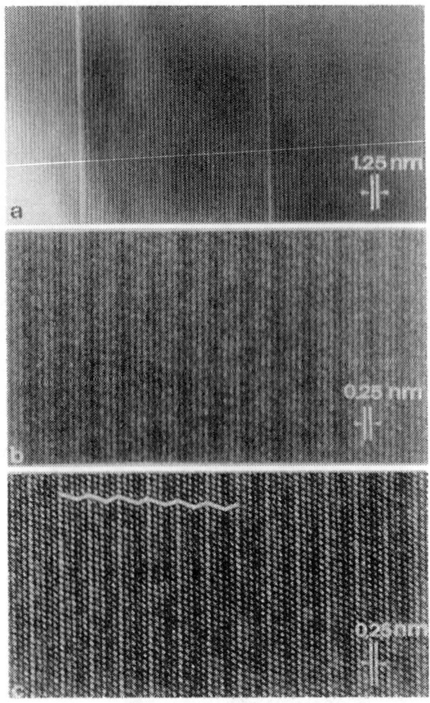

Figure 1.130 HREM image of the 15R polytype of SiC (abcbc...) (a) Only two successive spots in the 000ℓ row are used in the image formation; the period corresponding with the five layer lamellae is revealed. Two stacking different spacing. (b) A cluster of five spots in 000ℓ row is used to image the structure; individual atom layers are now visible within the five-layer lamellae. (c)Several parallel rows of spots are used. The stacking sequence can now be observed. [91]

network of dissociated dislocations all nodes are then dissociated; this is for instance the case in silicon (Figure 1.128) and in certain alloys (Ag-Sn). The two kinds of faults have opposite displacement vectors of the type a/3 [111] and can thus be distinguished by the characteristic fringe pattern which they produce when situated in inclined planes (see § 4.4.1). They can be identified directly in high resolution images, as in Figure 1.102 where intrinsic stacking faults are present between the partials.

A stacking fault of the type ... ababcacac ... in 2H wurtzite is imaged in Figure 1.129 by means of HREM.

Figure 1.130 is a lattice image of a crystal of the 15R polytype of SiC with stacking sequence abcbć ... . Lattice fringes with a spacing equal to the thickness of the five layered lamellae exhibit different lattice spacings at the level of two stacking faults. In Figure 1.130 all single atom layers are imaged as fringes. Finally in Figure 1.130 enough reflections were collected to produce atomic resolution of the 15R stacking.

A twin boundary imaged along a common atomic row is reproduced in Figure 1.131, which refers to $YTaO_4$. The structure is monoclinic; the twin plane is (001).

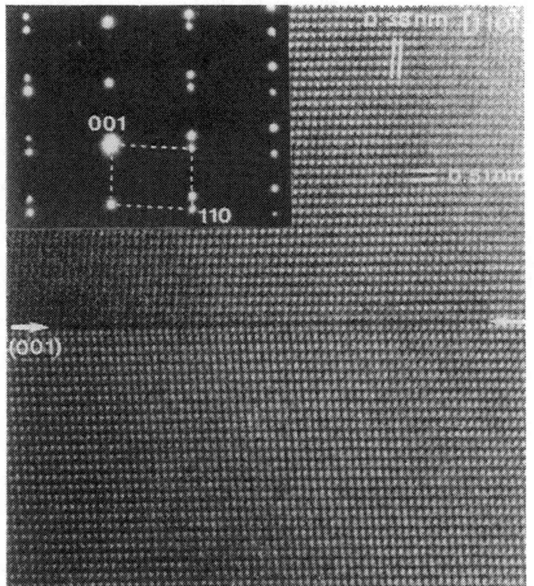

Figure 1.131 Twin interface along (001) in $YTaO_4$ imaged along a common lattice row (Courtesy G. Van Tendeloo).

1.47. Domain structures

Phase transformations are usually accompanied by a decrease in symmetry with decreasing temperature. As a result a single crystal of a higher symmetric phase becomes fragmented into domains of which the structures are related by the symmetry elements lost in the transition to the lower symmetry phase [17]. The lost rotation symmetry elements give rise to orientation variants of the low temperature phase of which the number is given by the ratio of the order of the point group of the high temperature phase and the order of the point group of the low temperature phase. The loss of translation symmetry gives rise to translation variants related by displacement vectors given by the lost lattice translations. Their number is determined by the ratio of the volumes of the primitive unit cells of the low and high temperature phases [17].

Orientation variants are separated by domain boundaries, whereas translation variants are separated by out-of-phase boundaries. The orientation of the domain boundaries is determined by the requirement that the strain energy should be a minimum. This will be the case for strain-free interfaces. As a result the orientation of certain interfaces (W) follows entirely from symmetry whereas others (W′) have orientations which depend on the lattice parameters of the two phases involved, at the transition temperature.

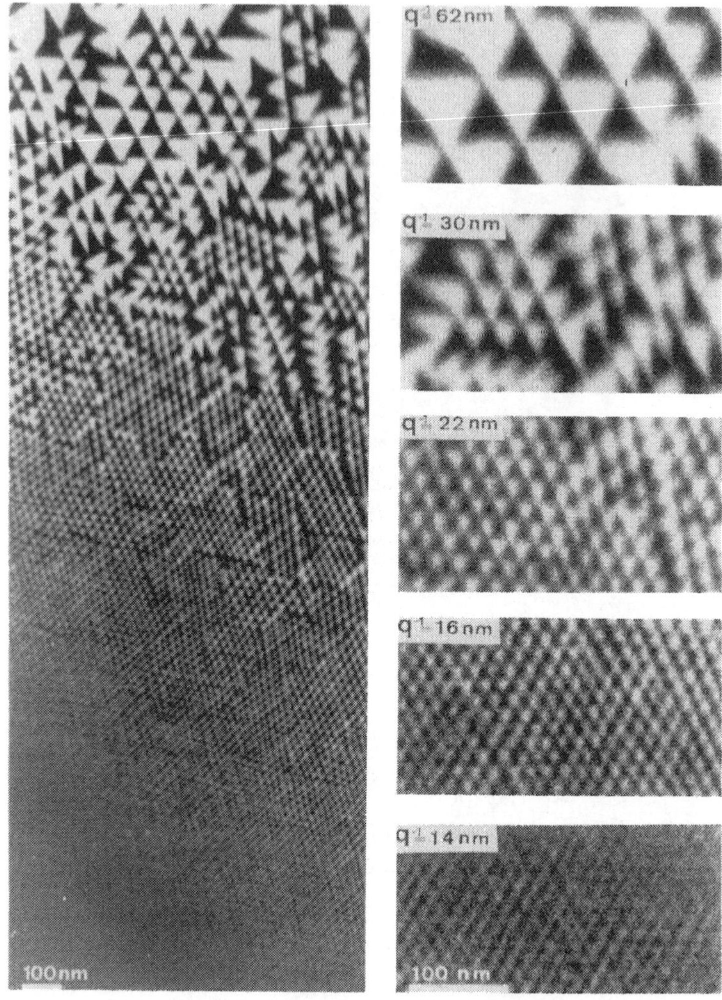

Figure 1.132 Domain fragmentation in quartz as a function of temperature. A temperature gradient is present across the specimen. At the highest temperature the incommensurately modulated phase is observed [16].

For example, in the α-β transition of quartz, referred to above, the α-phase has the point group 32 (order 6) and the β-phase the point group 622 (order 12). The number of orientation variants in the α-phase is thus $12 \div 6 = 2$ (α_1 and α_2; Dauphiné twins) and they are related by the lost 180° rotation about the threefold axis. There is no change in translation symmetry. Images of domain fragmented å-phase are shown in Figure 1.132. In the case of quartz the situation is actually somewhat more complicated by the occurrence of an intermediate incommensurate phase between α and β and

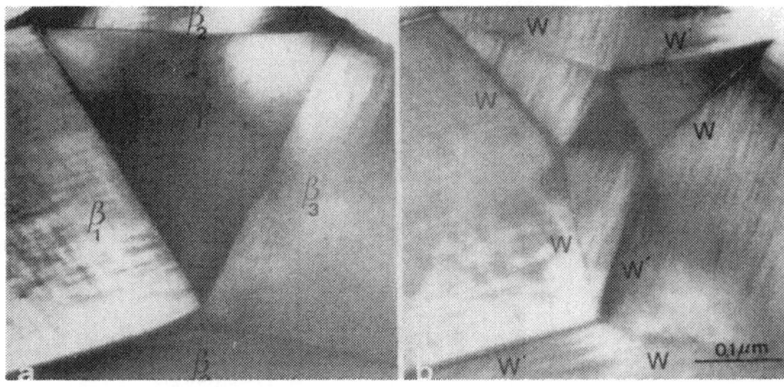

Figure 1.133 Domain pattern inlead orthovandate resulting from the $\gamma \rightarrow \beta$ phase transition. The central triangle of the star pattttern is still in the γ phase. The two photographs refer to the same area; in (b) the termperature was somewhat lowered with respect to that in (a) [92].

which is only stable in a narrow temperature interval ($\approx$ 1.5 K). This phase was discovered by diffraction contrast electron microscopy [16]. It consists of a regular texture of triangular prisms parallel to the c-axis, of α_1 and α_2 structure. The size of the triangular prisms decreases with increasing temperature in the vicinity of the transition temperature (Figure 1.132).

Quite striking domain structures were studied by diffraction contrast in the monoclinic room temperature phase of ferroelastic lead orthovanadate $\left(Pb_3(VO_4)_2\right)$ [92]. The structure is rhombohedral at high temperature (γ-phase) but on cooling it transforms at 120°C into a monoclinic structure (β–phase) which is stable at room temperature. The rhombohedral parent phase is fragmented in domain patterns which minimize the strain energy. They consist of combinations of completely symmetry determined walls (W) and of walls (W′) of which the strain-free orientation depends on the lattice parameters below and above the $\gamma \leftrightarrow \beta$ transition temperature (i.e. on the spontaneous strain tensor). The most striking configuration is the pattern of Figure 1.133; it contains three concentric "stars" of decreasing size. The pattern in (a) of Figure 1.133 contains a central triangle of metastable γ-phase, surrounded by areas consisting of three different variants of the β-phase. On cooling it transforms further "in situ" into the configuration (b), where the γ-triangle has become smaller and is rotated over 180°.

Similar patterns occur in other domain textures resulting from a phase transformation between parent and product phases

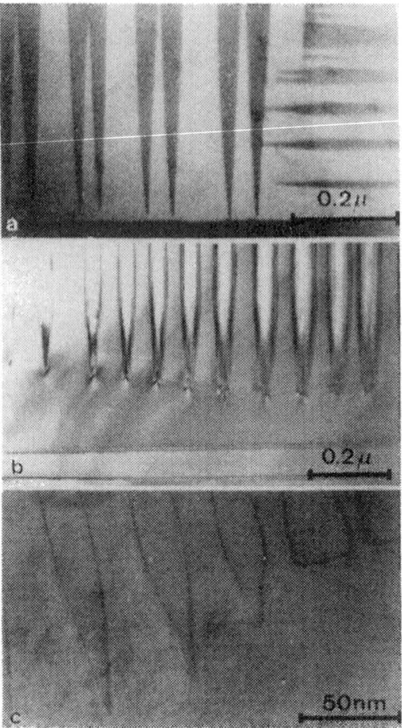

Figure 1.134 Orthorhombic twins in $YBa_2Cu_3O_{7-\delta}$ [93]. Revealed using three different contrast modes; (a) Domain contrast; (b) interface contrast; (c) High resolution imaging. Note the strain at the twin tips in (b).

belonging to the same point groups as γ and β lead orthovanadate respectively.

The compound $YBa_2Cu_3O_7$ is tetragonal at high temperature where the ... O-Cu-O-Cu ... chains in the CuO layers are disordered. Below the transition temperature, which depends on the oxygen content, the chains order in any given area along one out of two mutually perpendicular, equally probable, orientations, which then becomes the b_0 direction of the orthorhombic structure. The disorder-order transition thus produces two structural variants with their b_0 axis roughly perpendicular and which are twin related by a mirror operation with respect to (110) or $(1\bar{1}0)$. These two orientation variants are revealed, using different imaging modes, in Figure 1.134 [93].

The ordering of magnetic moments below the Néel temperature in anti-ferromagnetic materials, is usually accompanied by a structural phase transition. This leads to the formation of an anti-ferromagnetic domain structure of which the domain walls coincide with those due to the structural phase transition [94]. Such a combined transition occurs for instance in NiO, which has the sodiumchloride structure above the Néel point (525 K). Below this temperature the Ni-spins order in such a way that the spins in one of the families of (111) planes order ferromagnetically, the spin direction being parallel to these (111) planes, whereas successive (111) sheets contain oppositely oriented spins. As a consequence of magnetostriction the structure contracts along the <111> direction perpendicular to these sheets, and the lattice becomes rhombohedral ($\alpha = 90^\circ 4'$). The rhombohedral structures in adjacent anti-ferromagnetic domains contract along different <111> directions, and as a results such domains are separated by coherent twin boundaries with a very small twinning vector, which are imaged as δ-fringe patterns. Two such domain walls are shown in Figure 1.135 which is a bright field image of two parallel domain walls for which the δ-values are opposite in sign. This is reflected in the opposite nature of the edge fringes for the two boundaries.

1.48. The structure of ordered alloys

The structure determination of ordered alloys by means of X-ray diffraction is hampered by the occurrence of numerous orientation variants in "single crystals" of the ordered phase. As many as 12 orientation variants may occur, e.g. in Au_5Mn_2. X-ray diffraction is therefore mostly limited to the use of powder diffraction methods with their inherent limitations. Although electron diffraction, even when complemented by HREM, is usually far inferior to X-ray diffraction for the solution of structure determination problems it is on the contrary particularly convenient for the structure determination of

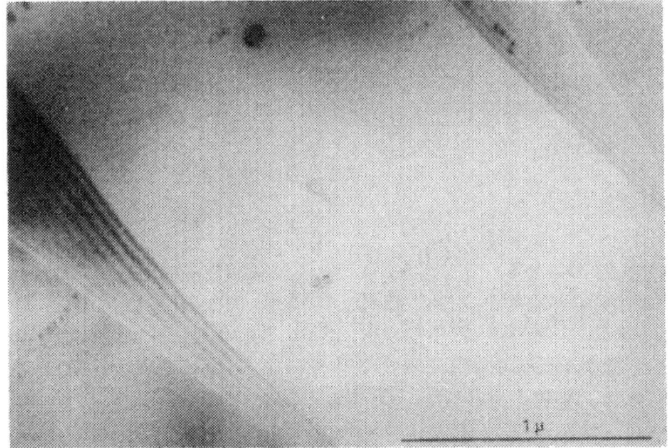

Figure 1.135 Two domain walls in antiferromagnetic nickel oxide; they are imaged as δ-fringe patterns [94].

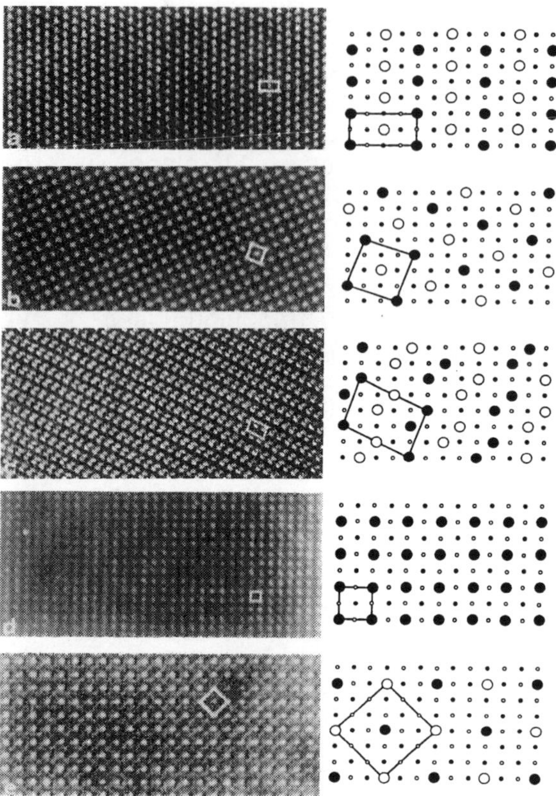

Figure 1.136 Ordered structures of a number of alloys as viewed in HREM along a cube direction of the FCC basic structure [95].
(a) Au_3Mn along [100]
(b) Au_4Mn along [001]
(c) Au_5Mn_2 along [010]
(d) Pt_3Ti along [100]
(e) Pt_8Ti along [001]
All images were made using the bright field superlattice imaging mode. The bright dots represent columns of minority atoms.

binary alloys, especially when also HREM is applied to the same specimen area [95].

The structures of binary alloys are usually superstructures based on a face-centered cubic lattice. It is therefore sufficient to determine the positions of the minority atoms to deduce the complete structure. The minority atoms define the superlattice and produce the superstructure spots in the diffraction pattern. These spots are located within the unit meshes of the FCC reciprocal lattice. When producing a high resolution image, along the cube direction, by selecting all superstructure spots within the square formed by {200} reflections, a sufficient number of Fourier components is excited to produce an adequate representation of the minority atom sublattice by bright dots for a suitably chosen focus. This imaging mode is called <u>bright field superlattice imaging</u>.

As an alternative, applicable to the simpler superstructures, one can select all superstructure spots within the square of FCC basic reflections 000, 200, 020, 220 and produce a <u>dark field superlattice image</u>. In this case the incident beam must be tilted so as to make sure that the centre of Ewald's sphere projects in the centre of the square formed by the four above mentioned reflections. Under these conditions the phase differences resulting from the angle dependent lens aberrations are minimized for the set of selected beams. In the particular case of the tetragonal Au_4Mn superstructure, viewed along [001], these aberrations are completely eliminated when applying this imaging mode [96] (see also § 1.37.2).

Figure 1.136 reproduces a number of bright field superlattice images of simple superstructures derived from a basic FCC lattice, viewed along a cube direction. In all images, the bright dots represent the minority atom columns. The relations between the images and the juxtaposed structure models is evident. In Figure 4.19 of the chapter on "Modulated Structures" a dark field

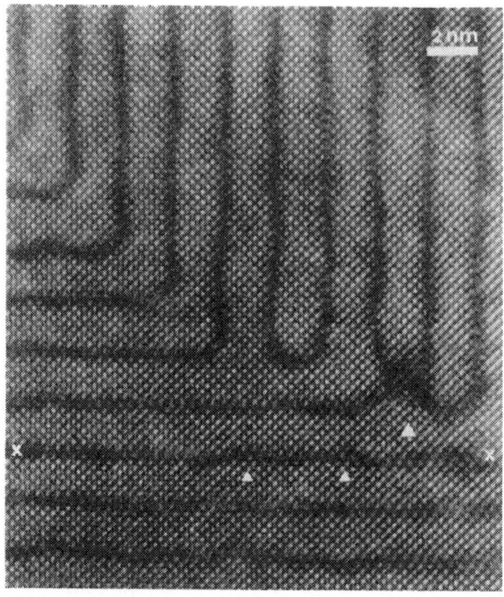

Figure 1.137 HREM image along [001] of the CuAu (II) structure. Note the somewhat wavy shape of the anti-phase boundaries {97}.

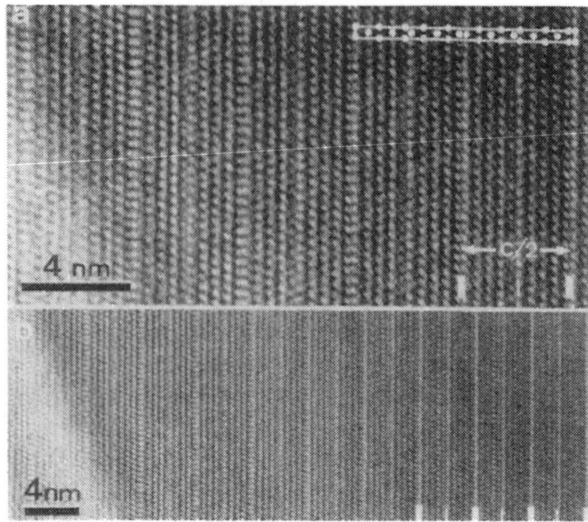

Figure 1.138 HREM image of the long period superstructure of the alloy Nb_5Ga_{13} derived from the DO_{22} structure and which contains alternatingly two different types of out-of phase boundaries [99].

superlattice image of a domain structure in Au_4Mn is reproduced. Two orientation variants with a common [001] axis, as well as a number of out-of-phase boundaries are revealed by the bright dot pattern, which indicates the positions of the manganese columns.

1.49. Long period alloy superstructures

Electron microscopy and electron diffraction have been particularly successful in the study of long period superstructures in general and in particular in alloys. Especially in this field X-ray diffraction and electron diffraction are complementary, the first being used to determine the basic structure, whilst the structure of the long period derivative is deduced from electron diffraction patterns or from HREM images.

The one-dimensional and two-dimensional long period structures derived from Au_4Mn and from $Au_{3+}Zn$ are discussed in the chapter on "Modulated structures".

The first long period alloy structure was discovered in 1925 in the equiatomic alloy CuAuII [98]. The period is about 10 basic unit cells long but its exact length depends on the composition and on alloying elements. The basic structure is a tetragonal superstructure of the FCC structure, it is derived from this structure by having alternating (001) atomic layers occupied by gold and by copper. The conservative anti-phase boundaries are situated in cube planes (100) or (010) and the corresponding displacement vectors are 1/2 [011] and 1/2 [101], i.e. a given (001) plane of the long period structure is alternatingly occupied by copper and by gold. A HREM image of the structure as viewed along a cube direction is reproduced in Figure 1.137. The interfaces are not atomically flat.

A unique long period alloy structure, with an exceptionally large c-parameter (c = 8.02 nm) occurs in Nb_5Ga_{13} which has a structure derived from the tetragonal DO_{22} structure $NbGa_3$ (Figure 1.138). The superstructure results from the introduction of periodic non-conservative Nb-rich out-of-phase boundaries on (001) planes of the DO_{22} structure. Whereas in all long period alloy structures all parallel out-of-phase boundaries are identical, i.e. all have the same displacement vector, this is not the case here; out-of-phase boundaries with two different displacement vectors 1/2 [011] and 1/2 [101] (with respect to the basic FCC structure) alternate, doubling in

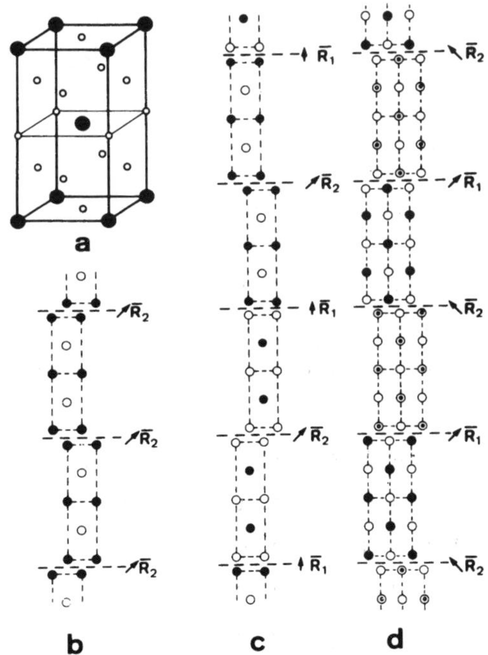

Figure 1.139 Model of the Nb_5Ga_{13} structure to be compared with Figure 1.138 [99].

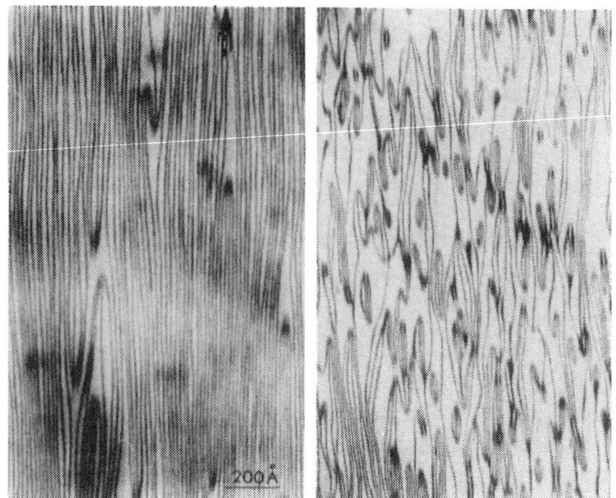

Figure 1.140 Fourfold discommensuration nodes in Ni_3Mo, revealed by diffraction contrast [100].

Figure 1.141 Diffraction contrast image of the first stage in the formation of a one-dimensional long period structure in Cu_3Pd. Note the "meandering" of the anti-phase boundaries {102}.

this way the repeat distance along the *c*-direction. A model of the structure, which can be compared with the HREM image of Figure 1.138 is shown in Figure 1.139; only Niobium atoms are represented [99].

The first observations of "discommensurations" and of "discommensuration nodes" were performed on the alloy $NI_{3+}Mo$ [100], using diffraction contrast, at a time when the term

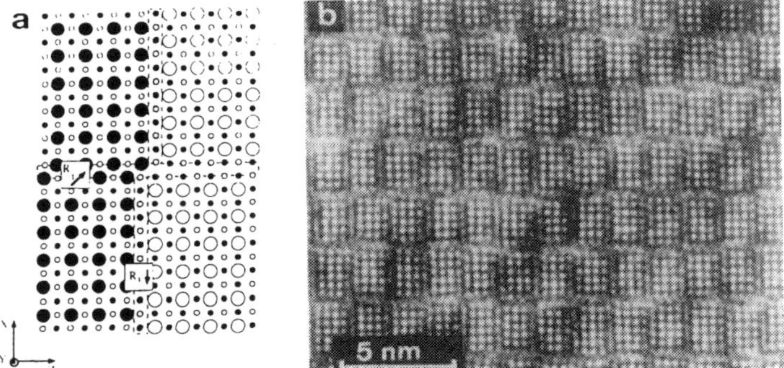

Figure 1.142 HREM image of the two-dimensional long period superstructure of Cu_3Pd compared with a model of the structure. The horizontal set of anti-phase boundaries is non-conservative, whereas the vertical set is conservative [102].

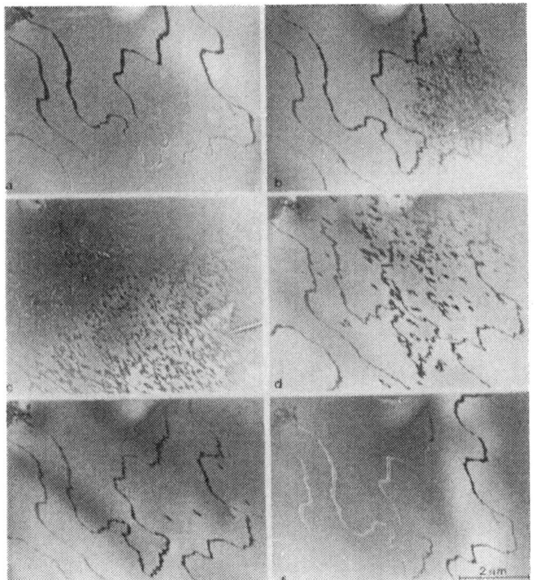

Figure 1.143 Evolution of anti-phase boundaries in anorthite ($CaAl_2Si_2O_8$) during a heating-cooling cycle from room temperature up to above 514 K. All images refer to the same area. Note the memory effect [104].

"discommensuration" had not yet been coined [101]. The interfaces of Figure 1.140 were described as "out-of-phase boundaries" with a displacement vector equal to one quarter of a lattice vector. Although in alloys there is no essential difference between out-of-phase boundaries and "discommensuration walls" the defects of Figure 1.140 would at present presumably be termed "discommensurations" by most authors.

Conservative anti-phase boundaries in the alloy Cu_3Pd with $L1_2$ structure revealed by diffraction contrast, are shown in Figure 1.141; they represent the first stage in the formation of a one-dimensional long period anti-phase boundary structure from the disordered phase. A number of non-conservative anti-phase boundaries become unstable and start "meandering", forming in this way parallel sets of conservative anti-phase boundaries [102].

The same alloy system also exhibits a two-dimensional long period superstructure, based on a rectangular grid of out-of-phase boundaries in the $L1_2$ basic structure. One family of anti-phase boundaries is conservative; the second one is non-conservative. A HREM image together with the corresponding model is shown in Figure 1.142. HREM studies have shed some light on the formation mechanism of such superstructures [103].

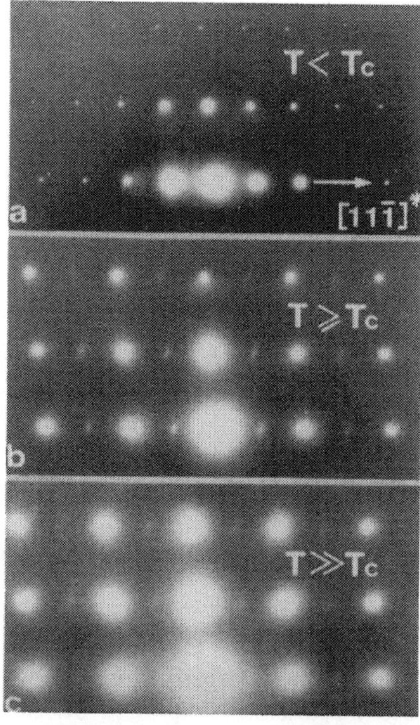

Figure 1.144 Evolution of the diffraction pattern during the same heating-cooling cycle as in Figure 1.143 [104] .

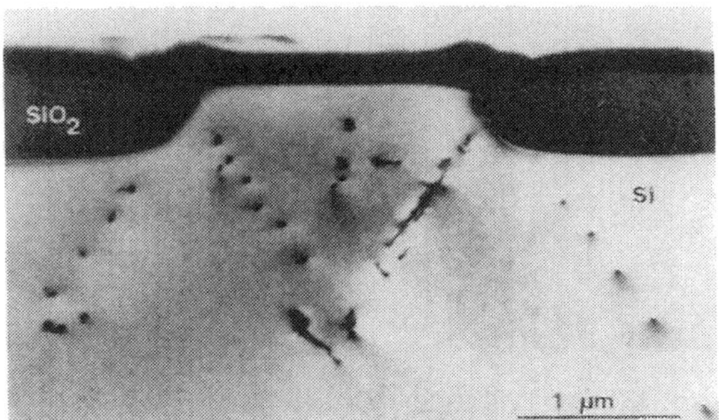

Figure 1.145 TEM image of a cross section of a field effect device. Dislocations are emitted from the edges of the constriction in the silicon oxide layer; the dislocations are seen end-on [105].

1.50. Minerals

Anorthite ($CaAl_2Si_2O_8$) is a complicated silicate which has a primitive triclinic Bravais lattice (space group $P\overline{1}$) at room temperature. Above T_c = 514 K the same unit cell becomes body-centered ($I\overline{1}$). This can be concluded from the diffraction pattern since the spots of the type h + k + l = odd gradually disappear above T_c. On cooling the crystal from the high temperature phase to room temperature it breaks up in two translation variants separated by very "raggy" anti-phase boundaries with a 1/2 [111] displacement vector. No orientation variants are formed. The domain boundaries are revealed by diffraction contrast dark field imaging in reflections for which h + k + l = odd. On heating above 514 K the boundaries disappear but on cooling they reappear at exactly the same place and with the same shape as before, i.e. there is a pronounced memory effect presumably due to impurity pinning. This is illustrated by the heating-cooling cycle in Figure 1.143, whereas the corresponding diffraction patterns along [101] are reproduced in Figure 1.144 [104].

1.51. Fabrication induced defects in semiconductors

Semiconductor single crystal "chips" often undergo a long sequence of fabrication steps (thermal treatment, oxidation, etching, ...) some of which can be accompanied by a deterioration of the crystal's physical properties and hence affect the performance of the final device. The micro-miniaturization of the electron devices makes detailed control of the crystal perfection strongly dependent on electron microscope techniques; both on high resolution images of

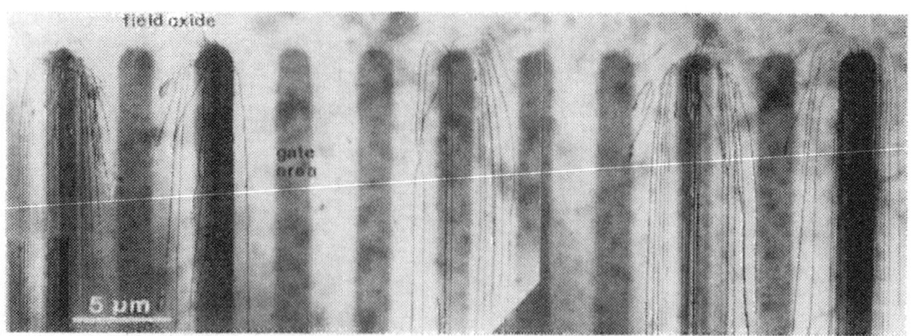

Figure 1.146 Finger shaped gate areas in a field oxide. Dislocations are generated along the edges; they are observed in a plane view. (Courtesy Vanhellemont)

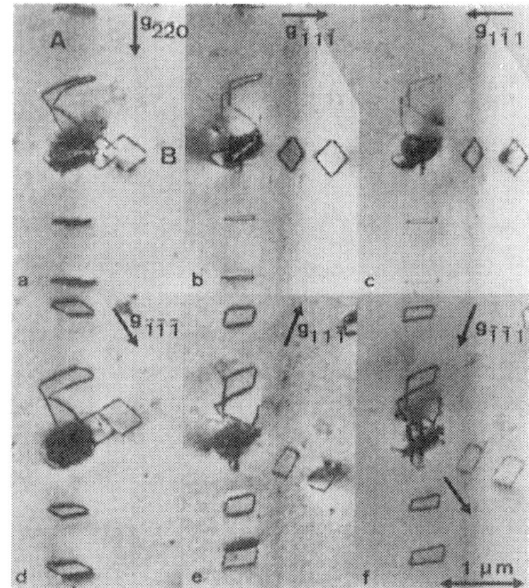

Figure 1.147 Prismatic punching around a precipitate particle in a silicon matrix. (Courtesy H. Bender)

cross-section specimens of devices and on high voltage electron microscopy for the study of "thick" specimens at low resolution and small magnification.

Figure 1.145 shows a TEM image of processions of dislocations observed end-on in a cross section view of a field-effect device. At the edge of the constriction in the silicon oxide layer sources have generated dislocations along the glide planes of maximum resolved shear stress, in order to relieve the stresses generated by the oxidation process. The dislocations apparently form "inverse" pile-ups, their spacing being smallest close to the source [105].

In Figure 1.146 "finger" shaped gate areas formed in a field oxide layer on a silicon chip have similarly generated stresses which were relieved by dislocation generation. In this case the dislocations are imaged in a plane view.

Oxide or other precipitate particles may put the surrounding silicon matrix under a compressive stress. This stress is often large enough to give rise to "prismatic punching" whereby discs of self interstitials surrounded by a loop of perfect dislocation are emitted. Such loops are glissile on a cylindrical surface of which the cross section is determined by the precipitate's shape and the direction of the generators by the Burgers vector of the dislocations, i.e. 1/2 <110> (Figure 1.147).

Sophisticated crystal growth techniques, such as molecular beam epitaxy, have recently made it possible to fabricate artificial periodic layer structures consisting of various combinations of semiconducting materials. Cross section samples, examined in electron microscopy, made it possible to study the precise geometry of the obtained products, in particular the quality of the interfaces and the thickness of of the successive layers. Since electron microscopy is destructive it is not likely to become a routine production control method, but it is important for the calibration of non-destructive, but less direct methods. In particular it allows to check the quality of the interfaces with atomic resolution (Figure 1.148).

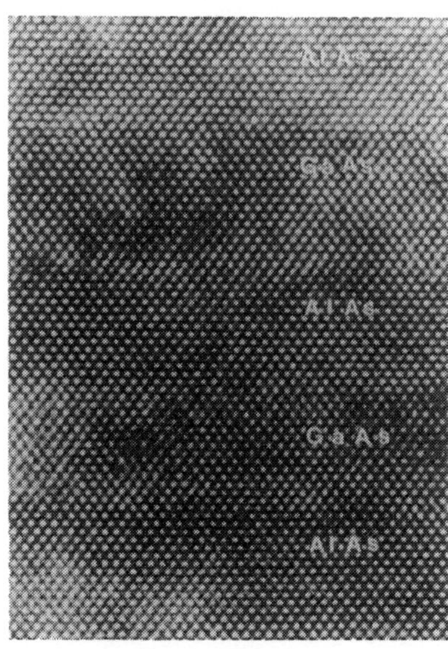

Figure 1.148 High resolution image of a cross section sample of molecular beam deposited layers of AlAs-GaAs.

Figure 1.149 Misfit dislocations along the interface between InSb and GaSb. The inset shows two dislocations; the supplementary halfplanes are formed in the GaAs layer (Courtesy W. Luytne)

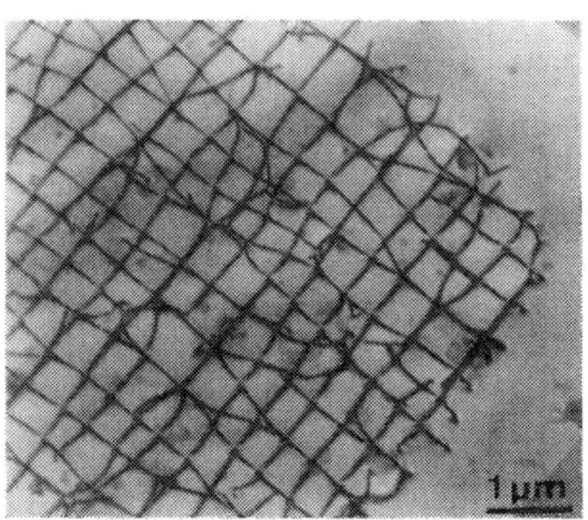

, **Figure 1.150** Network of interfacial dislocations in the contact plane between silicon and aluminium.

At the interfaces between different epitaxial layers, misfit dislocations are formed; in cases where the lattice parameters are not too different; these can be imaged directly in HREM, as for example in the interface between InSb and GaAs in Figure 1.149 from which it is clear that the supplementary halfplanes of the edge dislocations are formed in the GaAs layer of which the lattice parameter is smaller than that of the InSb layer.

Interfacial dislocations are often formed at the interface between the silicon substrate and metallic layers used as electrical contacts. In Figure 1.150 the networks of misfit dislocations between silicon and aluminium is observed in a plane view using diffraction contrast.

1.52. Superstructures due to non-stoichiometry

In the high T_c superconductors the oxygen content, as well as the way in which it is accommodated in the crystal is of considerable importance since it determines to a large extent the value of T_c and of other superconducting properties. In the $YBa_2Cu_3O_{7-\delta}$ compound oxygen deficiency is mainly associated with the CuO layers in the structure. These layers consist of chains of ... O-Cu-O-Cu-O ... parallel with the b_0 direction and separated by a_o. It is this feature of the structure that breaks the tetragonal symmetry and reduces it to orthorhombic. In oxygen deficient crystals there is a strong tendency to deplete complete chains in the CuO layer, rather than breaking up chains into short segments by oxygen vacancies. The average separation of chains then increases with increasing oxygen deficiency. In the particular case of $YBa_2Cu_3O_{6.5}$ the average separation of the chains is $2a_o$. This was first shown by means of electron microscopy [93][87]. Subsequently evidence was found for more complicated arrangements such as $3a_o$ and for mixed sequences of $2a_o$ and $3a_o$ [106]. Since oxygen is a low Z element it does not produce pronounced dot contrast, but nevertheless evidence could be obtained from images, as well as from the diffraction pattern (Figure 1.151) for the presence of the $2a_o$

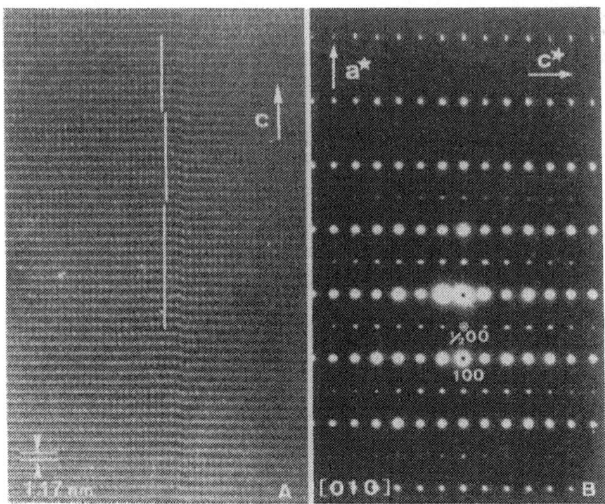

Figure 1.151 The vacancy ordered $2a_0$ structure in $YBa_2Cu_3O_{7-\delta}$ [105]; a HREM image; (b) Diffraction pattern.

structures and for the spatial arrangement of the chains in these structures.

In the compound 1-2-3½ similar $2a_o$ and $3a_o$ structures due to oxygen deficiency in the single CuO layers were found. The relaxation of the heavy atoms around the rows of vacancies allows to make the vacancy chains visible as in Figure 1.152 which visualizes the vacant chains as viewed end-on in the $3a_o$ structure in the 1-2-3½ phase [87].

Figure 1.152 The vacancy ordered $3a_0$ structure in the phase $Y_2Ba_4Cu_7O_{15}$. The high resolution images are compared with simulated images, based on a model that allows for relaxation of the barium ions around the oxygen vacancy rows. [87]

In $La_2CuO_{4-\delta}$ the oxygen deficiency is apparently accommodated along crystallographic shear planes formed by edge sharing CuO_6 octahedra. The shear planes have a strong tendency to be uniformly spaced giving rise to crystallographic shear structures. Such a shear structure is imaged at high resolution in Figure 1.153. The inset shows the well resolved atom columns in the thinnest part of the specimen [107].

The structure of $Ca_{0.85}CuO_2$ is remarkable by the way in which the complicated stoichiometry is accommodated in a rather simple structure [108]. The structure can be considered as consisting of an orthorhombic framework of ribbons of edge-sharing planar CuO_4 groups "stuffed" with calcium atoms. The ribbons are positioned in a centered arrangement when viewed along the ribbons, i.e. parallel with the b_0 direction. When viewed along the a_o direction, i.e. along the normal to the CuO_4 planes, the copper sublattice is centered as well. This framework gives rise to "tunnels" along the b_0 direction, consisting of interpenetrating deformed oxygen octahedra. The calcium atoms occupy positions along these tunnels. Coulomb repulsion tends to space them uniformly but on the other hand the centres of the oxygen octahedra are preferred sites for inserted atoms such as calcium. The actual configuration is a trade-off between a uniformly spaced array and an arrangement whereby along the tunnels arrays of five octahedral sites are occupied, regularly alternating with a vacant site, so as to realize a ratio Cu/Ca $\approx 6/5$. Imaging along a suitable zone electron microscopy made it possible to reveal the copper atoms of the framework simultaneously

Figure 1.153 HREM image of a shear structure attributed to oxygen deficiency in $La_2CuO_{4-\delta}$ [107].

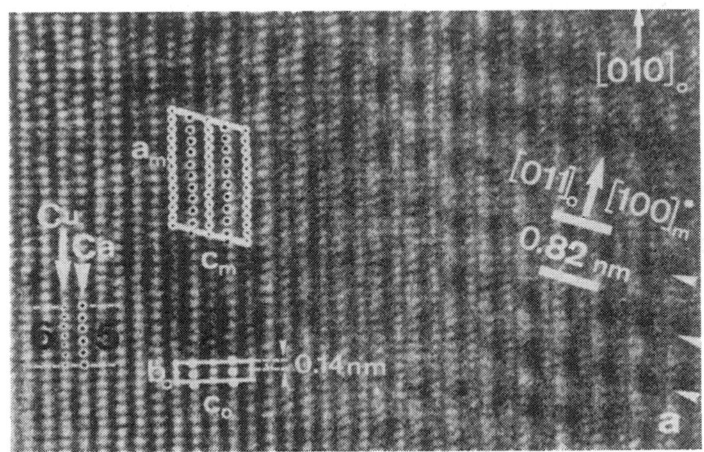

Figure 1.154 HREM image of the incommensurate structure in $Ca_{0.85}CuO_2$ [108]; $[010]_0$ zone) note the relation between the rows of calcium and copper atoms.

with the calcium atoms in the tunnels as parallel, alternating bright dot rows (Figure 1.154). It is then clear that seven dots along the copper row (i.e. six spacings b_0) correspond with six dots (i.e. five spacings) along the calcium rows. The superstructure has therefore a periodicity which is the smallest common multiple of the copper and the calcium spacing, i.e. $6b_0$, the average calcium spacing along the tunnels being $6/5\ b_0$.

1.53. Various applications

1.53.1. In-situ studies

The availability of cooling and heating specimen holders allows the "in-situ" study of the phenomena accompanying phase transitions. When going through a disorder-order transition different phases of the domain fragmentation can be followed. The creation and elimination of discommensuration walls is directly observable in dark field images made in clusters of incommensurate reflections.

Static images referring to the phase transitions in lead orthovanadate and in quartz are reproduced respectively in Figure 1.133 and in Figure 1.132. When performing such observations one should be aware of the effect of the electron beam on the specimen, which results in an increase in temperature depending on the thermal conductivity of the foil and which may also cause some radiation effects, which may interfere with the transition.

1.53.2 Radiation damage

Electron microscopy, in particular high voltage electron microscopy has extensively been used to study "in-situ" radiation effects as well as post irradiation defect configurations. The point defects, precipitates and small dislocation loop can be characterized using the methods mentioned above (§ 1.29.4).

1.53.3Radiation ordering

Some surprising results were found by "in-situ" studies of ordering alloys, exhibiting a short range order state, such as Ni_4Mo. When irradiated with 1 MV electrons at low temperature ordered Ni_4Mo becomes completely disordered. When irradiating in a temperature range below but close to the order-disorder transition temperature, the irradiation causes the alloy to order up to a certain degree. The order parameters can be determined by following the evolution of the intensity of the order diffraction spots. These phenomena result from the competition between the ordering effect due to radiation enhanced diffusion at the irradiation temperature and the disordering effect of the irradiation as a result of atomic collisions. In a certain temperature range the short range order state will be produced by irradiation. Certain alloy phases, which could not be ordered by thermal treatment, were found to order under electron irradiation (e.g. Ni_4W).

1.53.4. Magnetic domain structures

Magnetic domain walls, as well Néel walls as Bloch walls, can be studied, making use of Lorentz contrast. The magnetization vector changes in orientation and (or) sense across a domain wall. As a result the electron beam is differently deflected, because of the Lorentz force, in the domains on both sides of the wall. Under defocussed conditions the domain wall can thus be revealed either as a bright or as a dark line, depending on whether the change in the magnetization across the boundary, produces an excess or a deficiency of electrons behind the specimen of the position of the walls.

Acknowledgments

The authors are grateful to a number of colleagues for the use of photographs as illustrations in this survey: their names are mentioned either in the figure caption, in the text or in the reference appearing in the caption. A number of unpublished photographs was supplied by Prof. Dr. G. Van Tendeloo, to whom special thanks are due. We are grateful to our colleague Prof. Dr. J. Van Landuyt for his interest.

The authors are moreover grateful to the following colleagues for the use of diagrams or photographs from their papers: Aerts, E.; Art, A.; Bender, H.; Berghezan, A.; Booker, J.; Broddin, D.; Cockayne, D.J.H.; Coene, W.; Delavignette, P.; De Veirman, A.; Fourdeux, A.; Hirsch, P.; Head, A.K.; Howie, A.; Humble, P.; Krekels, T.; Kuypers, S.; Luyten, W.; Manolikas, C.; Meuter, J.W.; Milat, O.; Okabe, T.; Op de Beeck, M.; Reyes-Gasga, J.; Ruedl, E.; Silcox, J.; Siems, R.; Schryvers, D.; Takeda, M.; Van Landuyt, J.; Van Tendeloo, G.; Whelan, M.J.; Yasuda, K.; Zandbergen, H.

The photographs were skilfully printed for publication by F. Schallenberg and A. De Muynck. The manuscript was carefully typed by H. Evans and the line drawings prepared with care by M. Schrijnemakers.

REFERENCES

1. C.J. Humphreys and E.G. Bithell, *Electron Diffraction Theory*, Chapter 2, Volume I of this book.

2. D. Van Dyck (1986), "High Resolution Electron Microscopy", in *Microscopia elettronica in transmissione e technice di analisi di superfici nella scienza dei materiali*, Edizione ENEA, Roma.

3. R.D. Heidenreich (1949), *J. Appl. Phys.* **20**, 993.

4. S. Kikuchi (1928), *Japan J. Phys.* **5**, 23.

5. R. Gevers (1978), in *Diffraction and Imaging Techniques in Material Science*, Eds. S. Amelinckx, R. Gevers, J. Van Landuyt (North-Holland, Amsterdam-New York-Oxford), p. 9.

6. R. Gevers (1962), *Phil. Mag.* **7**, 59.
R. Gevers (1962), *Phil. Mag.* **7**, 651.
R. Gevers (1962), *Phil. Mag.* **7**, 1681.
R. Gevers (1963), *Phil. Mag.* **7**, 769.

7. P.B. Hirsch, A. Howie and M.J. Whelan (1962), *Phil. Mag.* [8] **7**, 2095.

8. G. Borrmann (1941), *Physik Z.* **42**, 157.
G. Borrmann (1950), *Z. Physik* **127**, 297.

9. C.G. Darwin (1914), *Phil. Mag.* **27**, 315, 675.

10. A. Howie, and M.J. Whelan (1960), *Proc. European Reg. Conf. on Electron Microscopy, Delft*, vol. 1, p. 194.
A. Howie, and M.J. Whelan (1961), *Proc. Roy. Soc.* **A263**, 217.
A. Howie, and M.J. Whelan (1962), *Proc. Roy. Soc.* **A267**, 206.

11. J.W. Menter (1956), *Proc. Roy. Soc.* **A236**, 119.

12. H. Yoshioka (1957), *J. Phys. Soc. Japan* **12**, 628.

13. S. Takagi (1962), *Acta Cryst.* **15**, 1311.

14. S. Amelinckx and J. Van Landuyt (1978), in *Diffraction and Images in Material Science*, Eds. S. Amelinckx, R. Gevers, J. Van Landuyt (North-Holland, Amsterdam-New York-Oxford), p. 107.

15. R. Gevers, J. Van Landuyt and S. Amelinckx (1965), *Phys. Stat. Sol.* **11**, 689.

16. G. Van Tendeloo, J. Van Landuyt and S. Amelinckx (1976), *Phys. Stat. Sol.* (a) **33**, 723.

17. G. Van Tendeloo, and S. Amelinckx (1974), *Acta Cryst.* **A30**, 431.

18. J. Van Landuyt, G. Van Tendeloo, S. Amelinckx and M.B. Walker (1985), *Phys. Rev.* **B31**, 2986.
G. Dolino, P. Bachheimer, B. Berge, C.M. Zeyen, G. Van Tendeloo, J. Van Landuyt and S. Amelinckx (1984), *J. Phys.* **45**, 901.

19. M. Meulemans, P. Delavignette, F. Garcia-Gonzales and S. Amelinckx (1970), *Mat. Res. Bull.* **5**, 1025.

20. C. Boulesteix, J. Van Landuyt and S. Amelinckx (1976), *Phys. Stat. Sol.* (a) **33**, 595.

21. C. Manolikas, J. Van Landuyt and S. Amelinckx (1979), *Phys. Stat. Sol.* (a) **53**, 327.

22. R. Serneels, M. Snijkers, P. Delavignette, R. Gevers and S. Amelinckx (1973), *Phys. Stat. Sol.* (b) **58**, 277.

23. M. Snijkers, R. Serneels, P. Delavignette, R. Gevers and S. Amelinckx (1972), *Cryst. Latt. Def.* **3**, 99.

24. J.D. Eshelby and A.N. Stroh (1951), *Phil. Mag.* [7], **42**, 1401.

25. R. Siems, P. Delavignette and S. Amelinckx (1962), *Phys. Stat. Sol.* **2**, 421.

26. M. Mannami (1962), *J. Phys. Soc. Japan* **17**, 1160.
H. Hashimoto and M. Mannami (1960), *Acta Cryst.* **13**, 363.

27. P. Delavignette, R. Trivedi, R. Gevers and S. Amelinckx (1966), *Phys. Stat. Sol.* **17**, 221.

28. M.F. Ashby and L.M. Brown (1963), *Phil. Mag.* **8**, 1083, 1649.

29. D.J.H. Cockayne, M.J. Jenkins and I.L.E. Ray (1971), *Phil. Mag.* **24**, 1383.
D.J.H. Cockayne, I.L.E. Ray and M.J. Whelan (1969), *Phil. Mag.* **20**, 1265.

30. R. De Ridder, and S. Amelinckx (1971), *Phys. Stat. Sol.* (b) **43**, 541.

31. A.K. Head (1967), *Aust. J. Phys.* **20**, 557.

32. P. Humble (1978), in *Diffraction and Imaging Techniques in Material Science*, Eds. S. Amelinckx, R. Gevers, J. Van Landuyt (North-Holland, Amsterdam-New York-Oxford), p. 315.

33. G.W. Groves and A. Kelly (1961), *Phil. Mag.* [8] **6**, 1527.

34. B. Edmondson and G.K. Williamson (1964), Proc. Joint Conf. on Inorganic and Intermetallic Crystals, Univ. of Birmingham (U.K.)., *Phil. Mag.* **9**, 277.

35. E. Ruedl, P. Delavignette and S. Amelinckx (1962), *Proc. I.A.E.A. Symposium on Radiation Damage in Solids and Reactor Materials*, Venice 1962, Vol. 1, p. 363.

36. R. Gevers (1963), *Phys. Stat. Sol.* **3**, 415.

37. C.J. Ball (1964), *Phil. Mag.* **9**, 541.

38. P. Humble (1969), *Aust. J. Phys.* **22**, 51.

39. H. Hashimoto, and M.J. Whelan (1963), J. Phys. Soc. Japan **18**, 1706.
H. Hashimoto, A. Howie and M.J. Whelan (1962), *Proc. Roy. Soc.* **A269**, 80.

40. R. Gevers, A. Art and S. Amelinckx (1963), *Phys. Stat. Sol.* **3**, 1563.

41. D.W. Pashley, J.W. Menter and G.A. Bassett (1957), *Nature* **179**, 752.

42. G.A. Bassett, J.W. Menter and D.W. Pashley (1958), *Proc. Roy. Soc.* **A246**, 345.

43. R. Gevers (1963), *Phys. Stat. Sol.* **3**, 2289.

44. O. Scherzer (1949), *J. Appl. Phys.* **20**, 20.

45. C.E. Shannon (1949), *Proc. IRE* **37**, 10.

46. D. Gabor (1965), *Laboratory Investigation* **14**, 2.

47. G. Fan and J.M. Cowley (1987), *Ultramicroscopy* **21**, 125.

48. M. Tournarie (1960), *Bull. Soc. Franc. Miner. Cryst.* **83**, 179.
M. Tournarie (1961), *Cr. Acad. Sci.* **252**.
M. Tournarie (1962), *J. Phys. Soc. Japan*, Suppl. BII, 98.
A. Howie and M.J. Whelan (1961), Proc. Roy. Soc. (London) Ser. **A263**, 217.
L. Sturkey (1957), *Acta crystallogr.* **10**, 858.
L. Sturkey (1962), *Proc. Phys. Soc.* **80**, 321.

49. K. Ishizuka and N. Uyeda (1977), *Acta Crystallogr.* **A33**, 740.

50. D. Van Dyck and W. Coene (1984), *Ultramicroscopy* **15**, 29.

51. V. Castano (1989), in *Computer Simulation of Electron Microscope Diffraction and Images*, Eds. W. Krakow and M. O'Keefe, The Minerals, Metals and Materials Society, Pennsylvania, p. 33.

52. S. Amelinckx, G. Van Tendeloo and J. Van Landuyt (1984), *Bull. Mater. Sci.*, Vol. **6**, nr. 3, 417.

53. D. Van Dyck, G. Van Tendeloo and S. Amelinckx (1982), *Ultramicroscopy* **10**, 263.

54. J. Lindhard (1965), *Mat. Fys. Medd. Dan. Vid. Selsk* **34**, 1.
A. Tamura and Y.K. Ohtsuki (1974), *Phys. Stat. Sol.* (b) **73**, 477.
A. Tamura and F. Kawamura (1976), *Phys. Stat. Sol.*
B. Buxton, J.E. Loveluck and J.W. Steeds (1978), *Phil. Mag.* **A3**, 259.
C.J. Humphries and J.C.H. Spence (1979), *Proc. 37th EMSA Meeting, Baton Rouge*, Louisiana, Claitor's Publ. Div., p. 554.

55. D. Van Dyck, J. Danckaert, W. Coene, E. Selderslaghs, D. Broddin, J. Van Landuyt and S. Amelinckx (1989), in *Computer Simulation of Electron Microscope Diffraction and Images*, Eds. W. Krakow and M. O'Keefe, The Minerals, Metals and Materials Society.

56. M.V. Berry and K.E. Mount (1972), *Rep. Progr. Phys.* **35**, 315.

57. K. Kambe, G. Lempfuhl and F. Fujimoto (1974), *Z. f. Naturforschung* **29a**, 1034.

58. D. Shindo and M. Hirabayashi (1988), *Acta Cryst.* **A44**, 954.

59. D. Van Dyck, J. Van Landuyt, F. Jellinek and S. Amelinckx (1979), *Ultramicroscopy* **4**, 467.

60. G. Van Tendeloo, R. Wolf, D. Van Dyck and S. Amelinckx (1978), *Physica Status Solidi* **(a) 47**, 105.
G. Van Tendeloo, R. Wolf, J. Van Landuyt and S. Amelinckx (1978), *Physica Status Solidi* **(a) 47**, 539.
R. Wolf, G. Van Tendeloo, J. Van Landuyt and S. Amelinckx (1978), *Physica Status Solidi* **(a) 48**, 39.
G. Van Tendeloo and S. Amelinckx (1978), *Physica Status Solidi* **(a) 49**, 337.
G. Van Tendeloo, R. De Ridder and S. Amelinckx (1978), *Physica Status Solidi* **(a) 49**, 655.
G. Van Tendeloo and S. Amelinckx (1979), *Physica Status Solidi* **(a) 51**, 141.
S. Amelinckx (1978-79), *Chemica Scripta* **14**, 197.
G. Van Tendeloo, R. Wolf, J. Van Landuyt and S. Amelinckx (1980), *Physica Status Solidi* **(a) 60**, 581.

61. G. Van Tendeloo and S. Amelinckx (1982), *Physica Status Solidi* **(a) 65**, 73, 431 (1981); **69**, 589 (1982); **71**, 185.

62. G. Van Tendeloo, J. Van Landuyt and S. Amelinckx (1982), *Physica Status Solidi* **(a) 70**, 145.
J. Van Landuyt, G. Van Tendeloo and S. Amelinckx (1982), *Physica Status Solidi* **(a) 70**, 407.
M. Van Sande, G. Van Tendeloo, S. Amelinckx and P. Airo (1979), *Physica Status Solidi* **(a) 54**, 499.
G. Van Tendeloo and S. Amelinckx (1982), *Physica Status Solidi* **(a) 69**, 103.

63. W. Coene, D. Van Dyck, G. Van Tendeloo and J. Van Landuyt (1985), *Phil. Mag.* **52**, No. 1, 127.

64. W. Coene, H. Bender and S. Amelinckx (1986), *Proc. XIth Int. Congr. on Electron Microscopy, Kyoto*, 1491.

65. D. Van Dyck, J. Van Landuyt, S. Amelinckx, Nguyen-Huy-Dung and C. Dagron (1976), *J. Sol. St. Chem.* **19**, 179.

66. H. Lichte (1991), *Adv. Opt. El. Microsc.* **12**, 25.

67. D. Van Dyck and M. Op de Beeck (1990), *Proc. XIIth Int. Congr. on Electron Microscopy, Seattle, U.S.A.*, Vol. 1, 26.

68. D. Van Dyck (1990), *Proc. XIIth Int. Congr. on Electron Microscopy, Seattle, U.S.A.*, Vol. 1, 64.

69. G.R. Anstis and M.A. O'Keefe, private communication (unpublished).

70. J. Frank (1973), *Optik* **38**, 519.
P.L. Fejes (1977), *Acta Cryst.* **A33**, 109.

71. M. Born and E. Wolf (1975), *Principles of Optics*, Pergamon Press, London, Chapter X.

72. S. Amelinckx (1979), in *Dislocation in Solids*, Vol. 2, Chapter 6, ed. F.R.N. Nabarro (North-Holland, Amsterdam), p. 68.

73. P. Delavignette, and S. Amelinckx (1962), *J. Nucl. Mater.* **5**, 17.

74. R. Siems, P. Delavignette and S. Amelinckx (1962), *Phys. Stat. Sol.* **2**, 636.
R. Siems, P. Delavignette and S. Amelinckx (1962), *Phys. Stat. Sol.* **2**, 421.

75. P. Delavignette, H.B. Kirkpatrick and S. Amelinckx (1961), *J. Appl. Phys.* **32**, 1098.

76. S. Amelinckx and P. Delavignette (1962), *J. Appl. Phys.* **33**, 1458.

77. J. Van Landuyt, G. Remaut and S. Amelinckx (1970), *Phys. Stat. Sol.* **41**, 271.

78. A. Art, R. Gevers and S. Amelinckx (1963), *Phys. Stat. Sol.* **3**, 967.

79. E. Ruedl, P. Delavignette and S. Amelinckx (1962), *J. Nucl. Mater.* **6**, 46.

80. A. Berghezan, A. Fourdeux and S. Amelinckx (1960), *Acta Met.* **9**, 464.

81. G. Van Tendeloo, M. Opdebeeck, S. Amelinckx, J. Bohr and W. Kretschmer (1991), *Europhys. Lett.* **15(3)**, 215.

82. C. Goessens, D. Schryvers, J. Van Landuyt, S. Amelinckx, A. Verbeek and R. De Keyzer (1991), *Journal of Crystal Growth* **110**, 930.

83. J. Bohr, L. Gråbaek, H.H. Anderson, A. Johansen, E. Johnson, L; Sarholt-Kirstensen, V. Surganov, I.K. Robinson, D. Broddin and G. Van Tendeloo (1991), in *Fundamental Aspects of Inert Gases in Solids*, Eds. S.E. Donnelly and J.H. Evans (Plenum Press, New York).

84. P.B. Hirsch, J. Silcox, R. Smallman and K. Westmacott (1958), *Phil. Mag.* [8] **3**, 897.

85. P.B. Hirsch and J. Silcox (1958) in *Growth and Perfection of Crystals*, Eds. R.H. Doremus et al, p. 262, Wiley, New York.

86. D. Schechtman, I. Blech, D. Gratias and J.W. Cahn (1984), *Phys. Rev. Letters* **53**, 1951.

87. S. Kuypers, G. Van Tendeloo, J. Van Landuyt, S. Amelinckx, H.W. Shu, S. Jaulmes, J. Flahaut and P. Laruelle (1988), *J. of Solid State Chemistry* **73**, 192.

88. T. Krekels, G. Van Tendeloo, S. Amelinckx, J. Karpinski, E. Kaldis and S. Rusiecki (1991), *Appl. Phys. Lett.* **59** (23), 3048.

89. S. Kuypers, G. Van Tendeloo, J. Van Landuyt and S. Amelinckx (1989), *Acta Cryst.* **A45**, 291.

90. E. Aerts, P. Delavignette, R. Siems and S. Amelinckx (1962), *J. Appl. Phys.* **33**, 3078.

91. W. Coene, H. Bender and S. Amelinckx (1985), *Phil. Mag.* **A52**, 369.

92. C. Manolikas, J. Van Landuyt and S. Amelinckx (1980), *Phys. Stat. Sol.* (a) **60**, 607.
C. Manolikas, and S. Amelinckx (1980), *Phys. Stat. Sol.* **61**, 179.

93. H.W. Zandbergen, G. Van Tendeloo, T. Okabe and S. Amelinckx (1987), *Phys. Stat. Sol.* (a) **103**, 45.

94. P. Delavignette and S. Amelinckx (1963), *Appl. Phys. Lett.* **2**, 236.

95. S. Amelinckx (1978-1979), *Chimica Scripta* **14**, 197.

96. G. Van Tendeloo, and S. Amelinckx (1978), *Phys. Stat. Sol.* (a) **49**, 337.

97. K. Yasuda, M. Nakagawa, G. Van Tendeloo and S. Amelinckx (1987), *Journal of Less-Common Metals* **135**, 169.

98. C.H. Johansson, and J.O. Linde (1925), *Ann. Physik* [4] **78**, 439.

99. M. Takeda, G. Van Tendeloo and S. Amelinckx (1988), *Acta Cryst.* **A44**, 938.

100. G. Van Tendeloo, and S. Amelinckx (1974), *Phys. Stat. Sol.* (a) **22**, 621.

101. W.L. McMillan (1976), *Phys. Rev.* **B14**, 1496.

102. D. Broddin, G. Van Tendeloo, J. Van Landuyt and S. Amelinckx (1989), *Phil. Mag.* **59**, Nr. 1, 47.

103. D. Broddin, G. Van Tendeloo and S. Amelinckx (1990), *J. Phys. Condens. Matter* **2**, 3459.

104. G. Van Tendeloo, S. Ghose and S. Amelinckx (1989), *Phys. Chem. Minerals* **16**, 311.

105. J. Vanhellemont and S. Amelinckx (1987), *J. Appl. Phys.* **61 (b)**, 2176.

106. J. Reyes-Gasga, T. Krekels, G. Van Tendeloo, J. Van Landuyt, S. Amelinckx, W.H.M. Bruggink and H. Verweij (1989), *Physica C* **159**, 831.

107. G. Van Tendeloo, and S. Amelinckx (1991), *Physica C* **176**, 575.

108. O. Milat, G. Van Tendeloo, S. Amelinckx, T.G.N. Babu and C. Greaves (1992), *Journal of Solid State Chemistry*, to be published.

General References

G1. S. Amelinckx (1964), *The Direct Observation of Dislocations*, Supplement 6 in "*Solid State Physics*", Eds. F. Seitz and D. Turnbull (Academic Press).

G2. *Diffraction and Imaging Techniques in Material Science* (1970, 1978), Eds. S. Amelinckx, R. Gevers and J. Van Landuyt (North-Holland Publishing Company, Amsterdam-New York-Oxford).

G3. *Dislocation in Solids* (1979), Ed. F.R.N. Nabarro (North-Holland Publishing Company, Amsterdam-New York-Oxford).

G4. P.B. Hirsch, R.B. Nicholson, A. Howie, D.W. Pashley and M.J. Whelan (1965), *Electron Microscopy of Thin Crystals*, (Butterworths, London).

G5. *Elektronenmikroskopie in der Festkörperphysik* (1982), Ed. H. Bethge and J. Heydenreich (Springer Verlag, Berlin-Heidelberg-New York).

G6. J.C.H. Spence (1981), *Experimental High Resolution Electron Microscopy. Monographs on the Physics and Chemistry of Materials*, Oxford Science Publications, (Clarendon Press, Oxford).

G7. G. Thomas (1962), *Transmission Electron Microscopy of Metals* (John Wiley and Sons Inc, New York).

G8. A.K. Head, P. Humble, L.M. Clarebrough, A.J. Morton and G.T. Forwood, (1973) *Computed Electron Micrographs and Defect Identification*, Eds. S. Amelinckx, R. Gevers, J. Nihoul, Vol. 7.

2
Disorder and Defect Scattering, Thermal Diffuse Scattering, Amorphous Material

J. Gjønnes

2.1 Introduction

Most properties and virtually all transformations in solids are strongly influenced by defects. Hence the characterization of defect structures has been a challenge to structure research ever since models of various crystalline defects were advanced in the 1930s. Optical microscopy, spectroscopy and in particular x-ray diffraction contributed significantly to insight in the nature of defects, notably the statistical description of short range order and the relation of crystal structures to transformations.

But it was through the application of electron microscopy techniques, at first by diffraction contrast imaging and selected area diffraction in the 1950s and 60s, that we were able to construct the present detailed and many-faceted picture of defects in inorganic solid matter: of dislocation configurations, of the nature and distribution of planar faults and of relations between short range and long range ordering in a variety of compounds. Schematic models and statistical pictures based on more circumstantial evidence could be confirmed, replaced or augmented by direct observations of defect arrangements and by diffraction information about local structure within individual crystal grains and precipitates. A further wealth of detail was offered in the 1970s by high resolution imaging of the projected atomic arrangement. The investigation of local structure - as distinct from the average structure normally studied with x-ray or neutron diffraction - became a prime domain of electron microscopy techniques.

The close relation between these techniques must be emphasized - and also the relation to the other diffraction methods. Electron diffraction is often used for surveying the qualitative features in disorder scattering, supplementing quantitative measurements by x-ray or neutron diffraction. It is also an essential tool for defining conditions for imaging defects in the electron microscope by high resolution or diffraction contrast. This wide utility of electron diffraction in connection with other techniques may to some extent have held up the development of the quantitative side of electron diffraction. This may change; instrumental development of electron optics, recording systems and energy filtering, as well as the increasing importance attached to local structure in solid state and materials science suggest increased emphasis on electron diffraction also as a quantitative technique in its own right.

Defects are visible in diffraction and microscopy because they scatter outside Bragg reflections, due to non-periodic components in the object. We may express this by writing the scattering potential as a sum of a periodic and a non-periodic part, ϕ_d:

$$\phi(\mathbf{r}) = \sum \Phi_g \exp[2\pi i \mathbf{gr}] + \phi_d(\mathbf{r}) \qquad (2.1)$$

assuming that we can define an average lattice. ϕ_d may include

. substitional disorder, i.e. variations in occupation of the atomic sites in the structure;

. displacement disorder, where atoms or groups of atoms are displaced (or rotated) from the average positions.

Displacements also occur as shifts $\mathbf{R}_n$ in position or orientation between different parts of the crystal; then an average lattice may not be defined. Cowley (1981) proposed the form

$$\phi(\mathbf{r}) = \sum_n \phi_n(\mathbf{r}) * \delta(\mathbf{r} - \mathbf{R}_n) \qquad (2.2)$$

for a set of blocks ϕ_n at positions determined by the shifts $\mathbf{R}_n$. The corresponding amplitude expression becomes very simple when there is just one type of block with scattering amplitude F(**q**), viz:

$$A(\mathbf{q}) = F(\mathbf{q})\sum_n \exp[2\pi i \mathbf{q}\mathbf{R}_n]$$

The scattering thus produced renders the defects visible in various techniques in the electron microscope, by diffraction contrast, in high resolution micrographs and in the selected area diffraction pattern by the diffuse scattering - which is more readily observed in electron diffraction than with x-rays or neutrons: stronger scattering means higher sensitivity. But quantitative determination of disorder in crystals, e.g. in terms of order parameters defined as statistical averages, has usually been left to intensity measurements of x-ray or neutron diffuse scattering from large single crystals. Such studies depend upon procedures which assume kinematical scattering and upon reliable background subtraction; conditions which are difficult to meet in electron diffraction where the pattern of diffuse scattering usually is modified by dynamical Bragg diffraction and multiple scattering.

Some progress has been made in overcoming these problems. Corrections for dynamical scattering have been developed. The inelastic background (including a substantial part of the multiple scattering) can be subtracted experimentally through energy filters. Even so, quantitative measurement of diffuse scattering with electrons may remain rare for some time; with most applications of electron diffraction based on its particular advantages in qualitative studies:

- Diffuse scattering from defects can be recorded from very small, selected specimen regions, down to a few nm in diameter.

- Diffraction information on defects and disorder may be correlated with high resolution electron microscope imaging from the same specimen area.

- The electron diffraction pattern approximates to a planar section of reciprocal space; complicated configurations of diffuse scattering can be readily visualized.

- Dynamical effects, though in general a disadvantage, may be exploited to obtain information about localization of defects within the unit cell.

Through these features electron diffraction has become a useful link beteen the direct observation of local atomic arrangement by high resolution imaging and the statistical picture of disorder in large volumes derived from x-ray or neutron studies.

2.2 Defects and their appearance in diffraction patterns: the qualitative aspect

Defects are revealed in diffraction patterns by a number of features: diffuse spots, streaks or curved lines; appearing between the Bragg spots or very close to the reflections from the average structure. The features may follow the repetition of the reciprocal lattice, be restricted to certain Brillouin zones or form a repetitive pattern with a larger reciprocal unit. These qualitative aspects of the scattering often include salient information about the defect structure. Let us first emphasize this side of electron diffraction, by referring to a few a broad categories.

Planar faults and diffuse streaks. Planar faults (Fig 1a-e) are among the most widely studied objects in electron microscopy and diffraction. Since the early 1960s an extensive literature exists on diffraction contrast from stacking faults in close-packed structures; see relevant chapters in textbooks by Hirsch et al (1977); Thomas and Goringe (1979); Reimer (1984); Spence (1988). Later the oxide structures described in terms of crystallographic shear planes served as a prime example of high resolution imaging, Cowley & Iijima (1971), see Spence (1988); Buseck, Cowley and Eyring (1989). In recent years faults in modulated structures have been studied in many laboratories, (see following chapter).

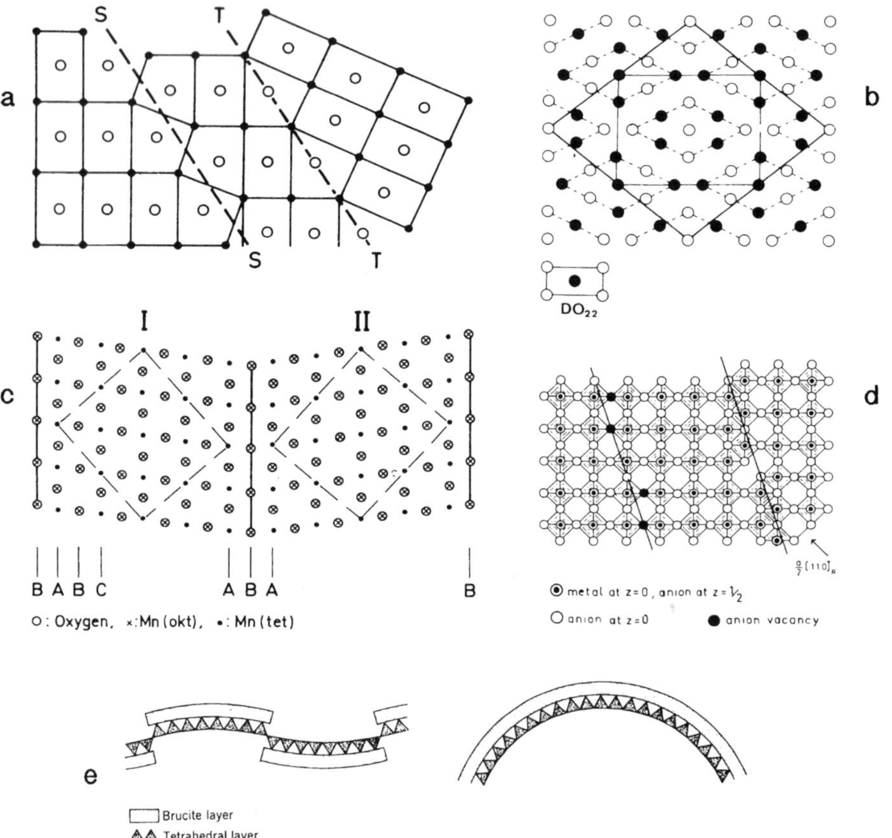

Fig. 1 Planar faults: a) stacking fault, S and twin, T in FCC lattice; b) twin model for a tetragonal spinel; c) antiphase structures in Au-Mn (Hiraga et al, 1980). d) shear defect in an oxide (after Eyring, 1970). e) curved lattices in antigorite and chrysotile (from Hurlibut and Klein, 1977)

In the diffraction pattern faults are recognized by streaks normal to the fault plane. For displacement faults, where the lattices at the two sides are displaced by a vector **R**, streaks will appear through reflections **g** for which **Rg** $\neq$ n - i.e.reflections from net planes which are shifted at the fault by a fraction of their spacing. These are also the reflections in which the fault is visible in diffraction contrast imaging. The observation of streaks (or of condition for diffraction contrast) thus allows determination of the fault vector **R**. If **R** lies in the fault plane (conservative fault), no streak is expected through the origin according to a kinematical argument. When a radial streak is observed, this is often due to Bragg diffraction interactions: kinematical diffuse scattering at **u** can be scattered by the reflection **h** into **u** + **h** Fujime, Watanabe and Ogawa (1964) showed that this multiple scattering effect will disappear when the incident beam lies in a symmetry plane containing the streak. When there are compositional variations associated with the faults, which may be the case when **R** has a component normal to the fault plane (Cowley, 1976), the radial streak will appear also in kinematical scattering.

Antiphase boundaries (APB) in superstructures have fault vectors **R** which are translation vectors in the subcell but not in the supercell. These may form a hierarchy of faults: APBs on (001) in the cubic $AuCu_3$ with **R** = [110]/2 can be ordered to form tetragonal superstructures with c=na, where n may be non-integer. In these structures APBs can again be formed, with fault vectors referred to the larger unit cell, see e.g. studies of the Au-Mn system, by the Sendai group (Hiraga et al, 1980; Terasaki et al, 1981) and by Amelinckx and his coworkers (van Tendeloo, de Ridder and Amelinckx, 1978).

Streaks which appear continous may on closer inspection be seen to include series of spots, revealing periodic faults. Different periodicities often appear in the same specimen forming a mixture of metastable superstructures. Such features are often overlooked or not resolved in diffraction patterns obtained from larger volumes. Fig 2c is an example: the diffraction pattern from a highly faulted specimen of the monoclinic intermetallic phase usually called Al_3Fe reveal faults on (001) with fault vector [100]/2. Note the weaker streaks through the origin and through spots with h=2n.

Transformation from a cubic to a tetragonal low-temperature phase frequently results in a microstructure consisting of fine twins on (110) planes. Fig 2a is a diffraction pattern from the tetragonal spinel Mn_3O_4. Several sets of periodic twins are seen in low magnification dark field micrographs taken around the twin spot positions, revealing an intergrowth of metastable superstructures. A classical example is the "disordered" form of SiC which produces x-rays patterns corresponding to a heavily faulted α-SiC. In SAD patterns from small regions (and in high resolution micrographs) the material is seen to consist of small domains with different long-period stackings (Shinozaki and Kinsman, 1978) - which illustrates that the definition of the state of order may depend upon the volume observed.

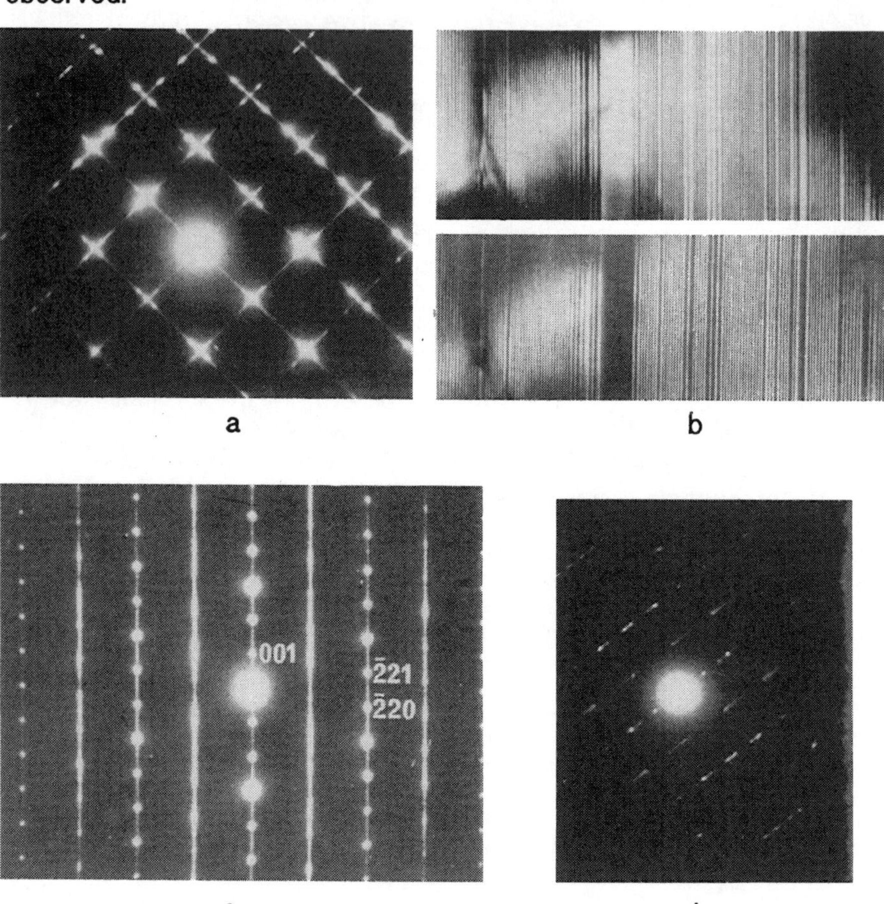

Fig 2 a) Diffraction pattern along [001] from two sets of (110) twins in tetragonal Mn_3O_4.b) two dark field pictures from two twin spots within the same set. c) Diffraction patterns from multiple faults in Al_3Fe. (Skjerpe, 1987) d) Diffraction pattern from chrysotile.

More detailed information about individual faults and their arrangement in disordered or ordered structures is obtained from images, especially in the HREM mode with the beam parallel to the fault plane, see e.g. Bourret, Rouvier and Penisson (1988); Smith (1989). But some features may evade the resolution or sensitivity of the microscope image, notably when the atomic displacements associated with the faults are small. Some microstructures in substituted 1:2:3 superconductors may serve as an example. With slight orthorhombic distortions diffuse [110] and [110] streaks can be related to a fine tweed-like structure on a nm-scale, interpreted as microtwins. In $YBa_2(Cu_{1-x}Fe_x)_3O_{7-\delta}$ Xu et al (1989) noted faint diffuse streaks even beyond the x-value .03 at which the b/a-ratio becomes unity according to powder diffraction measurements. In this range the tweed-like structure "was not clearly visible", but apparently some local deviations from the average tetragonal symmetry exist also in this composition range: a structural feature which may be called displacement disorder along the <110>-directions rather than faults.

Modulations: compositional waves, charge density waves.

Modulated structures, whether commensurate or incommensurate are not defect structures in the true sense, and are treated in another section (13). However, they may manifest themselves in the diffraction pattern in a similar way. The modulations frequently appear as a periodic stacking of planar defects, although basically described as relating to variations in composition or electronic structure (charge density waves). During spinodal decomposition the compositional variations may be accompanied by lattice strain. The resulting variations in local lattice spacing were determined by optical diffraction from high resolution micrographs in a study of the Au-Ni system by Sinclair, Gronsky & Thomas(1976). Modulations associated with charge density waves in transition metal chalcogenides have been studied extensively by Steeds and coworkers (Steeds et al. 1985) and by Amelinckx, (see 1).

Layer silicates: curved faults and bent lattices. Chemical and physical processes in rocks are reflected in microstructural features of minerals. Transformations due to temperature or pressure changes or by chemical reactions, such as leaching of ions, will leave intergrowths, modulations and intercalate structures. (For reviews of electron microscopy in mineralogy, see Wenk, 1976; Buseck, 1989). Disorders in the serpentine minerals have been the subject of diffraction and microscopy studies for some time. Curved layers are formed, due to the mismatch between the tetrahedral Si_2O_5 and the octahedral brucite layers forming the layer silicate. The bond stretching in the tetrahedral layer can be accommodated by bending the layers: with alternating curvatures forming a corrugated structure in the antigorite minerals or as rolls or scrolls in the chrysotiles, Fig 1e. Kinematical expressions for scattering by curved lattices were derived by Whittaker (1954-55). The characteristic diffraction features are seen in patterns from individual fibres, Fig 2e, - a more detailed picture is obtained from HREM images of cross sections (Yada, 1971). Some bending of the lattices may occur in many layer structures and affect also the intensities of the crystalline reflections through the deformation of the unit cell. This was considered by Cowley (1961), who proposed a bending correction formula for spot intensities in SAD patterns:

$$I(\mathbf{u}) = \sum_i W_i(\mathbf{u}) \exp[2\pi i \mathbf{u}\mathbf{r}_i] \exp[-\pi c^2 u^2 z^2]$$

where the intensity is calculated from different peaks $\mathbf{W}_i$ at $\mathbf{r}_i$ in the Patterson function. c is the bending in radians, z the component of $\mathbf{r}_i$.

<u>Substitutional disorder: diffuse spots.</u> The early investigations of sro in metallic solid solution with x-rays were classical diffraction studies of disorder in crystals. A statistical description by short range order parameters was derived from kinematical scattering. Electron diffraction from these alloys had an early impact through the study of longperiod superstructures. The description in terms of antiphase boundaries related to stoichiometry, electronic structure and transformations is seen as one of the major contributions of electron diffraction and electron microscopy to structure research.

Electron diffraction application to short-range order studies emerged more slowly, although Raether (1952) and later Marcinkowski and Swell (1963), Sato, Watanabe and Ogawa (1962), Watanabe and Fisher (1965) observed characteristic splitting of the diffuse maxima around the 100 and 110 positions in electron diffraction patterns from gold alloys. These quite sharp details, which later were confirmed by X-ray diffraction (Moss, 1966), suggested long range interactions, presumably through the conduction electrons as proposed by Krivoglaz (1969) and Moss (1969).Explanation in terms of Fermi surface interactions, similar to the Kohn anomaly in phonon scattering, fitted well with measurements carried out by Ohshima and Watanabe (1973). The separation of split spots for Cu-Pd (Fig 3a) and Cu-Pt alloys of different compositions could be interpreted with the electron/atom ratio as the only parameter, in a description in reciprocal space. Recently a combination of techniques: X-ray and electron diffraction, HREM and computer modelling, have been applied for a more precise direct space description

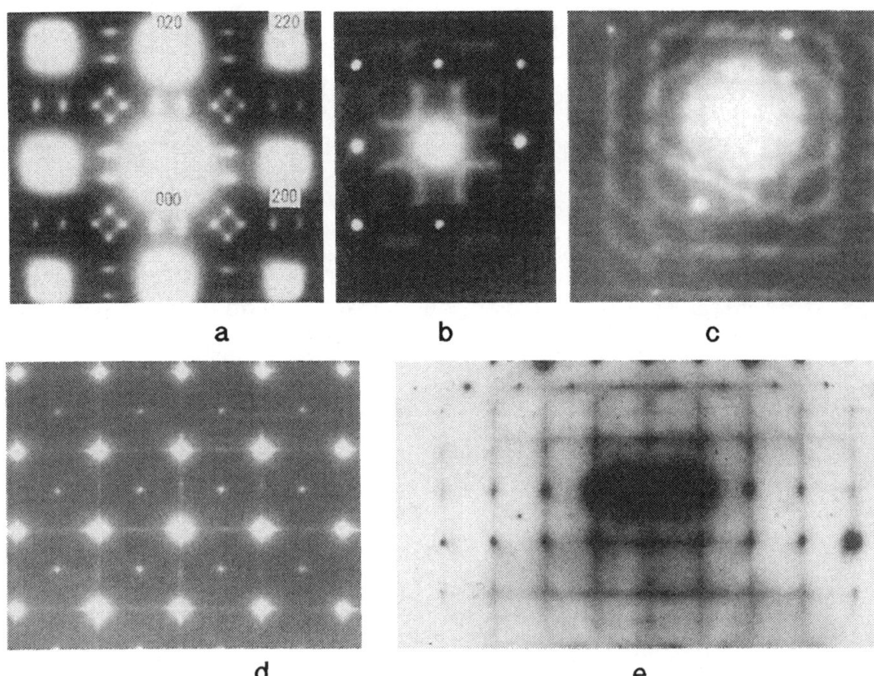

Fig. 3a) Fermi surface effects in sro scattering from Cu-Pd (Ohshima & Watanabe, 1973) b) diffuse scattering from $LiFeO_2$ near [100] projection. (van Tendeloo) c) High angle diffuse scattering from displacements in $VO_{1.23}$ [018] projection (Andersson, 1979) d) diffuse streaks from precipitates within Cu-rich phase in an Au-Cu-Ag alloy (Skjerpe et al, 1986) e) thermal diffuse streaks inn Ge (Honjo et al, 1964)

of the disordered state, (see Tanaka, Cowley and Ohshima (1987), van Tendeloo, Amelinckx and de Fontaine (1985)).

Many inorganic compounds with wide homogeneity range can be described in terms of point defects; vacancies or interstitials on one or more sites in the unit cell. These may form ordered superstructures - notably at low temperatures - or enter in short range ordered defect arrangements producing diffuse scattering. Lower oxides, carbides and nitrides of transition metals are examples of compounds with wide range of non-stoichiometry associated with high concentrations of defects on the metal or anion site. In the the defect rocksalt type monoxide of vanadium metal vacancies, metal intersitials at tetrahedral position and oxygen vacancies are involved in short range order in a phase which may be written $V_{1-x+t}O_{1-y}$

(Andersson, Gjønnes and Taftø, 1974). Several ordered superstructures are also found in this system (Andersson and Gjønnes, 1978). Other examples of defect rocksalt type structures with vacancy ordering are Ti-O (Castles, Cowley and Spargo, 1971), VC_{1-x} (Sauvage and Parthe, 1972), $LiFeO_2$ (Allpress, 1971). Vacancy ordering is found also in alloys with CsCl-type structure, e.g. $Al(Cu,Ni)_{1-x}$ (de Ridder et al. 1977a); sulfides $Me_{1-x}S$ based on a NiAs-type subcell; oxygen-deficient perovskites; metal-deficient spinels. Ordering of interstitial cations occur in many intercalation structure, e.g. transition metal dichalcogenides (de Ridder et al, 1976, 1977b).

The distribution of sro scattering within the reciprocal unit cell may contain quite sharp details, indicating correlation over long distances. Sometimes the details appear as thin shells of scattering within each Brillouin zone, as was first recognized by Allpress (1971) in patterns from $LiFeO_2$. (Fig 3b) Since then similar patterns have been reported from a number of alloys, oxides, carbides etc. They have attracted considerable interest and have been interpreted partly as transition states between local order and a superstructure, partly as an indication of long-range forces associated with the electronic structure. A representation in terms of sro parameters must include a large number of distances. Other lines of interpretation have therefore been explored, notably in terms of clusters, (see below).

Electron diffraction is a convenient and sensitive method for recording such detail - and the only one when single crystals suitable for x-rays or neutrons are difficult to produce. This is frequently the case when metastable phases or stages in a transformation are studied. Of particular importance is the application to non-homogeneous samples, not only with respect to specimen preparation, but also in relation to the structure itself. Disordered states which from x-ray investigations are ascribed to a single phase field may upon electron microscope investigations be seen to contain variations in compositions, symmetry and structure on a fine scale; see the model for $Fe_{1-x}O$ proposed by Andersson and Sletnes (1977).

Precipitates. Coherent precipitates formed from solid solutions are important microstructure elements in many alloys and ceramics. Selected area diffraction in combination with dark field microscopy is a well-established technique for the characterization of such material. For a textbook description, see e.g. Edington (1976). In alloy systems several precipitate particle types are often present, with different crystal structures and/or several orientation relationships. In the aged condition of an Al-Mg-Zn alloy four crystallographically different $MgZn_2$ precipitate orientations appeared with appreciable frequency in one particular state (Gjønnes and Simensen, 1970). With orientation multiplicities ranging from 4 to 8 this included 22 different reciprocal lattice orientations to the parent matrix. The dark field/SAD combination permitted identification of the lattices and description of their morphologies. Another example of diffraction from precipitates is shown in Fig 3e, an as-cast structure of an Au-Ag-Cu alloy. The solidification leads to phase separation in Ag- and Cu-rich lamellae. During cooling of the solid precipitation will occur within these lamellae, due to reduced solid solubility of Cu and Ag respectively. Diffraction patterns and dark field pictures show that the thin secondary lamellae within the Ag-rich regions possess an ordered tetragonal CuAu structure.

Point defects. Isolated point defects, i.e. vacancies, interstitials, foreign atoms, can be described in terms of the additional potential ϕ_d as producing an additional background term, (see Cowley, 1981). Due to the considerable background problem in electron diffraction and the influence of thickness variations etc, it has so far seemed impracticable to detect such scattering with electrons. A more realistic possibility may be to determine defect concentrations in compounds by accurate measurement of structure factors (Watanabe, Andersson, Gjønnes and Terasaki, 1974; Spence this volume). Another possibility is to detect the associated Huang scattering by the sophisticated critical voltage scattering technique shown by Sellar and Imeson (1988), (see below).

Thermal streaks and static displacement. Displacement of atoms, static or dynamical, relative to the average position in the periodic structure will produce diffuse scattering. The thermal diffuse scattering has been the subject of many diffraction studies with X-rays and especially with neutrons. The most extensive electron diffraction investigations may be those by Honjo and coworkers in the 1960s, see Honjo, Kodera and Kitamura (1964); Harada, Tanaka and Honjo (1966). From various substances they observed non-radial diffuse streaks in selected area diffraction patterns and found that these were in fact sections through walls of diffuse scattering in reciprocal space, normal to close-packed directions. At first they related the intensity distribution to thermal movement of strings of atoms - a picture which may appear different from the usual lattice wave description of thermal motion. Actually the streaks, and indeed the moving strings, can be well understood in terms of low frequency transverse lattice waves with polarization parallel to the string directions - a kind of soft modes. Rather similar patterns of diffuse scattering can be obtained with X-rays, as was noted quite early by Lonsdale and Smith (1941). In selected area electron diffraction patterns thermal streaks are often prominent and may obscure diffuse scattering from other sources. It is often assumed that electrons (and x-rays) are not very useful for the study of thermal scattering since one lacks the capability to resolve phonon energies, which makes neutron diffraction such a powerful technique. Even so, qualitative observations in electron diffraction can render detailed information about the spatial correlations of the atomic displacements and assist in model formulation. The facility for taking patterns at several temperatures is important, as shown in the study of pre-martensitic static displacments in In-Tl by Finlayson et.al. (1985). The pattern of diffuse scattering may change as fine-probe condition is approached, especially for static displacements. Bursill et. al. (1978) suggest that this can be used to distinguish between thermal and static displacements. It is expected that electron beam instrument of sufficient monochromacy to perform energy analysis in the range of 10^{-2}eV, corresponding to high-energy phonons, will be made in the future.

2.3. Experimental methods: parallel beam diffraction.

The standard way of observing diffuse scattering by electrons is with a parallel beam in the standard selected area diffraction mode - a technique which has not changed much since the introduction of reliable goniometer stages and beam tilt devices some 25 years ago. In marked contrast to the gas electron diffraction, rather little effort has until recently been devoted to intensity measurements in electron diffraction from solids: comparison with theoretical calculations have mostly been qualitative. There is scope for improvement along several lines, as discussed in other chapters. Digital microdensitometers can be used in combination with grey scales which are developped together with the diffraction photographs. Sequential recording by scanning systems, based on after-specimen deflection (ASD- or Grigson-coils: Grigson, 1962) which were introduced in electron diffraction nearly 30 years ago, have been used mainly for powders and amorphous patterns can be used for line scans of diffuse scattering, especially when combined with energy filtering for exclusion of the inelastic component of the scattering. They may be too slow (and thus sensitive to beam fluctuations and radiation damage) for two-dimensional patterns. For these, parallel recording systems with energy filtering may be needed. Additional problems arise when collection of three-dimensional intensity distributions of diffuse scattering is attempted.

Convergent beam diffraction. The convergent beam technique is mostly associated with perfect crystals but has been applied also to the study of faults and dislocations. A number of beautiful illustrations of such applications are presented in the book by Tanaka, Terauchi and Kaneyama (1988). The two-dimensional intensity distribution in CBED disks contains information which differs from what is obtained from the SAD-pattern. In the latter one records the scattering in a zone normal to the incident beam; in CBED the intensity variation in Bragg reflections is sampled along the direction parallel to the incident beam direction, i.e. as a function of the excitation error. This may include streaks in that direction, and offers the possibility of identifying faults parallel to the foil plane. Two effects are evident: reflections with $Rg \neq n$ will appear with a profile characteristic of the vertical position of the fault in the foil; the symmetry of the pattern will correspond to the faulted crystal. The latter effect was first shown by Johnson (1972): both effects are seen in the pattern Fig 4.

The effect on CBED profiles was treated by Tanaka, Terauchi and Kaneyama (1988) in relation to several CBED techniques: ZOLZ disks, LACBED, HOLZ lines. Experimental patterns were compared with calculations performed as two-beam rocking curves (see 2.3). The LACBED pattern has the advantage of combining diffraction and image information, HOLZ lines offer simpler profiles which are easily interpreted. With several reflections appearing in the CBED pattern, the fault vector can be obtained from one exposure. Based on detailed calculations of the Kossel line intensity profile the level of the fault in the foil can be determined. Defocussing of the probe can be used to locate

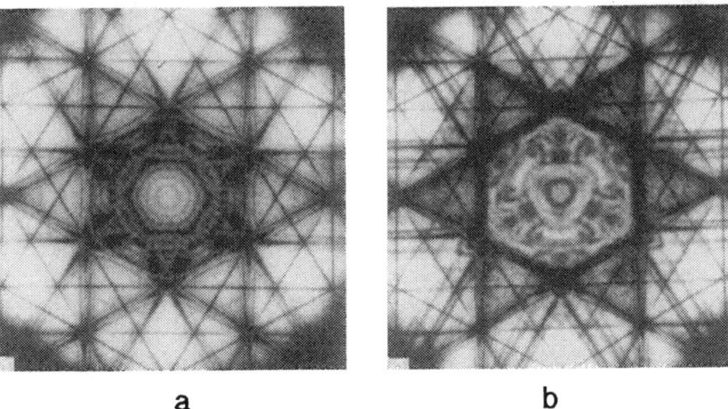

Fig 4. [0001] Tanaka patterns of graphite at 60kV from perfect crystal a) and crystal with stacking fault b).(Fung 1985)

the top and bottom of an oblique fault. Multiple faults as well can be analysed in this way; it is found that the CBED patterns from the heavily faulted layer structures are easier to interpret than SAD patterns when strong streaks intersect the Ewald sphere (Goodman, Olsen and Whitfield 1985; K. Gjønnes, 1985)

Determination of Burgers vectors from CBED patterns, first shown by Carpenter and Spence (1982) works on a similar principle, viz: the variation in lattice orientation across the dislocation. Using HOLZ lines, which are more sensitive to strains, the Burgers vector can be obtained uneqivocally from one pattern. Fung (1985) pointed out that the interpretation may be complicated by the split lines due to dynamic coupling between HOLZ reflecti o ns (Høier, 1972; Buxton, 1976), hence dynamical calculations of the profile and/or comparison with patterns from perfect regions should be made.

The sensitivity of HOLZ lines in CBED patterns to lattice orientation and unit cell dimensions (see chapter on CBED) can be used to detect local strains or lattice constant variations e.g. at coherent boundaries. From relative positions of HOLZ lines Twigg and Chu (1988) measured lattice mismatch in submicron heterostructures to an accuracy of 10^{-4} - essentially a local measurement of the average lattice. As was discussed by Treacy, Gibson & Howie (1985), the state of strain measured in the thinned specimen may be quite different from the original bulk material, hence caution is needed in the interpretation.

2.3 Theory: Kinematical intensity and Patterson methods

The kinematical description of the scattering has clear limitations in electron diffraction. Even weak components of the diffuse scattering may be strongly affected by Bragg reflection, as witnessed by effects such as Kikuchi lines. As serious is the quite strong multiple scattering especially of the inelastic component which is strongly peaked about the origin. This affects the level and slope of the steep background, and complicates all procedures for background subtraction. Even so, the kinematical theories will in many cases predict qualitative features in the scattering to sufficient accuracy and form a sound basis for the interpretation - if combined with an understanding of the basic effects of dynamical Bragg scattering.

Theoretical expressions based on kinematical theory can be developped along two lines: one follows the traditional approach from x-ray diffraction, based on intensity expressions related to a distribution of defects described in Patterson space. The other, which deals with the actual atomic configuration around a specific defect is used extensively for calculation of electron microscope contrast from defects. In electron diffraction we need both, since we are dealing with the distribution of diffuse intensity in SAD patterns from relatively large areas and with CBED from areas which may contain one or a few defects. Combination with modelling of atomic systems will shift the emphasis towards the direct space approach.

The x-ray approach is developped in considerable details in the books by Guinier (1963) and Warren (1969). Here we quote some typical results indicating only briefly the derivation. The expression for scattered intensity towards the direction $\mathbf{k}_o + \mathbf{u}$ can always be written in the form of a double sum over scattering units:

$$I(\mathbf{u}) = \sum \sum F_l \, F_{l'} \, \exp[2\pi i(\mathbf{r}_l - \mathbf{r}_{l'})\mathbf{u}] \qquad (2,3)$$

where the F_l may be defined according to the problem under study: sometimes representing the diffuse scattering only - i.e. the deviation ϕ_d from the average structure, cf equation (1); in other cases the F_l are taken as scattering factors for different layers in a stacking or expressed by occupation operators for short range order. The intensity can also be expressed as the Fourier transform of a Patterson type or correlation function,

$$I(\mathbf{u}) = \int P(\mathbf{r}) \, \exp[2\pi i \mathbf{r}\mathbf{u}]d\mathbf{r} \qquad (2,4)$$

In substitutional order the Patterson function P can be expressed in terms of order parameters, which are defined as correlations between occupation operators, σ_i(Flinn, 1956):

$$P_{sro} = \Sigma_j \, <\sigma_i\sigma_{i+j}> \, \delta(\mathbf{r}-\mathbf{R}_j) \qquad (2,5)$$

Displacement disorder is usually treated by the approximate expression

$$I(\mathbf{g}+\mathbf{s}) = \Sigma \, \Sigma \, |F|^2 \, <\exp[2\pi i(\delta_l - \delta_{l'})\mathbf{g}]> \, \exp[2\pi i \mathbf{R}_{ll'}\mathbf{s}] \qquad (2,6)$$

for the intensity in the vicinity of the reflection g. The first exponential in (6) may be called the correlation function for displacement short range order. Note that the relative displacement or strain $\delta_l - \delta_{l'}$ is paired with the reciprocal lattice vector g, whereas the distance in vector space $\mathbf{R}_{ll'}$ is paired with the deviation **s** from the reciprocal lattice point. The adaption of expressions like these to a large variety of disorders can be found in the literature cited.

Scattering from crystals with planar faults. A classical example is the treatment of stacking disorder in the close-packed FCC or HCP structures (Wilson, 1942). The close-packed planes can appear in three lateral positions A,B,C. One of these is chosen as origin in the x,y- plane - using hexagonal indices. The m'th plane will have scattering amplitude

$$A_m(h,k) = F_A \exp[(2\pi i/3)(2h+k)\delta_m] \exp[2\pi iml] \qquad (2,7)$$

where δ_m may take the values 0, -1, +1. Note that planes A,B,C scatter with the same phase for 2h+k=3n; diffuse streaks are expected only for 2h+k ≠ 3n∓1. The intensity is obtained as a double sum

$$I(h,k,l) = |F_A(h,k)|^2 \sum_m \sum_{m'} \exp[(2\pi i/3)(2h+k)(\delta_m - \delta_{m'})]\exp[2\pi i(m-m')l] \qquad (2,8)$$

which can be written in terms of probabilities associated with each distance n between planes:

$$I(h,k,l) = |F_A(h,k)|^2 \sum_n (\Sigma_\delta P^\delta(n) \cos[(2\pi/3)(2h+k)\delta]) \exp[2\pi inl]$$

$$= |F_A(h,k)|^2 \sum_n \langle\exp[i\Phi(n)]\rangle \exp[2\pi inl] \qquad (2,9)$$

Measurement of intensity along the diffuse streak allows in principle the determination of probabilities, P(n), by taking the Fourier transform

$\int I(h,k,l) \exp[-2\pi inl]\, dl$ of (2,9) - and by considering

$$\langle\exp[i\Phi(n)]\rangle$$

$$= P^0(n) + P^+(n)\exp[2\pi i(2h+k)/3] + P^-(n)\exp[-2\pi i(2h+k)/3] \qquad (2,10)$$

where we may assume $P^+(n) = P^-(n)$. Alternatively the P's can be expressed in terms of one or two "continuing probabilities", which for several models allows the summation over m to be carried out. Examples are given in the literature; for generalizations, see Jagodzinski (1949, 1954) and Kakinoki and Kimura (1952) who extended a matrix formulations introduced by Hendricks and Teller (1942) to a set of different layers, i,j,.. with a continuing probability P_{ij} that a layer j will follow the layer i. Assuming P_{ij} to depend only on the preceding layer, i, the intensity expression can be summed into an expression which is written in matrix form e.g.

$$I(u) = N\,\mathbf{B}_o + \sum_{m=1}^{N-m} (N-m)\mathbf{B}_m + \text{c.c.}\,, \qquad \mathbf{B}_m = \text{spur}\;\mathbf{VFQ}_m$$

where the matrices **VF** and **Q** are defined as $(\mathbf{VF})_{ij} = \mathbf{F}_i p_j \mathbf{F}_j$ with p_j a priori probability of a layer j; $(\mathbf{Q})_{ij} = \mathbf{P}_{ij}\exp[-\phi_j]$ with ϕ_j the phase shift due to the thickness of the layer j. Other formulations are given in the papers by Kakinoki and Kimura. Jagodzinski (1954) introduced the term reichweite, s, in order to include cases when the continuing probability depends not only on the layer i but on the combination within s preceding layers. In these treatments and in the books by Warren (1969) and Guinier (1963) a number of examples can be found. Different possibilites of stacking closed packed planes were illustrated by Jagodzinski (1949) as a probability tree, Fig 5a.

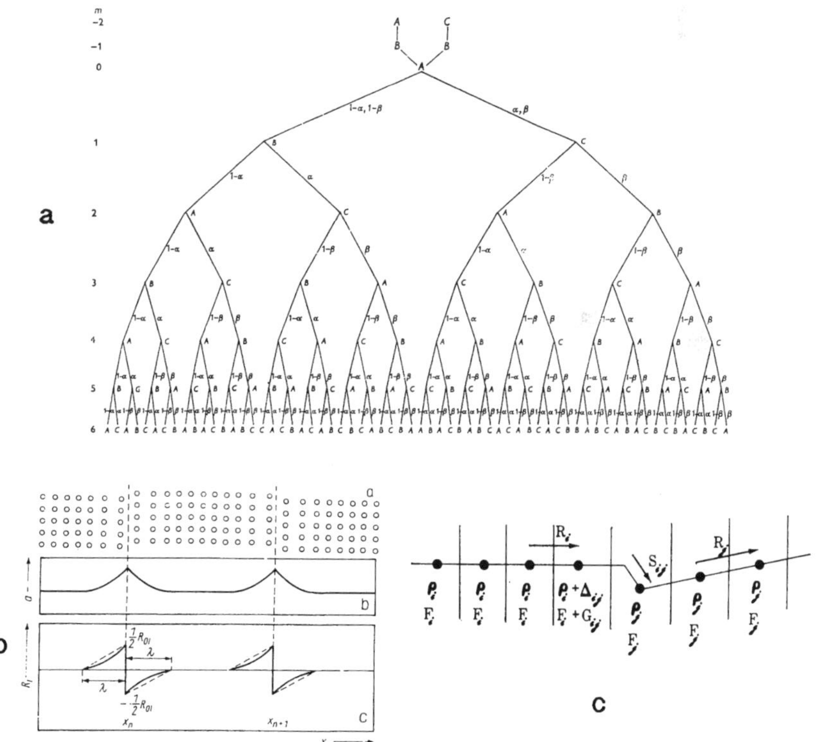

Fig5. Fault models: a) probability tree, (Jagodzinski, 1949) b) relaxations around a fault, (van Dyck et al 1980) c) fault description by Cowley (1976)

Cowley (1976) presented the intensity expression in a somewhat different way. Starting from a Patterson function he arranged the terms according to interlayer vectors: scattering from independent layers, interference between nearest neighbour layers, next neighbours and so on. He also included the modification Δ_{ij} of the content ρ_i in layer i resulting from a different preceding layer j as well as a displacement S_{ij} which is added to the interlayer distance r_i between i-layers, see Fig 5c. When S_{ij} has a component perpendicular to the layers material must be subtracted or added, usually resulting in change of composition, i.e. a non-conservative fault. The modification of the structure arising from faults was also treated by van Dyck et al, (1980) by a relaxation in the boundary region, i.e. as an increment to the fault vector **R**, see Fig 5b.

In the CBED patterns taken from a small area, diffraction is obtained from a single fault or the faults occuring in a cross section through the foil. The intensity calculation then is the same as the treatment developped originally by Hashimoto,

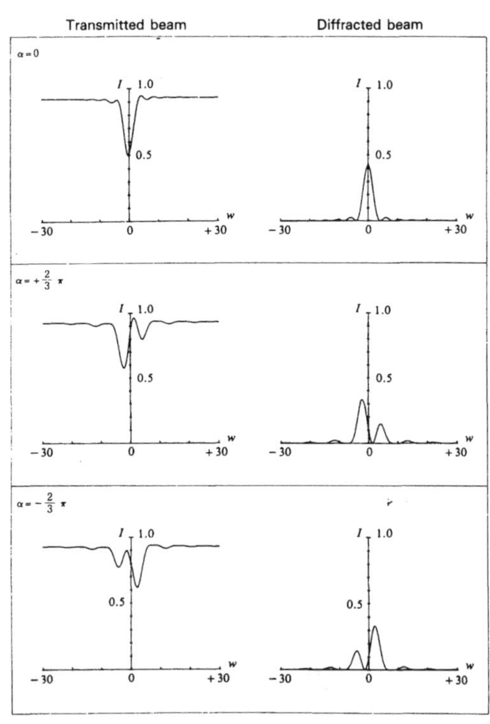

Fig 6 Calculated rocking curves for different values of the parameter $\alpha = \mathbf{g}\delta$ for thickness $t = \xi/4.25$ at the middle of the fault, $t_1 = t/2$. (from Tanaka, Terauchi and Kaneyama, 1988).

Howie and Whelan (1962) for diffraction contrast. Assuming that the CBED illumination can be considered as perfectly incoherent (for a discussion of this point see Spence and Carpenter (1986) and Cowley, this volume), the intensity profile is given by the calculated rocking curve. For a fault parallel to the foil and roughly normal to the incident beam the rocking curve is readily calculated from two-beam approximation using the well-known matrix expression found in textbooks, (see e.g. Reimer, 1984):

$$\begin{vmatrix} \Phi_o(t) \\ \Phi_g(t) \end{vmatrix} = \begin{vmatrix} S_{oo}(t-z) & S_{og}(t-z) \\ S_{go}(t-z) & S_{gg}(t-z) \end{vmatrix} \begin{vmatrix} \Phi_o(z) \\ \Phi_g(z) \end{vmatrix}$$

which may be used even for an oblique fault, assuming the column approximation (see chapter 3), and allowing for some thickness spread due to the finite lateral extension of the probe. For reproducing the fringe pattern, the two-beam expression can be used even when other beams are excited, provided that an effective Fourier potential is introduced. Extension to several faults is straight forward, involving more matrix multiplications. For HOLZ lines the corresponding kinematical expression will often be adequate, viz:

$$I_g(s) = U_g^2 \{[2\sin(\pi Ls/2)\cos(\pi Ls/2 + \alpha/2)]^2 + 4\sin^2(\pi sz)\sin(\alpha/2)\cos(\pi Ls+\alpha/2)\} \qquad (2,12)$$

Substitutional short range order; Cowley-Warren parameters, Borrie-Sparks expression. Scattering from short range order in binary alloys was treated by Cowley (1950) in terms of order parameters α_j for each interatomic distance R_j. He showed how an expression for the short range order scattering can be extracted from a general intensity expression, for the disordered binary alloy A-B, with atomic fractions m_A and m_B. Later authors (Clapp & Moss, 1966) introduced the so-called Flinn (1956) operators σ_j ($\sigma_j = m_B$ if the position j is occupied by an A-atom, $\sigma_j = -m_A$ when it is occupied by a B-atom). The amplitude expression for the diffuse scattering becomes

$$A(\mathbf{u}) = [f_A(m\mathbf{u}) - f_B(\mathbf{u})]\sum_l \sigma_l \exp[2\pi i m \mathbf{R}_l \mathbf{u}] \qquad (2,13)$$

and the corresponding intensity expression

$$I_{sro}(\mathbf{u}) = \sum_j \alpha_j |f_A - f_B| \exp[2\pi i \mathbf{R}_j \mathbf{u}] \qquad (2,14)$$

in terms of the Cowley-Warren sro parameters :

$$\alpha_j = \langle\sigma_l\sigma_{l+j}\rangle / m_A m_B$$

which can be obtained from measurement of diffuse scattering in one Brillouin zone. Extension to several atom species and to ordering on different sites follow similar lines, but becomes quickly quite involved. Hayakawa and Cohen (1975), Hashimoto (1987)

The different atom species involved will usually have different radii. Hence they may be displaced from the average sites referred to the ideal substructure. This "size effect" can be included through an additional exponential factor containing the atomic displacements δ_l in the intensity expression for short range order, as shown by Borie & Sparks (1971). Starting with the expression

$$I(\mathbf{u}) = \sum_p\sum_q f_p f_q \exp[2\pi i\mathbf{u}(\mathbf{r}_p-\mathbf{r}_q)]\exp[2\pi i\mathbf{u}(\delta_p-\delta_q)] \qquad (2,15)$$

and expanding the last exponential in a power series, they obtained additional intensity contributions in terms of displacement parameters, which describe the relative displacement of atoms at sites separated by a vector $\mathbf{r}_j$. Hayakawa and Cohen (1975) extended the treatment to several sites in the structure and wrote the short range order scattering in the form

$$I_d/I_{LM} = \Sigma_{lmn}\alpha_{lmn}A_{lmn} + \Sigma_1^3 Q^p h^p + \Sigma_1^3 R^p h^p + \Sigma_1^3 S^{p,p+1} \qquad (2,16)$$

where

$$Q^p = i\Sigma_{lmn}\gamma^p A_{lmn};\ R^p = \Sigma_{lmn}\delta^p A_{lmnn};\ S^{p,p+1} = \Sigma_{lmn}\varepsilon^{p,p+1}A_{lmn};$$

$A_{lmn} = \exp[2\pi i(h_1 l+h_2 m+h_3 n)]$ and $\gamma,\delta,\varepsilon$ are displacement parameters averaged over different sites. Simplified expressions involving only nearest neighbour distance increments are quoted e.g. by Warren (1969). The "size effect"-contributions vary through the reciprocal space with the components h_p of the reciprocal vector $\mathbf{u}$, but the factors $Q^p, R^p, S^{p,p+1}$ have the same periodicity in reciprocal space as the first sum. From measurements of the diffuse intensity in different Brillouin zones, the contributions can be separated, *i.e.* the parameters, α,τ,δ, and ε can be determined for each distance j. It may be noted that the atomic displacements will affect also the average structure through a Debye-Waller type factor and that the R and S-terms above correspond to so-called Huang scattering.

Expansions in order parameters include the total information contained in the diffracted intensity. In order to proceed beyond this description in correlation or vector space additional information or assumptions must be introduced. One approach to this problem is the so-called cluster expansion developped in Amelinckxs group (de Ridder et al., 1976, 1977; van Dyck et al. 1977, 1980). This applies to the frequently observed case where the diffuse scattering is concentrated on a shell in the Brillouin zone. They describe this locus by the equation f(u) = 0 which is then represented by a Fourier series:

$$f(\mathbf{u}) = \sum_k \omega_k \exp[2\pi i \mathbf{u}\mathbf{r}_k] \qquad (2,17)$$

On the other hand the diffuse scattering is given by the standard amplitude expression in terms of occupation operators σ. Hence the identity $A(\mathbf{u})f(\mathbf{u}) = 0$, or in terms of the Fourier components:

$$\sum_k \omega_k \sigma_{j+k} = 0 \, , \qquad (2,18)$$

valid for all j. The power of this relation, where the ω_k can be experimentally determined, may be appreciated by reference to the much more complicated relation (5) based directly on the α's. (18) is now used as a "cluster relation", *i.e.* relating the occupation operators within a cluster consisting of sites involved in the ordering. It follows that the cluster should have the macroscopic composition. Examples of clusters are shown in Fig7 a-f, they should not be confused with defect clusters which appear as structural entities in short- or long-range order, Fig 8. Note that the octahedron 7a corresponds to the equation

$$(\cos \pi h + \cos \pi k + \cos \pi l) - \tau(\cos \pi h \cos \pi k \cos \pi l) = 0 \qquad (2,19)$$

where h, k, l are reciprocal coordinates. This was noted by Sauvage and Parthe (1974) for the distribution of diffuse scattering from several carbides of transition metals, VC_{1-x}, TaC_{1-x}, NbC_{1-x}, TiC_{1-x}; around composition corresponding to T_6C_5. The octahedral cluster is taken to consist of one vacancy and five carbon atoms. The clusters are seen as elements in a transition state between short range order and long range order, which may be envisaged as small ordered domains separated by planar boundaries. (de Ridder, van Tendeloo, van Dyck and Amelinckx, 1976).

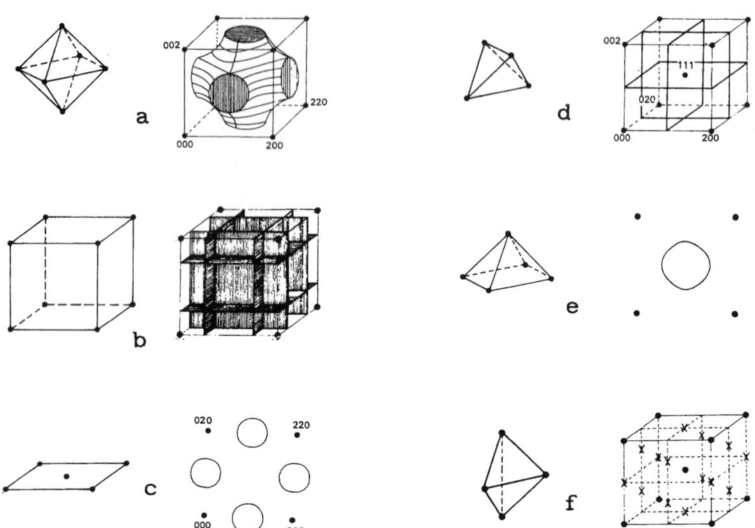

Fig. 7. Different cluster types with their corresponding diffuse intensity distributions : a) octahedron, b) cube, c) centered square, d) tetrahedron, e) square pyramid, f) irregular tetrahedron, after Amelinckx, de Ridder and van Tendeloo (1976).

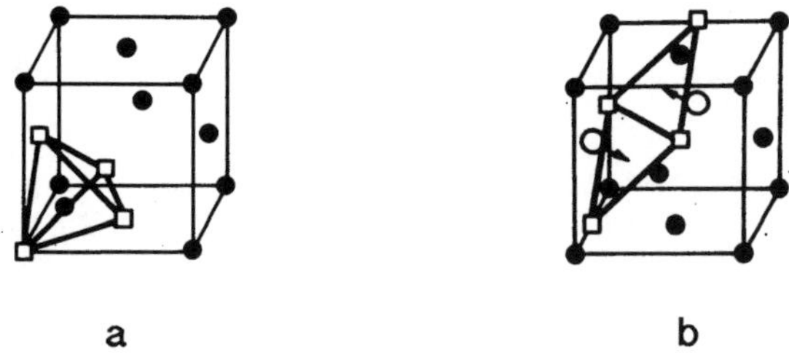

Fig. 8. Defect clusters in the V-O system: a) Tetrahedron of metal vacancies around inter- stitial metal atom in V_{1-x}. b) Sheet of oxygen vacancies with displaced metal atoms in VO_{1-y}. Filled circles: metal atoms; squares: vacancies; open circles : displaced metal atoms.

Displacement disorder or strains. Atomic displacements in crystals may originate from various sources with spatial correlations differing widely in character. At one end large relative displacements which vary over short distances, e.g. the "size-effect" associated with substitutional order. These can produce diffuse scattering over appreciable parts of the Brillouin zone.

At the other end we have the small elastic strains extending over long distances from dislocations, coherent precipitates and coherent boundaries. These are not conveniently related to an average lattice and are better described as local variations of lattice orientation and parameters and studied by imaging techniques or by CBED, see 2.2.

Much of the diffraction information relating to atomic displacements is contained in scattering close to the Bragg spots and difficult to record. This is so partly because crystals may be bent, but also because there is usually strong inelastic scattering close to the Bragg spots. Filtering of the inelastic scattering in diffraction patterns may substantially improve the situation. Further gains can be obtained by exploiting dynamical effects for suppressing/emphasizing different components in the scattering (see below)

Thermal diffuse scattering. This a special case of displacement, described in terms of lattice vibrations q (see e.g. Slater (1958)

$$\delta_l = \sum_q \sum_r A^r_q \; W^r_q \; \exp[2\pi i q R_l + i\alpha^r_q - \omega^r_q t]$$

where the sums are over all vibration modes, r, for each wave vector q within the Brillouin zone in an expression of type (6). The resulting intensity expression can be expanded in zero-phonon, one-phonon etc terms. The zero order term is Bragg reflections with the Debye-Waller factor $\exp[-M_l]$ on each atom. The first order term can be written

$$I_1(u) = \sum_q \sum_r \left| \sum_l f_l \exp[-M_l] \; A^r_q(j) W_q \; \exp[2\pi i g R_l] \right|^2 \delta(u-g+l) \qquad (2,20)$$

A better approximation to this one-phonon contribution term is given by Slater, op cit. Expressions for the second order term are given by Warren (1969) and Walker (1956). The two-phonon term near the Bragg spots called Huang type scattering is difficult to observe, see however Sellar and Imeson (1988).

Diffraction, reflection and refraction at interfaces

In the treatment of diffraction from faults, bicrystals etc, it is often assumed that two perfect crystals are joined together at a sharp interface. In more sophisticated treatment the relaxation of atom planes or atom positions at the interface is considered, see van Dyck et al. (1987). In addition to these effects there are reflection/refraction effects which are seen in the diffraction pattern as a faint streak extending from the central spot, as shown by Taftø (1986) who pointed out that this could be used to determine the sign of the differences between mean inner potential of different materials. When a Bragg reflection is strongly excited the average potential U seen by the incident electron will be modified according to

$$U_0^{eff} = U - 2\,k\gamma^j + \Sigma_g |C_g^j|^2 2ks_g$$

for the Bloch wave j (Taftø and Gjønnes, 1988). By varying the excitation error of the strong reflection, the effective potential difference can be adjusted to the potential at the other side of the interface - suggesting this as a method for quantitative determination of mean inner potential in e.g. the components in heterostructures.

The angular intensity variation in the streak may be estimated by a simple phase object calculation based on a slight modification of the standard procedure explained by Cowley (1981). Let the incident beam be a planar wave entering parallel to the interface. Take the projection of the potential along a direction bisecting the angle between the incident and diffracted wave. For diffraction into the angle 2θ the projected potential will have the lateral variation as shown with the scattered amplitude given by

$$A(s) = \int \exp[i(\sigma U/2b)x]\, \exp[2\pi isx]dx = \frac{\sin[(\sigma U+2\pi s)]}{\sigma + 2\pi s} \qquad (2,21)$$

2.5.Theory. Adaption to electron diffraction: dynamical and multiple scattering

Modification of the diffuse scattering due to Bragg diffraction is apparent even in quite thin crystals. In addition to scattering from the direct beam $\mathbf{k}_0$ into directions $\mathbf{k}_0+\mathbf{u}$ each Bragg beam $\mathbf{k}_0+\mathbf{g}$ will also be subject to diffuse scattering in $\mathbf{k}_0+\mathbf{g}+\mathbf{u}$. Similarly will diffuse scattering in directions $\mathbf{k}_0+\mathbf{u}$ near Bragg condition for a reflection h be scattered into an excess Kikuchi line appearing at $\mathbf{k}_0+\mathbf{u}+\mathbf{h}$, leaving a deficient line at $\mathbf{k}_0+\mathbf{u}$. The effects of multiple diffuse scattering may be as serious and even more complicated, by producing appreciable changes in the slope and level of the largely inelastic background.

A comprehensive and quantitative theory for diffuse scattering of electrons should ideally include dynamical Bragg diffraction effects and multiple diffuse scattering - which presents a formidable task. In practice considerable simplifications are always introduced. The Bragg diffraction part has been considered by several authors, the first fairly complete theory was presented by Kainuma (1954) who gave expressions for Kikuchi line contrast at points $\mathbf{u} + \mathbf{g}$ in the diffraction pattern in terms of scattering factors which may be written

$$|\Phi_d(\mathbf{u})|^2; \quad |\Phi(\mathbf{u}+\mathbf{g})|^2 \quad \text{and} \quad |\Phi_d(\mathbf{u})\Phi^*(\mathbf{u}+\mathbf{g})| \qquad (2,22)$$

The theories, which are first order in the diffuse scattering, can be visualized as Bragg scattering of the incident beam 0 into reflections g, followed by diffuse scattering into a set of beams $\mathbf{u}+\mathbf{g}$ at the level z and then by Bragg scattering between these beams. The amplitude contributions from all levels dz (see Fig. 9) must then be added:

$$A(\mathbf{u}+\mathbf{g}) = \sum_h \sum_f \int S_{gf}(t-z;\mathbf{k}_0+\mathbf{g})\Phi(\mathbf{u}+\mathbf{f}-\mathbf{g})S_{ho}(z;\mathbf{k}_0)dz \qquad (2,23)$$

In the intensity expression

$$I(\mathbf{u}+\mathbf{g}) = \int \sum_h \sum_{h'} \sum_f \sum_{f'} S_{gf'}(II)\, S^*_{gf}(II) \langle\Phi(\mathbf{u}+\mathbf{f}-\mathbf{h})\, \Phi^*(\mathbf{u}+\mathbf{f}'-\mathbf{h}')\rangle S^*_{h'o}(I)S_{ho}(I)dz \qquad (2,24)$$

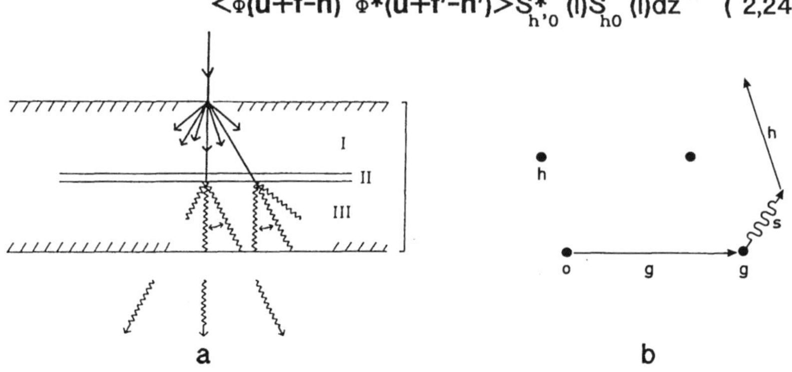

Fig 9 a) Model for calculation of single diffuse scattering with Bragg interactions. b) Schematic representation of amplitude term in eq.(23).

where the interference between diffuse scattering at different levels is neglected we have introduced the abbreviated arguments I and II for the upper and lower part of the crystal.

In order to perform the integration over z, the Bragg scattering matrices S can be expanded in terms of Bloch waves. Resulting expressions for the two beam case as given by Gjønnes (1966), can be interpreted as intraband and interband scattering of the Bloch waves by the potential Φ.

The complete expression (24) has only occasionally been used in structure studies, see however Andersson, Gjønnes and Taftø (1974). Further approximations may often be warranted, e.g. neglecting the $\Phi(\mathbf{u} + \mathbf{g})$ for $\mathbf{g} \neq \mathbf{0}$ for the inelastic scattering, or neglecting the Bragg scattering of the incident beam by omitting terms with $\mathbf{h} \neq 0$ in equation (24). In studies of diffuse scattering from defects and thermal motion two types of terms should be noted. Consider the factors $\langle\Phi(\mathbf{u+h})\Phi(\mathbf{u+h'})\rangle$ for different values of h,h': Those with h=h' result only in a redistribution of intensity between corresponding points in the Brillouin zones, preserving the total intensity. Those with h≠h' lead to enhancement or reduction of total diffuse intensity - and hence to increased/reduced absorption from the Bragg beams and to enhanced/ reduced intensity of secondary radiation (photons, inelastically scattered electrons) i.e. anomalous absorption and channelling effects (ALCHEMI).

Channelling effects will be present also in scattering from defects, and may in principle provide additional informations. Gjønnes and Høier (1971) expressed such information in terms of the Fourier transform R of the generalized Kikuchi line form factor i.e.

$$Q(\mathbf{u},\mathbf{h}) \equiv \langle \Phi(\mathbf{u})\, \Phi(\mathbf{u} + \mathbf{h})\rangle\, u^2\, (\mathbf{u} + \mathbf{h})^2 = \mathcal{F}(R(\mathbf{r},\mathbf{q})) \qquad (2,25)$$

which includes information about correlation between sources of diffuse scattering and about their position within the projected unit cell. (The factor $\mathbf{u}^2(\mathbf{u} + \mathbf{h})^2$ removes the Coulomb factor in Φ). Several special cases can be derived from (25). The section $Q(\mathbf{u},0)$ represents the kinematical intensity, hence $\int R(\mathbf{r},\mathbf{q})\, d\mathbf{q}$ is the Patterson function. The integral of $Q(\mathbf{u},\mathbf{h})$ in the plane gives the anomalous absorption (Yoshioka, 1957) which is related to the distribution $R(\mathbf{r},0)$ of scattering centres across the unit cell. Channelling effects in disorder scattering can be seen, but applications analogous to ALCHEMI by spectroscopic signals appear difficult, since these will rely on quantitative and comparable measurements of diffuse scattering at different orientations of the incident beam.

However, there are significant effects which can be recognized and understood in terms of the scattering factors. For substitutional order at one site in the unit cell we obtain

$$< \Phi(u)\ \Phi(u + h)> = |\Phi(u)|^2\ [f_A(|u-h|) - f_B(|u-h|)]/[f_A(u) - f_B(u)] \qquad (2,26)$$

It is seen that Q(u, **h**) here does not contain any new information in this case: the ordering involves only one site, which is known. When several sites are involved in the ordering, this factor becomes less trivial, scattering factors for ordering on different sites will then include factors $\exp[2\pi i r_m h]$ and thus position information about the sites involved in the ordering.

Diffuse scattering from displacements, e.g. by lattice waves, is affected differently. We can express the scattering factor in terms of phonons:

$$< \Phi(u)\ \Phi(u-h)> = G_j(u,q)\ G_j(u-h,q) \qquad (2,27)$$

Since the scattering amplitude for phonons and indeed for all displacement contributions is antisymmetrical about the origin, $G_j(u,q)$ and $G_j(u-h,q)$ will have opposite sign and lead to reduction of the diffuse scattering from displacements especially in the low angle region, see Fig 3c. The same applies to static displacements - which is responsible for the very weak size effect scattering seen in many electron diffraction patterns, see Watanabe and Fisher (1965). Further out in the pattern, the effect may be opposite, since diffuse scattering from strong beams may appear with the same sign, enhancing the scattering from displacement. This effect was used by Andersson (1979) in his study of sro in the VO-system.

The studies referred above allow several conclusions about the effect of Bragg scattering on defect scattering: The predominant effect is a redistribution of diffuse scattering between corresponding points in the different Brillouin zones. This has its minimum for a symmetrical pattern. The interference terms, $h \neq h'$ in (22) affect the sro substitutional and displacement components differently. In the central part of the pattern diffuse scattering from substitutional order will dominate, since it is kinematically stronger and is favoured by the dynamical Bragg interactions. Further out in the pattern the displacement component is higher both through the increased δu and because of dynamical interactions. Since electron diffraction patterns extend to higher values of the scattering variable, u, than can be attained with x-rays, they may offer a better possibility of separating displacement and substitutional components - if a workable procedure for separating other background components can be found.

The main obstacle to quantitive studies by sro scattering of electrons is multiple diffuse scattering, especially of the inelastic component, and its modification by dynamical Bragg scattering. Some progress has been made towards a workable multiple scattering formulation: Høier (1973) developped a theory which includes Bragg scattering only in the final segment of a sequence of scattering processes, and explained qualitative effects in Kikuchi patterns based on essentially two beam treatment:

$$I(u+g)$$

$$= \sum_j |C_g^j|^2 \left\{ A_1^j \sum_{g \neq g'} \sum F_1(u,g,g') C_g^j C_{g'}^j + \sum_n \sum_g A_{ng}^j F_n(u,g) |C_g^j|^2 \right\} \quad (2,28)$$

where the summation is over Bloch waves, j, intermediate reflections g,g′ and multiple scattering order n. It is conceivable that analytical expressions based on this form may reproduce the background. The task would be simplified to a considerable degree by removing the inelastic scattering by a filter.

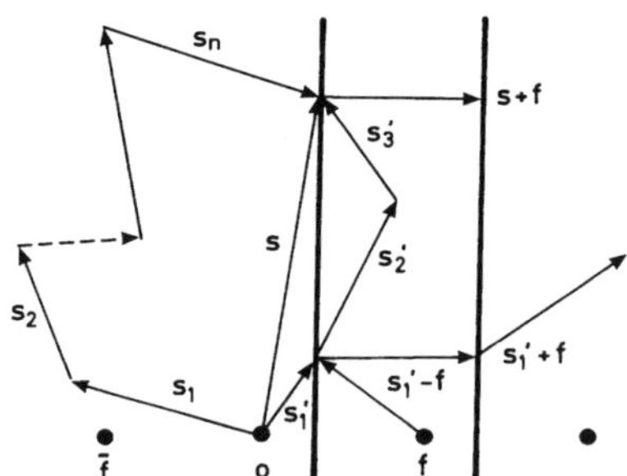

Fig. 10 Schematic representation of an intensity term in (2,28) (Høier, 1973)

Let us end this section by sketching two approximate approaches to dynamical corrections for the diffuse scattering, assuming thin specimensand/or small scattering angles. For the general case of a periodic, average structure plus a non-periodic part (2,1), we can also write the phase object approximation in the form:

$$\Psi(r) = \exp[-\sigma\phi(r)] = \exp[-i\sigma\phi(r)](1-i\sigma\Delta\phi(r)+ \ldots) \quad (2,29)$$

The Bragg reflections will then be given by the periodic part, viz:

$$<\exp[-i\sigma\phi(r)]> = \exp[-i\sigma\phi(r)] \quad \exp[-\sigma^2<\Delta\phi^2(r)/2>] \quad (2,30)$$

- where the last exponential is an absorption factor.
From the leading term in the expression for diffuse scattering i.e $-i\sigma\Delta\phi(r)\ \exp[-i\sigma\phi(r)]$, the intensity becomes

$$I_d(u) = |\Phi_0(u)|^2 * \Phi_{av}^{\ 2}(u)| \quad (2,31)$$

i.e. a convolution of the kinematical diffuse scattering amplitude with the Bragg reflections. Note that diffuse scattering which is periodic in reciprocal space will be modified only to a limited extent. The shape of diffuse maxima will be preserved; this is exploited in the cluster interpretation of diffuse scattering surfaces.

Another approach may be used for faults normal to the foil surface. Using the expression (2) for the kinematical case, a column approximation may be introduced for scattering within each block, leading to the very simple result

$$I(u) = |\Phi_0(u)|^2 |D(u)|^2 \quad (2,32)$$

which preserves the features of the kinematical case, when Ψ has the same symmetry as the kinematical Φ, but may deviate from this when the beam is tilted away from the zone axis. The calculation may be improved by calculation of scattering from periodic structures made up from such blocks.

2.6 Amorphous matter

Electron diffraction has important advantages for the study of amorphous substances. Small samples can be used, in particular thin films - the form in which amorphous material is often obtained and used. Measurements can be made to quite large values of the diffraction variable; $s = \sin\theta/\lambda \approx 3$ is easily reached, whereas the upper limit for e.g. MoK -rays is 1. The strong scattering leads to some problems. Although dynamical scattering in the sense it is encountered in crystals may be negligible, multiple scattering must be considered. Early experiments were carried out in diffraction apparatus of the type used for gases, i.e. without lenses after the specimen and with a rotating sector. The procedures developed in that field for handling photographic intensities were used; the interpretation followed the theories developed for x-ray scattering from liquids and amorphous solids. Multiple scattering constituted a main problem, to some extent also the lack of adequate models for the amorphous state.

Since then there has been a revival of the field, spurred by experimental development and by the interest in amorphous film, notably in microeletronics. Electron microscopes equipped with deflection coils for radial scan of the diffraction pattern across a detector are used for recording the intensity, essentially the method developed by Grigson (1962). The inelastic scattering is removed by energy filtering in a retarding potential or by a commercial energy selecting spectrometer (Cockayne and McKenzie, 1988).

Interpretation of the scattering is usually based on the radial distribution function, which is a spherically symmetrical Patterson function. Following standard theory used in x-ray and neutron studies of non-crystalline matter or gas electron diffraction we may write for the total scattered intensity:

$$I(u) = \sum_{i \neq j}\sum f_i f_j \frac{\sin uR_{ij}}{uR_{ij}} + \sum_j f_j^2 + \sum_j S_j / s^4 \qquad (2,33)$$

where $f_{i,j}$ are the atomic scattering factors for electrons, R_{ij} the interatomic distances and S_j is the incoherent scattering factor for the inelastic scattering, which is removed by the energy filtering.

The double sum is often called the molecular scattering I_M. In the case of one kind of atoms this can be written

$$I_M = f^2 \int \sigma(r) \frac{\sin ur}{ur} r^2 dr \qquad (2,34)$$

where $\sigma_M(r)$, the radial distribution function for atoms, can be obtained by a simple Fourier transformation of the intensity:

$$\sigma_M(r) = \int \frac{I_M(u)}{f^2(u)} \frac{\sin ur}{ur} u^2 du$$

The subscript M implies that we have subtracted the contribution to $\sigma(r)$ from the average density - and thus the small angle scattering from I(u).

When different atom species are involved several procedures have been developed. Some are based on average atomic scattering factors

$$\langle f \rangle = \Sigma_j N_j f_j / N$$

(Cockayne, 1988; Warren 1969) Other procedures are based on the so-called normal curves for each type of atomic pair ij. Such procedures - and indeed any quantitative interpretation of continous scattering curves - depend upon background subtraction. This is particularly important if the scattering is to be normalized for determination of the coordination through the weight of peaks in the σ-curve. This calls for knowledge of thickness and some procedure to correct also for muliple scattering. Even for quite thin films multiple scattering can affect the results, since the scattering at large angles will receive considerable contribution from second or higher order scattering.

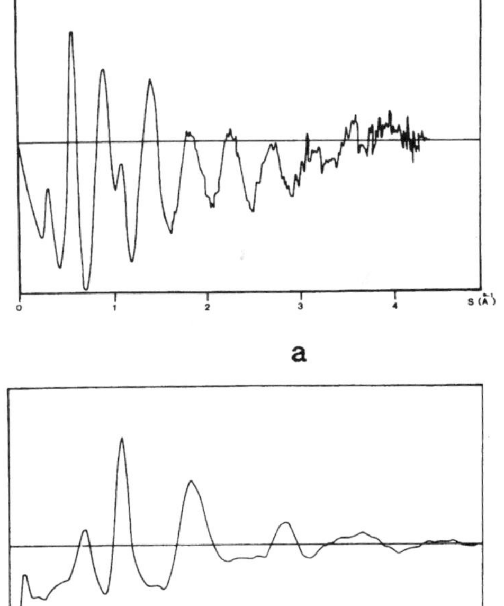

Fig 11 Example of reduced intensity curve as function of the scattering variable sin θ/λ (a) and of the corresponding radial distribution curve from hydrogenated silicon (from Liu et al, 1988)

Expressions suitable for calculating multiple scattering by a series of convolutions were formulated quite early by Goudsmit and Saunderson (1940), Moliere (1948). The derivation runs as follows: if single scattering follows the normalized angular distribution $i_1(2\theta)$, double scattering will have the normalized distribution

$$i_2(2\theta) = i_1(2\theta)*i_1(2\theta)$$

and similarly for triple scattering, with the convolution integrals taken on the unit sphere, that is as integration over angles, 2θ and the azimuth β. If i_1 is expanded in Legendre polynomials of 2θ :

$$i_1(2\theta) = \Sigma(2l+1)h_1P_1(\cos 2\theta)/4\pi$$

the convolution integral can be expressed by powers of h_1 and in fact summed to yield

$$I(2\theta) = \Sigma(2l+1)H_1P_1(\cos 2\theta)/4\pi$$

where $$H_1 = \exp[-\alpha t(1-h_1)]$$

with α as the cross section for first order scattering. This expression was first derived by Goudsmit and Saunderson (1940) and introduced in a correction procedure for multiple scattering in electron diffraction by Gjønnes (1959). In the angular region encountered in high energy electron diffraction the Legendre-expansion can be approximated by a Fourier-Bessel series. In recent work the energy loss is included, see Johnson and Isaacson (1988): note that the multiple inelastic scattering, notably the plasmon loss, can be used to determine thickness. The need for multiple scattering corrections is considerably reduced by energy filtering, but should definitely be included in order to achieve proper normalization of I_M. Experience from gas electron diffraction and from x-ray and neutron work shows the importance of proper attention to background subtraction. The facility of energy filtering is of considerable help in reducing tha background problem - but the sequential recording involved in scanning may increase the noise compared with the photographic method, which should not be dismissed. There are still relatively few electron diffraction studies of amorphous material - and considerable scope for improvement. Low temperature specimen stages will increase the selection of materials which can be studied.

The radial distribution description of the amorphous structure in Patterson space is a statistical average. Several attempts have been made in order to obtain a more detailed picture, either from high resolution imaging or by microdiffraction from very small areas. As the illuminated area becomes very small, considerable variations in the diffraction patterns, which are no longer symmetrical is found. The salient question is to what extent these variations should be interpreted as statistical fluctuations or representing small microdomains.

REFERENCES

Allpress, J.G. (1971) J. Mater. Sci. **6**, 313-318.
Andersson, B. (1979) Acta Cryst. **A35**, 718-727.
Andersson, B., Gjønnes, J. and Taftø, J. (1974) Acta Cryst. **A30**, 216-224.
Andersson, B., Gjønnes, J. and Forouhi, R. (1978) J. Less-Common Metals **61**, 273-291.
Andersson, B. and Sletnes, J. (1977) Acta Cryst. **A33**, 268-276.
Billingham, J., Bell, P.S. and Lewis, M.H. (1972) Acta Cryst. **A28**, 602-606.
Borrie, B. and Sparks, C.J. (1971) Acta Cryst. **A27**, 198-201.
Bourret, A., Rouviere, J.L and Penisson, J.M. (1988) Acta Cryst. **A44**, 838-846.
Bursill, L.A., Netherway, D.J. and Grey, I.E. (1978) Nature **272**, 405-410.
Buseck, P.R., Cowley, J.M and Eyring, L. (1988) 'High Resolution Transmission Electron Microscopy and Associated Techniques' New York: Oxford.
Buxton, G.F. (1976) Proc. Roy. Soc. London **A300**, 335-361.
Carpenter, R.W and Spence, J.C.H. (1982) Acta Cryst. **A38**, 55-61.
Castles, J.R., Cowley, J.M. and Spargo, A.E.C. (1971) Acta Cryst **A27**, 376-383.
Clapp, P.C. and Moss, S.C. (1966) Phys. Rev. **142**, 418-427.
Cockayne, D.J.H. and McKenzie, D.R. (1988) Acta Cryst **A44**, 870-877.
Cowley, J.M. (1950) J. appl. Phys. **21**, 24-30.
Cowley, J.M. (1961) Acta Cryst. **14**, 920-926.
Cowley, J.M. (1976) Acta Cryst. **A29**, 83-87.
Cowley, J.M. (1973) Acta Cryst. **A29**, 537-546.
Cowley, J.M. (1981) Diffraction Physics. New York: North-Holland.
Cowley, J.M. and Iijima, S. (1972) Z. Naturforsch **27A**, 445-451.
de Ridder, R., van Tendeloo, G. and Amelinckx , S. (1976)Acta Cryst. **A32**, 216-224.
de Ridder, R., van Tendeloo, G. and Amelinckx , S. (1977a) Pys. Stat. Solidii **A43**, 133-139.
de Ridder, R., van Tendeloo, G. van Dyck, D. and Amelinckx , S. (1976) Pys. Stat. Solidii **A37**, 591-606.
de Ridder, R., van Tendeloo, G. van Dyck, D. and Amelinckx, S. (1976) Pys. Stat. Solidii **A38**, 663-674.
de Ridder, R., van Dyck, D., van Tendeloo, G. and Amelinckx, S. (1977b) Pys. Stat. Solidii **A40**, 669-683.
de Ridder, R., van Tendeloo, G. van Dyck, D. and Amelinckx , S. (1977) Pys. Stat. Solidii **A41**, 555-561.
Edington, J.W. (1976) 'Practical Electron Microscopy in Materials Science' vol **4**. London: MacMillan.
Eyring, L.(1970) ed: 'The Chemistry of Extended Defects in Non-Metallic Solids' Amsterdam:North-Holland.

Finlayson ,T.R., Goodman, P., Olsen, A., Norman, P. and Wilkins, S.W. (1985) Acta Cryst. **B40**, 555-560.
Flinn, P.A. (1956) Phys. Rev. **104**, 350-356.
Fung, K.K. (1985) Ultramicroscopy **17**, 81-86.
Fujime, S., Watanabe, D. and Ogawa, S. (1964) J.Phys. Soc. Japan **19**, 711-716.
Gjønnes, J. (1959) Acta Cryst. **12**, 976-980.
Gjønnes,J. and Høier, R. (1971) Acta Cryst. **A27**, 166-174.
Gjønnes, J. and Simensen, C. (1970) Acta Met. **18**, 881-890.
Gjønnes, J. (1966) Acta Cryst. **20**, 240-249.
Gjønnes, K. (1985) Ultramicroscopy, **17**, 133-140.
Gleiter, H. (1983) in Cahn, R.W. and Haasen, P. Physical Metallurgy. New York: North-Holland.
Goodman, P., Olsen, A. and Whitfield, H.J. (1985) Acta Cryst. **B41**, 292-298.
Goudsmit, S. and Saunderson, J.L. (1940) Phys. Rev. **57**, 24-29.
Grigson,C.W.R. (1962) J. Electronics and Control **12**, 209-232.
Guinier, A. (1963) 'X-Ray Diffraction'. San Fransisco: Freeman.
Harada, J., Tanaka, M. and Honjo, G. (1966) J. Phys. Soc. Japan **21**, 968-972.
Hashimoto, H., Howie, A. and Whelan, M.J. (1962) Proc. Roy. Soc. **A269**, 80-103.
Hashimoto, S. (1987) J. Appl. Cryst. **20**, 182-186.
Hayakawa, M. and Cohen, J.B. (1975) Acta Cryst. **A31**, 635-645.
Hendricks, S.B. and Teller, E. (1942) J. Phys. Chem. **10**, 147-167.
Hiraga, K., Shindo,, D. Hirabayashi, M., Terasaki, O. and Watanabe, D. (1980) Acta Cryst **B36**, 2550-2554.
Hirsch, P.B., Howie, A., Nicholson, R.B., Pashley, D.W. and Whelan, M.J.(1977) Electron Microscopy of Thin Crystals. Florida: Robert E.Krieger.
Honjo,G., Kodera,S. and Kitamura, N. (1964) J.Phys. Soc. Japan **19**, 351-367.
Hurlibut, C.S. and Klein, C. (1977) Manual of Mineralogy. New York: Wiley.
Høier, R. (1972) Phys. Stat. Solidii **A11**, 596-610.
Høier, R. (1973) Acta Cryst. **A29**, 663-672.
Jagodzinski, H. (1949) Acta Cryst. **2**, 201-207; 208-214.
Jagodzinzki, H. (1954) Acta Cryst. **7**, 17-25.
Johnson, A. (1972) Acta Cryst. **A28**, 89-91.
Kainuma, Y. (1954) Acta Cryst.**8**, 247-257.
Kakinoki, J. and Kimura, Y. (1952) J.Phys. Soc. Japan **7**, 30-35.
Krivoglaz, M.A. (1969) 'Theory of X-ray and Thermal Neutron Scattering by Real Crystals' New York: Plenum.
Liu, Z.Q., McKenzie, D.R., Cockayne, D.J.H. and Dwarte, D.M. (1988) Phil. Mag. **57**, 753-761.
Lonsdale, K. and Smith, H. (1941) Proc. Roy. Soc. London. **A179**, 8-50.
Marcinkowski, M.J. and Zwell, L. (1963) Acta Met. **11**, 373-390.
Moliere, G. (1948) Z. Naturforsch. **3a**, 78-97.

Moss, S.C. (1966) in Cohen, J.B and Hilliard, J.E.(ed): 'Local Atomic Arrangement Studied by X-Ray Diffraction' pp 95-122.
Moss, S.C. (1969) Phys. Rev. Lett. **22**, 1108-1111.
Ohshima, K. and Watanabe, D. (1973) Acta Cryst. **A29**, 520-526.
Raether, H. (1952) Angew. Phys. **4**, 53-59.
Reimer, L. (1984) 'Transmission Electron Microscopy'. New York: Springer-Verlag.
Sato, H., Watanabe, D. and Ogawa, S. (1962) J. Phys. Soc. Japan **17**, 1647-1651.
Sauvage, M. and Parthe, E. (1972) Acta Cryst. **A28**, 607- 616.
Sauvage, M. and Parthe, E. (1974) Acta Cryst. **A30**, 239-246.
Sellar, J.R. and Imeson, D. (1988) Acta Cryst **A44**, 768-771.
Shinozaki, S. and Kinsman, K.R. (1978) Acta Met. **26**, 769-776.
Sinclair, R., Gronsky, R. and Thomas, G. (1976) Acta Met. **24**, 789-796.
Skjerpe, P. (1987) Metallurgical Transactions 189-200.
Skjerpe, P., Gjønnes, J., Sørbrøden, E. and Herø, H. (1986) J. Mat. Sci. **21**, 3986-3992.
Slater, J.C. (1958) Rev.Mod.Phys. **30**, 197-227.
Smith, D.J. (1988) in P.R. Buseck, J.M. Cowley and L. Eyring ... (ed) 'High Resolution Transmission Electron Microscopy and Associated Tecniques' New York: Oxford.
Spence, J.C.H. (1988) 'Experimental High Resolution Electron Microscopy'. 2.ed. New York: Oxford University Press.
Spence, J.C.H. and Carpenter, R.W. (1986) in D.C. Joy, A.D. Romig and J. I. Goldstein (ed) 'Prinicples of Analytical Electron Microscopy New York: Plenum.
Steeds, J.W., Bird, D.M., Eaglesham, D.J., McKernan, S., Vincent, R. and Withers, R.L. (1985) Ultramicroscopy **18**, 97-110.
Taftø, J., Jones, R.H. and Heald, S.M. (1986) J. Appl. Phys. **60**, 4316-4318.
Taftø, J. and Gjønnes, J. (1988) Acta Cryst. **A44**, 833-837.
Tanaka, M., Terauchi, M. and Kaneyama, T. (1988) 'Convergent-Beam Electron Diffraction II'. Tokyo: JEOL.
Tanaka, N. and Cowley, J.M. (1987) Acta Cryst. **A43**, 337-346.
Tanaka, N., Cowley, J.M. and Ohshima, K. (1987) Acta Cryst. **B43**, 41-48.
Terasaki O., Watanabe, D., Hiraga, K., Shindo, D. and Hirabayshi, M. (1981) J. Appl. Cryst. **14**, 392-400.
Thomas, G. and Goringe, M.J. (1979) 'Transmission Electron Microscopy of Materials'. New York: Wiley.
Treacy, M.M.J., Gibson, J.M. and Howie, A. (1985) Phil. Mag. **A51**, 389-417.
Twigg, M.E. and Chu, S.N.G. (1988) Ultramicroscopy **26**, 51-58.
van Dyck, D., de Ridder, R., van Tendeloo, G. and Amelinckx, S. (1977) Pys. Stat. Solidii **A43**, 541-553.
van Dyck, D., de Ridder, R. and Amelinckx , S. (1980) Pys. Stat. Solidii **A59**, 513-530.

van Tendeloo, G., de Ridder, R. and Amelinckx, S. (1978) Phys. Status Solidii **A49,** 655-666.

van Tendeloo, G., Amelinckx, S. and de Fontaine, D. (1985) Acta Cryst. **B41**, 281-292.

Walker, C.B. (1956) Phys. Rev. **103**, 547-557.

Warren, B.E. (1969) 'X-Ray Diffraction' Reading, Mass: Addison-Wesley.

Watanabe, D., Andersson, B., Gjønnes, J. and Terasaki, O. (1974) Acta Cryst. **A30**, 772-776.

Watanabe, D. and Fisher, P.M.J. (1965) J.Phys.Soc. Japan **20**, 2170-2179.

Wenk, H.-R. (ed) (1976) 'Electron Microscopy in Mineralogy' Berlin: Springer.

Whittaker, E.J.W. (1954) Acta Cryst. **7**, 827-837.

Whittaker, E.J.W. (1955) Acta Cryst **8**, 261-265, 265-271, 726-729.

Wilson, A.J.C. (1942) Proc. Roy. Soc. **A180,** 277-285.

Xu, Y., Suenaga, M., Taftø, J. Sabatini, R.L. and Moodenbaugh, A.R. (1989) Phys. Rev. **B**. **39**, 6667-6680.

Yada. K. (1971) Acta Cryst. **A27**, 659-664.

Yoshioka, H.(1957) J. Phys. Soc. Japan. **12**, 618-626.

3
RHEED and REM

K. Yagi

3.1 Introduction

A brief historical survey of reflection high energy electron diffraction (RHEED) has been included in Volume 1, section 1.6. Until the nineteen fifties, RHEED was used as a unique method for the study of surface processes such as the initial oxidation and corrosion of metals and alloys with crystallographic information (Cates, 1933) and the growth of thin films on crystalline surfaces (Uyeda, 1942). In the same period, progress was made in basic research on diffraction phenomena such as the refraction effect (Yamaguti, 1935) Kikuchi patterns (Kikuchi, 1928), surface wave resonance (Kikuchi and Nakagawa, 1933: Miyake et al., 1954) and asymmetry in RHEED patterns (Uyeda and Miyake, 1957).

However RHEED became less widely used after the development of transmission electron microscopy (TEM) (Hirsch et al., 1965) and scanning electron microscopy (SEM) because these techniques can give direct information in real space on the nature of surfaces and the microstructure of thin films.

In the nineteen seventies, the situation changed again. Ultra-high vacuum (UHV) RHEED was developed (Menadue, 1972; Ino, 1977) as a technique for characterizing structures of clean and adsorbate surfaces which was equivalent to low energy electron diffraction (LEED) (Germer and Hartman, 1960) but more convenient in its handling. One simple advantage of RHEED over LEED is that the method does not need much open space around the specimen. It requires only small apertures (subtending small solid angles) in the glancing-angle directions to allow the incident and exit beams to pass. This means that the technique can be incorporated readily with other methods for surface characterization and with devices for thin film formation or crystal growth as a means for characterizing the surface structure of the specimen. One recent and typical example is the observation of oscillations of the RHEED intensities during the layer-by-layer growth of thin films deposited by molecular beam epitaxy (MBE) as discovered by Neave et al. (1983) and analyzed theoretically by Kawamura and Maksym (1985).

Another important development to be mentioned here is reflection electron microscopy (REM). The method was invented just after the development of TEM by Ruska (1933). At that time and until the nineteen sixties conventional transmission electron microscopes (CTEMs) were used and the electrons used for imaging were those scattered diffusely in the forward direction from the surfaces with a glancing angle of incidence. However this gave little more than silhouettes of the surface irregularities (Menter, 1952-3; Watanabe,

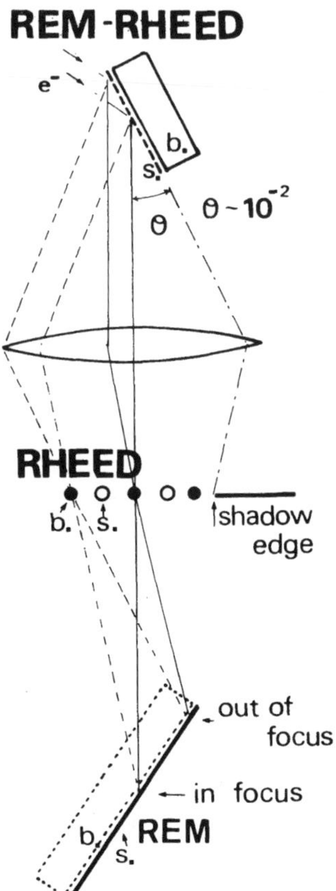

Figure 3.1 A schematic illustration of electron optics for image formation in REM with use of CTEM. The specimen crystal and the incident beam are aligned so that the imaging beam is parallel to the optic axes of the objective lens. Note foreshortening of images and over- and under-focus conditions of them.

1957) and the technique did not attract much attention because SEM gave similar images with less restriction on the imaging conditions.

Later, the electrons Bragg-reflected from crystal surfaces were used for imaging (Halliday and Newman, 1960; Nielsen and Cowley, 1976) to give crystallographic information on the surfaces in much the same way as dark-field TEM does in relation with transmission electron diffraction (TED) (Hirsch et al, 1965). Figure 3.1 illustrates the formation of a REM image in a CTEM instrument. The RHEED pattern consists of diffracted beams from the bulk (b) and surface (s) of the specimen. The image (REM) is formed by selecting one or several of these beams by use of an objective aperture. The REM images, thus formed, are foreshortened. Hence specimens with flat surfaces are needed. It is also noted that images are in focus only along one line perpendicular to the optic axis of the microscope. Thus, Fresnel diffraction effects (see Fresnel fringes in Figs. 3.13 and 14), reflecting local phase changes of diffracted waves, are important in the image analysis. REM, however, was not used widely until the end of the nineteen seventies. This is mainly due to the fact that the observed surfaces were not clean and flat enough to give meaningful information on the surface structure.

A breakthrough came when the technique was combined with UHV TEM and clean specimen preparation techniques (Osakabe et al. 1980a, b and 1981a, b). Images, thus obtained, clearly showed surface structure at a monolayer level such as steps and domain structures on the clean and adsorbate surfaces as will be shown in a later section.

A method which is equivalent to REM as a consequence of reciprocity theory is scanning REM (SREM), where a micro- or nano-beam is scanned over the surface with a glancing angle of incidence and one of the diffracted spots in the RHEED pattern is used as signal for the scanning images . If a dedicated scanning transmission electron microscopy (STEM) instrument is used, with accelerating voltage 100kV and a reasonably good vacuum (10^{-7}Pa or better), the imaging conditions and the resolution obtainable are much the same as for REM in a TEM instrument (Cowley, 1988 a,b). On the other hand a number of investigators have used lower-energy systems in the medium-energy range of 10-30keV, built specifically for surfaces studies under UHV conditions (Cowley et al., 1975; Ichikawa et al. 1984; Bennett and Johnson, 1988). In the latter cases the basic apparatuses are similar to ordinary SEM instruments and are more easily constructed for UHV conditions than CTEMs because the SREM instruments do not have complicated structures around the specimen. Moreover, with a large space around the specimen it is easy to introduce the attachments which are necessary for in-situ experiments and surface chemical analysis. Foreshortening can be corrected electrically, although the resolution can not be improved. A large space around the specimen (with a long working distance for the objective lens), in turn, means that the resolution is worse than CTEM and dedicated STEM in general.

On the present topics, a recent book, "Proceedings of NATO Advanced Research Workshop on Reflection High Energy Electron Diffraction and Reflection Electron Imaging of Surfaces" (Larsen and Dobson, 1988), contains recent results and is a very useful guide to the present status of the topics. For REM, review papers by the present author (Yagi, 1980, 1987, 1989a, 1989b; Yagi et al., 1982) should be consulted.

3.2 Instrumentation

3.2.1 UHV Performance

Although the ordinary high vacuum of electron microscopes is sufficient for the observation of RHEED and REM, under these conditions the applications are restricted to only a limited set of problems. Therefore, a UHV performance is necessary for the wide applications to characterize structures of surfaces and to observe surface dynamic processes.

In the case of RHEED it is rather easy to design for UHV conditions or easy to incorporate it into ordinary UHV surface analyzing systems. Figure 3.2 schematically shows a RHEED apparatus made by us in 1977. Usually a heated tungsten hair-pin filament is used as an electron source. A field emission gun (FEG) is used in SREM apparatus (Cowley et al., 1975; Cowley, 1982a, b; Ichikawa et al. 1984). The electrons from the filament are accelerated to 10-30keV and hit the specimen surface with a glancing angle with the help of a condenser lens and beam deflector coils. The glancing and azimuthal angles are controlled by a specimen goinometer. Diffracted electrons hit a fluorescent screen on a glass plate and the RHEED pattern is observed through the glass.

An example of the RHEED pattern from a clean Si(111)7x7 surface is shown in Fig. 3.3 (a), which was taken with use of the RHEED apparatus shown in Fig. 3.2. A schematic illustration of the RHEED pattern with two dimensional indices is shown in (b). It was taken at 30kV with an incident azimuthal direction nearly parallel to the [211] direction. Zero-th (O'-A), 1st (B-C) and 2nd Laue zones (F) from the fundamental lattice are seen. Between the

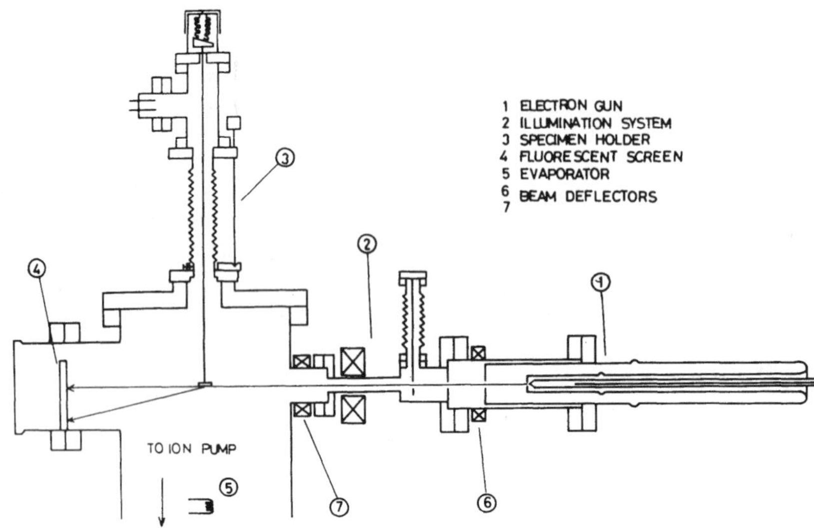

Figure 3.2 A schematic illustration of a simple RHEED apparatus.

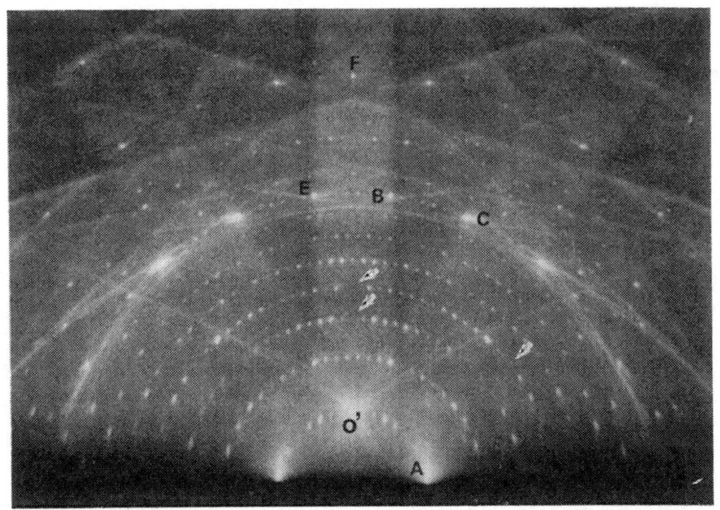

Figure 3.3 (a) A typical RHEED pattern from a SI(111)7x7 surface taken at 30kV. The specular spots 0', reflections from the bulk structure (A,B,C,D,F,) and reflections from the reconstructed superlattice structure between them are seen. The incident beam direction is [2 1 1].

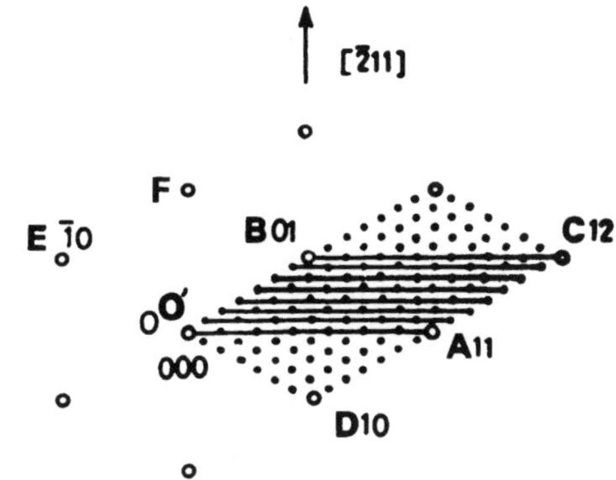

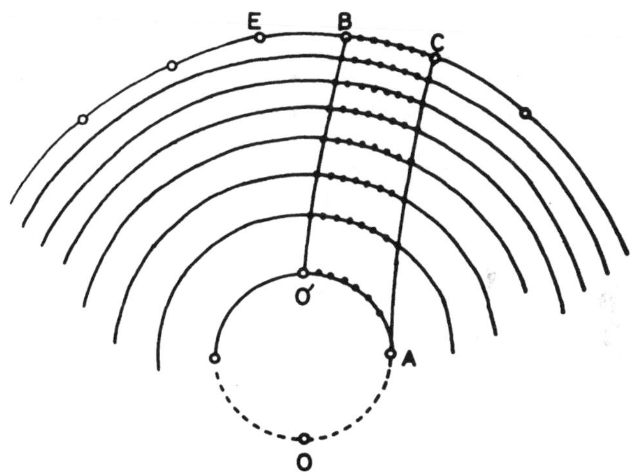

Figure 3.3 (b) Schematic illustrations of the pattern with two dimensional indices (h k).

fundamental reflections on each zone and between the zones, six superlattice reflections and six n/7th Laue zones from the 7x7 reconstructed structure, respectively, are seen. It is noted that some of the spots are missing as indicated by arrows, which reflects the characteristics of the reconstructed structure (Ino, 1980a, b; Horio and Ichimiya, 1989). Due to the grazing-angle condition, the reciprocal lattice is severely distorted in the RHEED pattern as seen in (b) and it is not easy to see the geometry of the reciprocal lattice from such patterns. Therefore, Ino (1977) invented a spherical glass fluorescent screen, whose center of curvature is the specimen position, to observe undistorted reciprocal lattices of surface structure (Ino, 1988).

Figure 3.3(b) also shows Kikuchi-lines, Kikuchi envelopes and Kikuchi bands .

Contrary to the case of RHEED, it is not so easy to build a UHV-CTEM instrument without loss of the amenities of commercial CTEM machines. We have developed three UHV microscopes (Takayanagi et al., 1978; Honjo et al., 1980; Takayanagi et al., 1986) and now several UHV-CTEMs have been developed specially for in-situ surface and interface studies and for fine particle studies (Wilson and Petroff, (1983); Heinemann and Poppa, 1986; Swann et al. 1987; Metois et al., 1989; Kubosoe et al., 1989). Some of them are commercialized.

3.2.2 Attachments

The important attachments in RHEED and REM include those for recording RHEED patterns, measuring the intensity of individual reflected beam and obtaining REM images. The intensity measurement is important for getting rocking curves (see Fig. 3.8) and RHEED oscillations during MBE growth and for getting images in the SREM instruments. A TV camera system with a microcomputer analyzer (Bolger and Larsen, 1986; Nakahara and Ichimiya, 1989), a photo-multiplier with an optical fiber coupling (Ichikawa et al., 1984, Sakamoto et al., 1985) or with a lens (Putike et al. 1988), which is connected to a fluorescent screen, or a channeltron (Cowley et al., 1975) are generally used. SREM needs a detector with a small solid angle subtended from the specimen, because this, by reciprocity, corresponds to the convergence of the incident beam illumination in REM (Yagi, 1980; Ichikawa et al., 1984). The image contrast for REM is sensitive to the illumination angle (Bragg condition. See discussions on Fig. 3.15). The TV camera system is also important in REM (Osakabe et al., 1981b) in a CTEM instrument to record dynamic processes of the surfaces together with ordinary photographic film recording. An image plate system (Mori et al., 1988) may also be useful in REM to record not only images but also RHEED patterns quantitatively.

Another important attachment is the specimen holder. In the case of RHEED the techniques developed for surface studies can be used. However, to get RHEED patterns and REM images using a

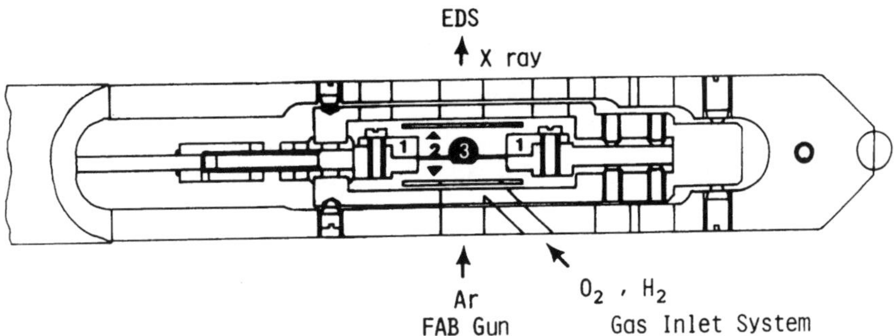

Figure 3.4 A cross section of a side entry single tilt REM specimen holder. A bulk crystal can be clamped and resistively heated. The holder has holes for various specimen treatments and analysis.

CTEM, a special specimen holder, which can clamp a bulk specimen and control glancing and azimuthal angles, is needed in addition to UHV performance. Fig. 3.4 shows a side-entry single-tilt REM specimen holder (Ogawa et al., 1987). It has electrodes (1) for resistive heating of a specimen (3), radiation shields (2) and holes for in-situ experiments and analysis. A double tilt REM holder has also been developed (Tanishiro et al., 1982). A beam tilting facility which is common in CTEM instruments also controls these angles in order to adjust the imaging diffracted beam to be parallel to the optic axis of the objective lens.

In both types of apparatus, RHEED and CTEM, it is necessary to have specimen cleaning facilities such as are incorporated in ordinary surface analysis equipment. Flash heating of the specimens, ion sputtering and annealing and in-situ deposition are typical methods.

Devices for in-situ treatment such as an evaporator (or MBE cell), a gas inlet device, specimen heating and cooling facilities, depending on the purpose of each project, are essential in surface experiments. A specimen preparation chamber is also useful for ex-situ studies. Surface chemical analyzing equipments such as X-ray spectroscopy (Ino et al., 1980; Hasegawa et al., 1985), Auger electron spectroscopy (Ichikawa et al., 1984; Horio and Ichimiya, 1985) and electron energy loss spectroscopy (EELS) are also important to increase the potential of the methods. These are common subjects in RHEED, REM and SREM.

Another point to mention here is that electron holography in REM was recently developed by Osakabe et al., (1988 and 1989) and Banzhof et al. (1988). The technique can reveal details of relative phase changes of reflected electrons around steps and dislocations.

3.3 Basic diffraction problems in RHEED and REM

Diffraction theory for the Laue case has been well developed for both perfect and imperfect crystals (see other chapters of this book and also Cowley (1981)). However, for the Bragg case theory has not been so well developed. Before going into detail it is helpful to describe some characteristics in RHEED.

3.3.1 Refraction

The refraction effect is very important in RHEED. The small positive deviation, of say 10^{-4}, of the refractive index from 1 is not important in TED but it gives rise to notable effects in RHEED. The glancing angle θ_V of an incident beam on the crystal surface from the vacuum is smaller than the θ_{cryst} of the refracted beam on the crystal net plane inside the crystal as shown in Fig. 3.5 (a). This means that θ_{cryst} is always larger than θ_C, the critical angle for total internal reflection. Therefore, the Ewald sphere construction after a correction for refraction is like that shown in (b) where an incidence beam nearly parallel to the surface is assumed. It is evident that with this azimuthal orientation we can only measure intensities of parts of the(h k) reciprocal rods indicated by thick lines by changing the glancing angle θ_V. By changing the azimuthal orientation other parts of the rods indicated by thin lines can be measured. However, the dotted parts cannot be measured by RHEED. The situation is similar in LEED and is quite different from those in TED and X-ray diffraction from surface structures. In the latter cases, the diffraction from the surface structures is kinematic and intensities corresponding to the projection of the surface structure on a plane parallel to the surface can be measured, i.e., diffraction with zero momentum transfer normal to the surface. The notable fact in Fig. 3.5(b) is that the Ewald sphere always cuts the reciprocal rods with finite angles in the regions indicated by thick lines. This means that for a flat surface or large domains of surface structures, the rods are thin and the surface gives sharp spots in RHEED patterns.

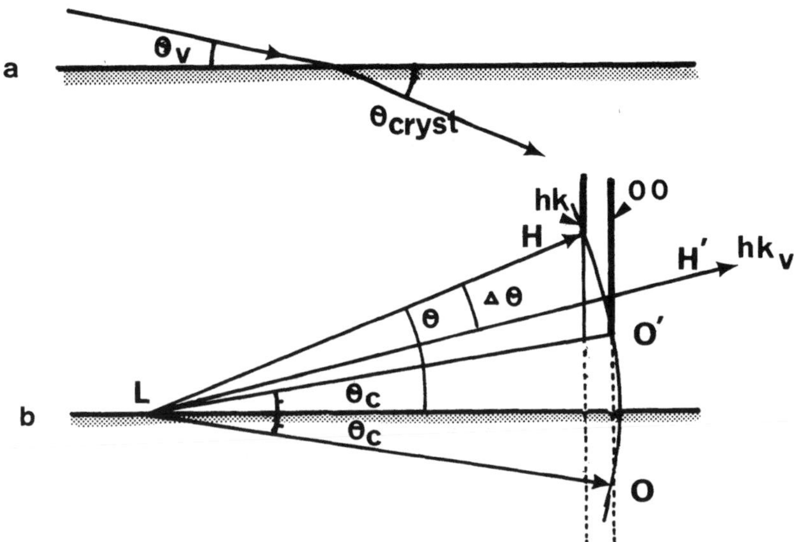

Figure 3.5 Schematic illustrations of the refraction effect in RHEED. (a) illustrates that the incident beam direction for the Ewald sphere construction is not that in the vacuum. (b) shows that the Ewald sphere cuts the reciprocal rods with finite angles in the reciprocal lattice space where we can observe in the patterns.

Because of the refraction, the (h k) reflection which is in the direction LH inside the crystal is in the direction LH' in the vacuum. The difference in angle ($\Delta\theta$) depends on the amount of the momentum transfer normal to the surface. From such a dependence a mean inner potential can be determined from the analysis of positions of horizontal Kikuchi lines (Yamaguchi, 1936) or from a split of a Bragg reflected beam from a crystal edge (Yamamoto and Spence, 1983).

The refraction gives rise to distortions of RHEED patterns; near the shadow edge, Laue zones are distorted from circles and diffraction disks in convergent beam RHEED (CBRHEED) are not circular (Shannon et al., 1985) and oblique Kikuchi lines are not straight.

3.3.2. RHEED patterns and surface topography

Another point to mention is the relation between RHEED patterns and surface topography. Fig. 3.6 shows 4 typical cases of the surface structure. In (a) the surfaces is dirty and undulated. The RHEED patterns are not clear and have strong background. The corresponding REM images are dim and shows undulations, which are seen as black and white horizontal lines by foreshortening.

In (b), the case of clean and flat surfaces, the RHEED pattern shows sharp spots as mentioned before. The REM images are clear and show atomic steps and boundaries of the surfaces, such as domain boundaries and out-of-phase boundaries of the surface structure. When the surface is covered by disordered adsorbates of monolayer thickness, the REM images do not change except for changes due to destruction of the surface structure which may occur during adsorption. This is because the adsorbates give rise to very weak diffuse scattering and do not affect Bragg reflected imaging beams. As a result there are no changes, for example, in step images.

When the surface has large protrusions or three dimensional islands, a case shown in (d), TED spots due to transmission through the protrusions or islands overlap with the pattern from the flat parts of the surface. REM images due to those spots shows bright images of the protrusions or islands.

When the surface has undulations with angles smaller than the Bragg angles, the RHEED pattern shows streaks (a case shown in (c)). This is due to the fact that coherently reflecting regions are small, i.e. terraces are narrow along the beam direction, and (h k) rods are thick enough to give streaks. The corresponding REM images show narrowly spaced steps or bright and dark contrast due to the undulations without well-defined step images. The bright and dark contrast, which is mainly determined by geometrical factors has been discussed (Nielsen and Cowley, 1976. See also recent experiments by Hsu and Lehmpfuhl (1989)). Sometimes it has been stated, through a misunderstanding, that steaks in RHEED patterns are evidence for atomically flat surfaces. The streaks are due to thick rods! It should be noted here that thick rods may also be caused by small two dimensional islands on atomically flat surfaces. The reasons for this are: 1) the islands reflect electrons incoherently, 2) even in the case of coherent growth of islands there is a phase difference between reflected waves from the substrate and islands

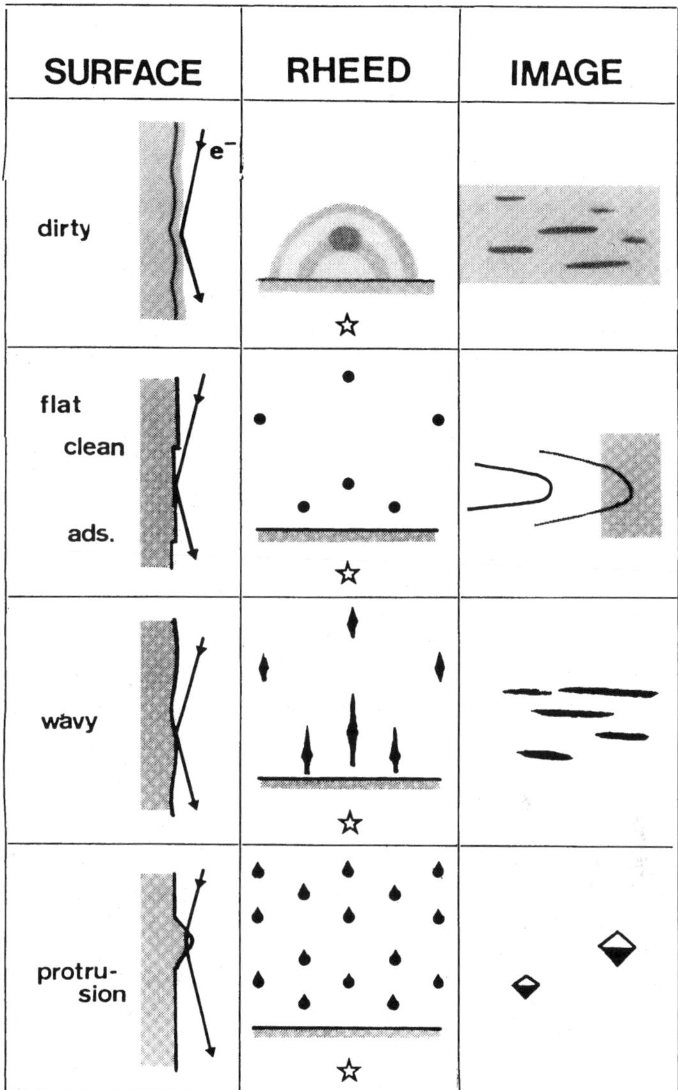

Figure 3.6 Schematic illustrations of relations among surface topography, RHEED and REM image. (a) a dirty and wavy surface, (b) a flat and clean or disordered adsorbate surface, (c) a clean but wavy surface (or with many small monolayer islands) and (d) a surface with protrusions.

and 3) steps around the islands reflect electrons incoherently. These problems are closely related to the origins of RHEED intensity oscillation during MBE growth and the origins of image contrast of surface steps in REM (see discussions on Figs. 3.12 and 3.13).

3.3.3 RHEED calculations

Now we return to the RHEED from perfect, flat surfaces. Generally dynamically diffraction theories should be applied. This is because incident and reflected beams cross many surface atoms under the glancing angle conditions. However, in special cases such as the consideration of RHEED intensities of superlattice reflections from reconstructed structures of surfaces, the kinematical scattering theory has sometimes been adopted under certain reflection conditions with relatively reasonable results. Ino's finding, in structure analysis of the Si(111) 7x7 structure assuming the kinematical scattering, that there are scatterers having a 2a spacing (where a is the lattice constant of the unreconstructed two dimensional (111) surface lattice) clearly indicates validity of the kinematical approximation to some extent (Ino, 1980a,b). The scatterers with 2a spacing were later proved to be the adatoms and dimers of the DAS model (Takayanagi et al. 1985), nowadays accepted. Very recently Horio and Ichimiya (1989) reexamned the structure by RHEED by making kinematical calculations of intensities from the 7x7 structure as a function of the incident angle on the basis of the DAS structure. Comparison with their observations showed the fine structural details. However, applicability is limited and we need full dynamical calculations.

The most simple dynamical theory is the two beam case, similar to that used for X-ray diffraction in the Bragg case (see, for example, Newkirk and Mallett (1967)). Although its results are not useful for practical problems, it generally gives us basic concepts in RHEED such as refraction and Bragg reflection, total reflection originating from the dynamical effect, the Bragg width, standing waves parallel to the surface and the penetration depth of the electrons.

Many beam theory has been devised in various ways. Some approaches, however, assume surfaces to be truncations of bulk structures, and the Bethe's theory is applied to get Bloch waves to be connected to the waves in the vacuum (Moon, 1972; Colella, 1972; Colella and Menadue, 1972). However, real surfaces are relaxed or reconstructed and these methods cannot be applicable. Therefore, in numerical methods, crystals are sliced parallel to the surface (Fig. 3.7(a)) and transmission and diffraction of transmitted and reflected waves in successive slices are calculated (Maksym and Beeby, 1981;

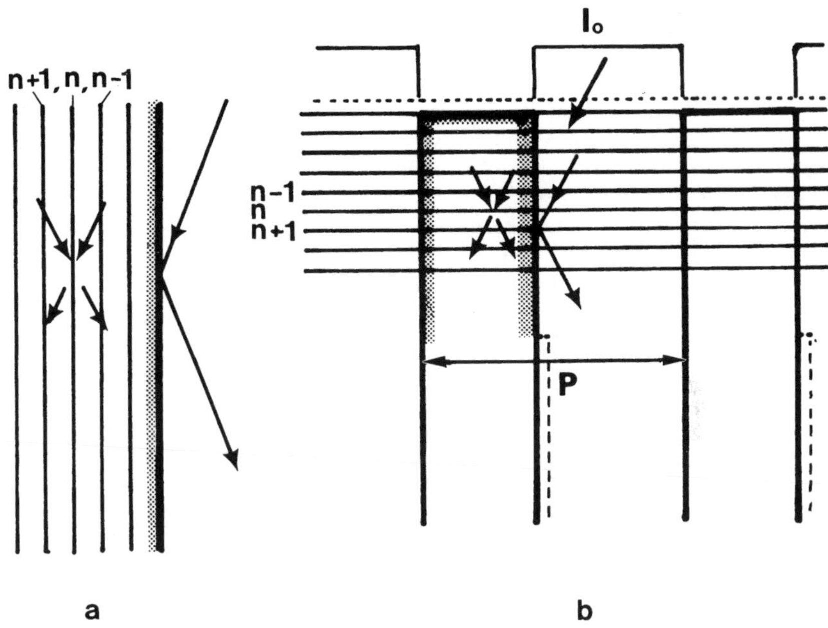

Figure 3.7 Two types of multislice calculations of RHEED intensities. (a) slices are parallel to the surface as in LEED calculation. (b) slices are perpendicular to the surface and calculation methods for TEM and TED are applied. Defects such as steps as schematically illustrated by dotted lines can be easily included in the calculation by the method (b).

Ichimiya, 1983; Tong et al., 1988; Smith and Lynch, 1988). Inelastic scattering is included phenomenologically by adding imaginary potentials in the calculations. It was found that the theory developed by Ichimiya (1983), gives similar (but slightly different) rocking curves from those calculated by the theory developed by the Maksym and Beeby (1981) (Kawamura et al., 1988a).

On the other hand, Peng and Cowley (1986) applied the multislice calculation as used in transmission mode(see chapter 4 of Volume 1) to RHEED problems as shown in Fig. 3.7(b). In this case the slices are perpendicular to the surface and the periodic continuation technique is used. The advantage of this technique is in that it can be applied to any surface structures including not only reconstructed structures but also defects such as steps (see dotted lines in Fig. 3.7 (b)), out-of-phase boundaries, and domain boundaries. Furthermore, the calculation is in real space and it can tell how the surface wave field is formed or disturbed by the defects mentioned above. Very recently, Anstis (1989) modified the technique using an idea developed for the calculation of image contrast of defects in TEM (Howie and Basinski, 1968). The main problem in this sort of calculation is that the calculation must be carried out over a long distance parallel to the surface (along the beam direction: more than 70nm in the case of GaAs (110) surface (Wang et al., 1989)) to get to a steady state wave-field, free from the effects of the boundary condition at the entrance surface. Also to know the effect of the defect, another long-distance calculation is needed and to do it the continuation period (in the case of Peng and Cowley's approach) should be large enough to allow that all the surface under consideration is illuminated by the incident beam.

As an example of a calculation, the specular beam intensity as a function of glancing angle, θ, (the so-called rocking curve) for a Pt(111) surface is shown in Fig. 3.8 (Ohse et al. 1988). The incident beam is along the $[1\bar{1}0]$ direction and 13 beams were included in the calculation. It is evident that periodic intensity peaks, which are expected from a simple theory, are absent due to many-beam dynamical effects.

Surface wave resonance (SWR), which was first noted by Kikuchi and Nakagawa (1933) as an anomalous intensity enhancement of the specular (0 0) beam under certain diffraction conditions, is one of the important and useful phenomena in RHEED and REM. Noting that the SWR occurs when an oblique Kikuchi line crosses the specular beam ((0 0) beam), Miyake et al. (1954) found that the SWR condition is to excite Bragg reflections parallel to the surface. Fig. 3.9 shows this geometrical consideration, where O and S correspond to the direct and specular beams, respectively. The Kikuchi lines (h_1,h_2,h_3) oblique to the shadow edge (SE, (001) plane) are indicated by K_h'. Under this condition another Kikuchi line K_h' $(h_1,h_2,\bar{h}_3)$ symmetric to SE, crosses the spot O provided that the surface is a mirror plane of the crystal lattice. Thus, the incident beam is at the Bragg condition for the reflection and experimentally a strong enhancement occurs when this Bragg diffracted spot G' is close to SE, which means that a diffracted beam runs nearly parallel to the surface (surface channeling: See also chapter 10 of Volume 1).

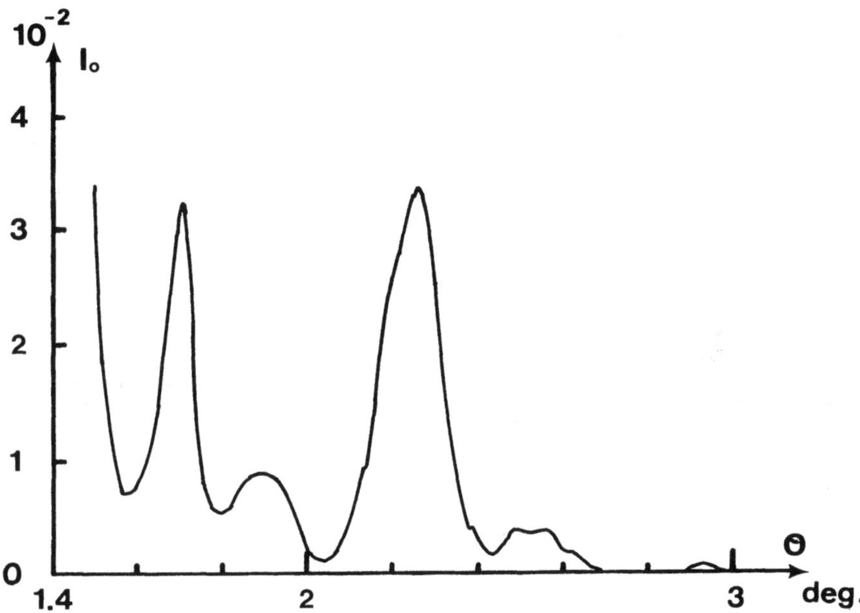

Figure 3.8 A rocking curve (specular beam intensity as a function of glancing angle) from Pt (111) surface with the [1 1 0] azimuth direction for 100kV electrons.

Calculations showed that under the SWR condition the wave fields are strongly localized to a few surface atomic layers with a modification in directions parallel to the surface (Horio and Ichimiya, 1985; Marten and Meyer-Ehmsen, 1985; Peng and Cowley, 1986). The surface waves travel for long distances (Kambe, 1988; Wang et al., 1989). Enhancement of the intensities of surface structure reflections in RHEED and of emission of secondary electrons including Auger electrons (Marten, 1988) under SWR conditions has been observed. Similar enhancement of LEED patterns is also noted when the Ewald sphere is tangential to a reciprocal rod (McRae, 1971; Miyake and Hayakawa, 1970).

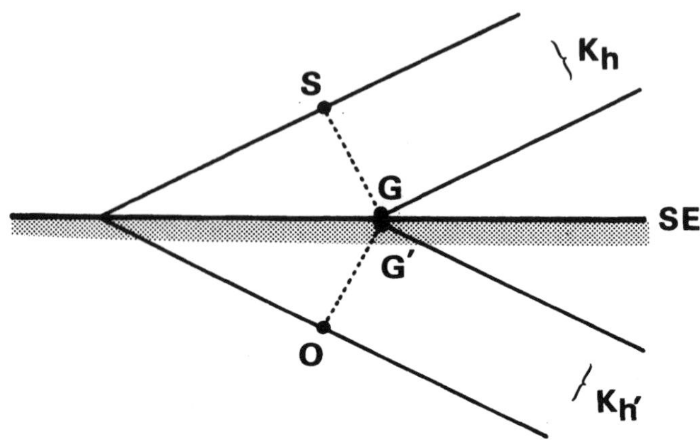

Figure 3.9 A schematic illustration of a SWR condition. A oblique Kikuchi line K_h crosses the specular reflection S, which means that the incident beam 0 satifies the Bragg condition for G' reflection. An intensity enhancement of the specular beam is noted when the G' is close to the shadow edge.

Calculation of the wave field is very important in some cases for the structural analysis, especially for determinations of positions of host and foreign atoms. Depending on the Bragg condition and structure, a wave field is formed and it determines the relative excitations of core electrons of the atoms in question, and hence the relative signals of Auger electrons, characteristic X-rays and core loss edges in EELS. This kind of experiment has a relatively long history (Miyake et al., 1968; Anderson and Howie, 1975; Woodruff, 1975: Baines et al., 1975; Chang, 1977; Ichimiya and Takeuchi, 1983, Marten, 1988) and the phenomena are analogous to ALCHEMI in TEM-TED (see chapter 10 Volume 1) and the standing wave method in X-ray diffraction (for example see Materlik et al., 1985).

CBRHEED has been explored experimentally and theoretically (Ichimiya et al., 1980; Shannon et al., 1985; Lehmpfuhl and Dowell, 1986; Eades and Shannon, 1988; Lehmpfuhl and Uchida, 1988; Smith, 1988; Smith and Lynch, 1988). However, its usefulness is not clear at present, because the experimental works including symmetry studies (Eades and Shannon, 1988) were not made on clean and well defined surfaces and calculations (Smith, 1988) were on truncated bulk surfaces. Drastic changes of reflected intensities with change of Bragg condition and contrast reversals for domains with a slight change of incident beam direction (see Fig. 3.15) seem to suggest usefulness of CBRHEED so long as the strong inelastic scattering in RHEED does not mask the fine details.

3.3.4 Inelastic scattering

Phonon, plasmon (surface and bulk plasmons, see section 1.3 Volume 1) and single electron excitations are important in RHEED and REM as in TEM and TED (see chapters 2 and 10, Volume 1). Diffuse scattering in RHEED pattern was discussed by Miyake (1962) and the effect of thermal diffuse scattering on RHEED patterns has been discussed by Peng and Cowley (1989). Phonon scattering causes loss of coherency of diffracted waves and is considered to be a cause of loss of REM image contrast at high temperatures (Ogawa et al. 1987; Yagi et al., 1988).

Figure 3.10 shows a low-energy-loss region of an EELS spectrum from a clean Si(111) surface at 80kV under the excitation of the 333 reflection (Krivanek et al., 1983). Surface plasmon excitation and its multiples are seen. A mean collision number ,$<n>$ is 1.8 and agrees with a theoretical prediction $<n>=e^2/8\varepsilon_0 hv_\perp$. Here, e, ε_0 and e have their usual meaning and $v_\perp$ is the normal component of the electron velocity. $v_\perp$ depends mainly on the order of surface reflection but not strongly on the incident electron energy. Therefore, to reduce chromatic aberration for high resolution, HVEM is advantageous. The classical theory of electron energy loss of electrons under glancing angle conditions has been discussed and compared with experimental results (Howie, 1981, 1983; Cowley, 1982a; Bleloch et al., 1989).

In Fig. 3.10, the bulk plasmon loss peaks are absent, indicating shallow penetration into the bulk of the electrons which form the specular beam. On the other hand, an analysis of Kikuchi lines showed a relatively strong bulk plasmon loss peak, indicating that Kikuchi lines are due to electrons generated deep in the crystal (Krivanek et al. 1983). With an increase of the glancing angle, the penetration increases and bulk plasmon loss peaks appeared even for the specular beam.

Single electon excitation loss edges in EELS are somewhat smeared by strong surface plasmon excitation and at first it was

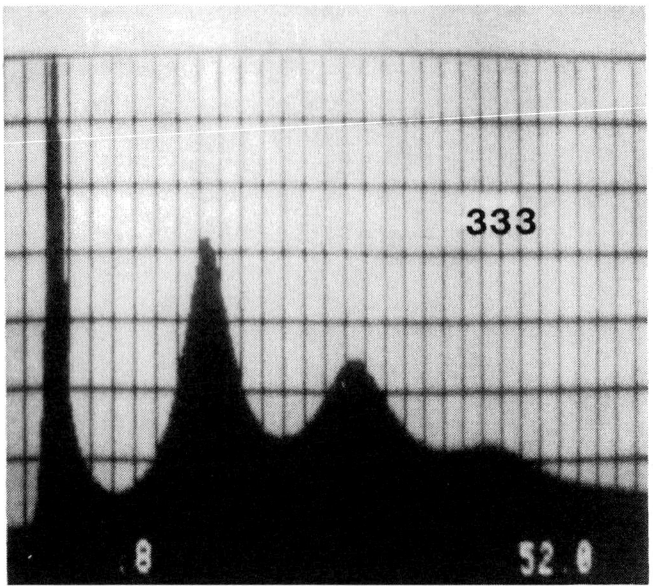

Figure 3.10 A low loss region of an EELS of the (333) reflection from a Si(111) surface at 80kV. Surface plasmon loss and its multiples are seen.

considered to be not useful for surface chemical analysis (Krivanek et al., 1983). Later, however, using higher incident energy and exploring the high energy loss region under a special diffraction condition such as SWR, clear core-loss edges were noted (Wang et al., 1987; Wang and Cowley, 1988a,b; Wang and Egerton, 1988). Relative changes of signals from Ga and As in a GaAs surface have been noted for changes of the Bragg reflection conditions, giving changes in the wavefield in the crystal surface (Wang et al., 1987).

3.3.5 Image contrast problems in REM

Low resolution images.

Image contrast analysis is one of the most urgent problems in REM. Geometrical ray diagram considerations on images from the wavy surfaces by Nielsen and Cowley (1976) and Hsu and Lehmpfuhl (1989) should be extended to more quantitative analysis for practical problems such as the quantitative analysis of undulations of MBE grown surfaces (Shimizu and Muto, 1987. See also discussions on Fig. 3.6).

The first dynamical calculations of REM image contrast were given by Shuman (1977) who analyzed image contrast of dislocations and stacking faults. He used a column approximation, where reflected intensities from a part of a distorted surface are assumed to be those from a perfect surface whose lattice orientation is identical to that part of the real surface. Osakabe et al. (1981a) analyzed their REM images of atomic steps and dislocations following Shuman's method and the importance of Fresnel contrast as well as Bragg contrast was reported.

High resolution images

High resolution REM using two or more diffracted beams in a RHEED pattern as shown in Fig. 3.11 is one important direction for the development of REM. Usually the specular beam is included in the OL aperture, indicated by a circle, together with surface structure reflections indicated by small spots, to record lattice fringes of a superlattice. When we also include spots from the fundamental reflection from the surface (indicated by larger spots), fringes due the fundamental surface lattice together with those from the superlattice can be recorded (Koike et al., 1989).

Lattice fringes from the Si(111)7x7 surface (2.3 nm in spacing) were first taken by Shimizu et al. (1985) to identify oxidized areas. Situations which complicate lattice fringes in REM compared to those in the TEM case, such as a large defocus effect (see Figs. 3.1 and 3.18), the beam alignment (which affects fringe direction), a beam divergence effect (which confines the observation of lattice fringes to a certain focus range), an effect of distortion of the reciprocal lattice in RHEED patterns (note in Fig. 3.11 that the imaging spots are not on a straight line and the foreshortening factor differs for each imaging beam), were discussed by Tanishiro et al. (1986).

Another problem to be mentioned here is whether the lattice fringes reflect a surface structure projected along the beam direction to some extent. Generally, the answer is no. There are several reasons. 1) RHEED intensities are not kinematic and structure-related fringes are not expected. 2) Momentum transfer normal to the surface differs for different reflections on the zero-th Laue zone. Therefore, the meaning of interference fringes of these reflections is complicated even under the kinematical diffraction conditions. 3) Imaging reflections are not on a straight line across the specular

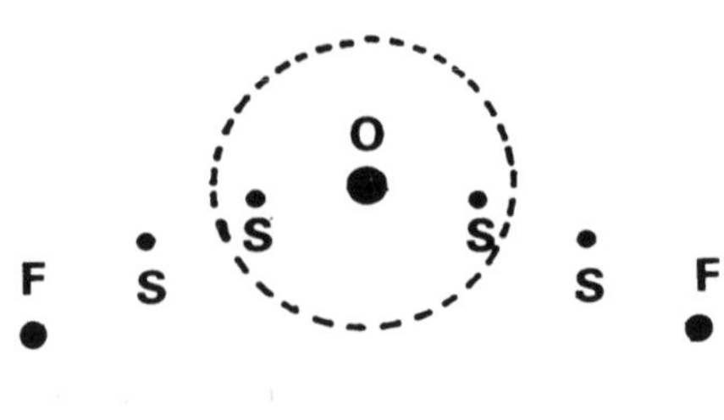

Figure 3.11 A schematic illustration of a geometrical condition for imaging of lattice fringes in the reflection mode. Specular beam 0 is assumed to be parallel to the optic axes of the microscope. F and S are fundamental and superlattice reflections, respectively.

beam as seen in Fig. 3.11. Therefore, the foreshortening factors for the imaging beams differ, and interference is not due to beams from the same surface position in out-of-focus regions.

Resolution of REM is mainly limited by an energy spread due to excitation of surface plasmons as mentioned before. Therefore, higher resolution is expected for high voltage electron microscopes (HVEM, see Figs. 3.19 and 3.20). Dynamical computational approaches to lattice fringes have not been done and are problems for the future.

3.4 Some Experimental Results

3.4.1 Analysis of RHEED intensities

Structure analysis

Structure analysis assuming kinematical diffraction has been done for Si(111)7x7 structure as mentioned before (Ino, 1980a, b; Horio and Ichimiya, 1989) with success to some extent. The case of

Si(111) $\sqrt{3} \times \sqrt{3}$-Ag was also studied (Horio and Ichimiya, 1983). The structure analysis, made by comparing the observed rocking curves with those from dynamical calculations, has been carried out mainly by the Nagoya group; Si(111)'7x7'-H (Ichimiya and Mizuno, 1987), Si(111)-Alkali metals (Muzuno and Ichimya, 1988), GaAs (001) 2x4 (Knibb and Maksym, 1988), Si(111) "5x5"-Si (Nakahara and Ichimiya, 1989), Si (111) $\sqrt{3} \times \sqrt{3}$-Ag (Ichimiya et al., 1989), Si(001)2x1 (Kawamura et al., 1988b). Ichimiya (1987b) found that in the case of Si (111), the rocking curves can be analyzed by a one-beam approximation for an azimuthal orientation of 7.2° from the $[11\bar{2}]$direction, where interlayer correlations or distances are obtained.

RHEED oscillations

RHEED oscillations, which is the phenomenon that the reflected intensity oscillates with a periodicity corresponding to the deposition of a monolayer during MBE growth, was first reported by Neave et al. (1983). It can be considered simply as follows. Monolayer islands formed on flat terraces give diffuse streaks around diffraction spots and the Bragg reflected intensity decreases. After one monolayer is deposited, flat surfaces are regained and the intensity recovers to some extent. Together with this geometrical factor, including the anisotropic shape of the nucleated islands, changes of the surface structures after each monolayer completion, for example, from 2x1 to 1x2 structures in the case of Si(001), affects the oscillations (Sakamoto et al, 1986). Therefore, the oscillation depends on the azimuthal orientation, the glancing angle and the type of reflection measured and the situation is rather complex (see a review paper by Joyce et al. (1988)).

Theoretical approaches were first carried out by Kawamura et al. (1984), who calculated rocking curves from the stepped Al (001) surface. Later similar calculations were used to show the oscillation by Kawamura and Maksym (1985). Calculations have been extended to surfaces which are more close to the experimental conditions such as bulk-terminated Si(001) surfaces (Kawamura et al., 1987) and dimer-form reconstructed Si(001) surfaces. Figure 3.12 shows an example of the calculated oscillation in the case of Si (001) (Kawamura, 1988). On Si(001), the surface atom arrangement rotates by 90° when a monolayer is deposited. Since the RHEED intensity observations were done with a fixed azimuthal orientation, the calculations of specular beam intensity should be carried out for surfaces of different step density of the two types of the surface as shown in Fig. 3.11 (a) and (b). The results are combined in (c) to get oscillations at two glancing angles 1.58° and 0.88°. Although calculations based on a more realistic model are necessary for good agreement with the experimental results, it is clear that such calculations are quite useful for basic understanding of the RHEED oscillations (for an alternative approach, see Ichimiya, 1987a).

From these calculations a correlation between the oscillation and the step density was apparent. In Monte Carlo simulations of MBE growth processes, the step density was calculated from the simulated growth pictures and was qualitatively compared with the observed growth parameter dependences of RHEED oscillations (Clark and Vvedensky, 1987a, b).

It is worth noting here that REM or SREM studies of MBE processes are very useful not only to understand growth processes but also to understand the diffraction effects from the growing surfaces. SREM studies by Ichikawa and Doi (1987, 1988) and Doi and Ichikawa (1989) showed the MBE processes in more realistic ways. It should be pointed out that similar intensity oscillations were observed by LEED during MBE growth (Henzler, 1988).

3.4.2 REM study of surfaces

Characterization of surface structures

Step images

Figure 3.13 reproduces a REM image and a RHEED pattern from a Si(111)7x7 surface (Osakabe et al. 1980a) taken with use of a modified JEM 100B UHV electron microscope (Takayanagi et al. 1978). The image is illustrated in such a way that the imaging electrons run downward so that the image is foreshortened in the vertical direction as indicated by the magnification scales. The upper part of the figure is in over-focus conditions and the lower part is in under-focus conditions (see also Fig. 3.1). At the center, the image is in-focus and there the images of surface atomic steps, seen as a wood print, are sharp. In over- and under-focus regions, bright (B) and dark (D) Fresnel fringes are seen along the steps. The bright and dark characteristics depend on the sense of the steps (step up or step down along the beam direction) and the sign of the defocus. From these rules (Table 3.1), the sense of the steps is determined (Osakabe et al., 1981a). For absolute determination of the sense of steps observations of motions of the steps during sublimation , or

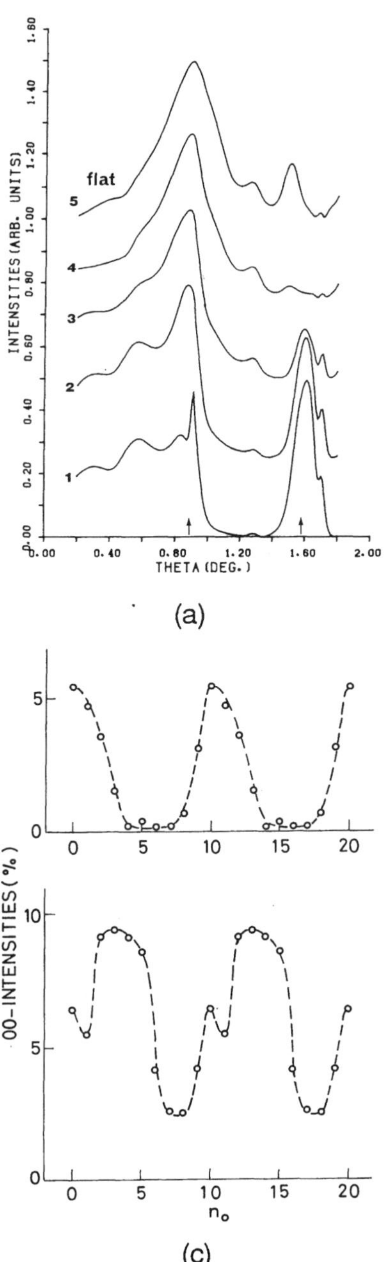

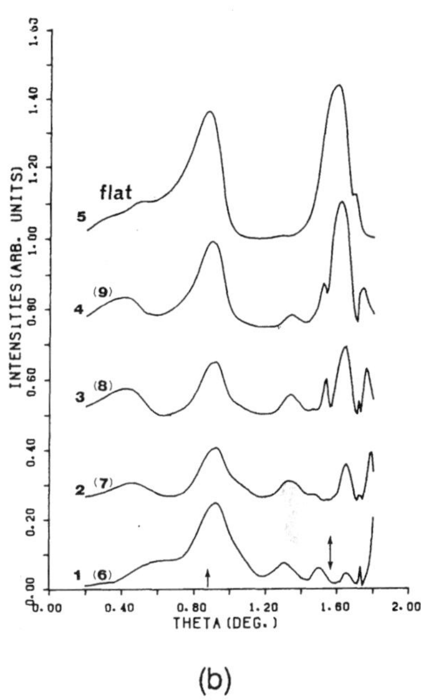

Figure 3.12 Calculated intensity variations of the specular beam from stepped Si(001) surfaces as a function of the glancing angle. (a) and (b) are for the [110] and $[1\bar{1}0]$ azimuthal directions, respectively. The number for each curve is the topmost atomic array on the surface. Two curves in (c) show intensity variations during growth (from 1 to 10 in (a) and (b)) at glancing angles of 1.58° and 0.88° (indicated by arrows in (a) and (b)). (Courtesy Dr. T. Kawamura).

considerations on the spherical form of the sample crystals (Hsu, 1983; Hsu and Cowley, 1983) are needed.

Table 3.1
Characteristics of image contrast of steps

	over-focus	under-focus	$\theta>\theta_B$	$\theta<\theta_B$
step up (along the beam)	bright* dark	dark* bright	dark	bright
step down	dark bright	bright dark	bright	dark

θ_B: Bragg angle (or a angle corresponding to a sharp peak in a rocking curve in a many beam condition).

*bright and dark Fresnel diffraction contrast along the beam direction

For the explanation of the characteristic nature of the contrast of steps, Osakabe et al., (1981a) assumed slight lattice distortion at the steps. On the other hand, Peng and Cowley (1986), Lehmpfuhl and Uchida (1986) and Uchida and Lehmpfuhl (1987) interpreted them simply by phase differences between the waves reflected from upper and lower terraces and changes of beam directions caused by refraction and reflection at the step edges. A quantitative measurements of phase difference was recently given by a holographic technique (Osakabe et al., 1988, Banzhof et al. 1988 and see Fig. 3.14). Effects of SWR on the image contrast of steps have been discussed by Uchida et al. (1984), Lehmpfuhl and Uchida (1988) and Hsu and Peng (1987). A double image of steps was attributed to SWR conditions (Uchida and Lehmpfuhl, 1987; Wang, 1988). It is noted, however, that a double image of a strain field is also expected under the many-beam condition REM as in the case of TEM (Hirsch et al., 1965). Wang et al. (1989) observed an extra spot in the RHEED patterns from the steps using a nanometer-size probe in a SREM instrument under SWR conditions which were confirmed by the form of the dark field image.

At the place indicated by an arrow in Fig. 3.13, one step is terminating and a dark horizontal line is seen. The dark line is a foreshortened image of a screw dislocation terminated at the surface, i.e. the distorted area around the dislocation is slightly out of Bragg

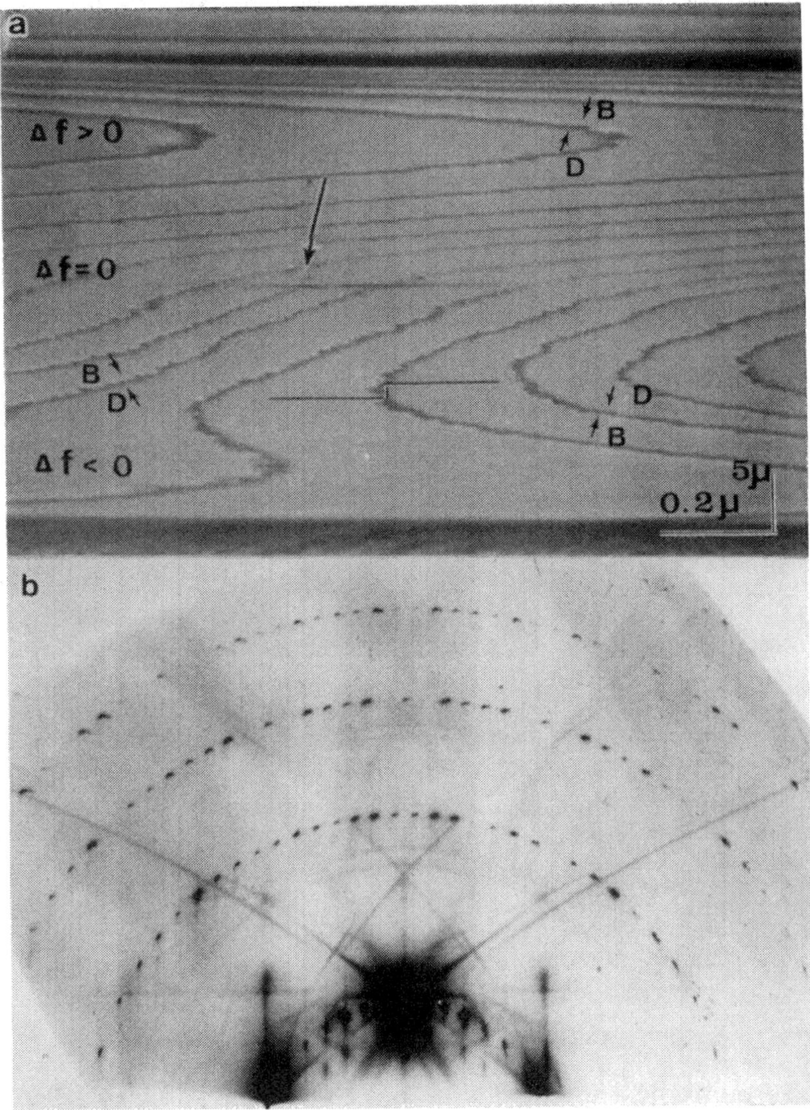

Figure 3.13 A REM image and a corresponding RHEED pattern from a Si(111) 7x7 surface. In the foreshortened image of (a) steps of wavy lines and bright (B) and Dark (D) Fresnel fringes in over- and under-focus regions are noted. A big arrow in (a) indicates an image of a screw type dislocation.

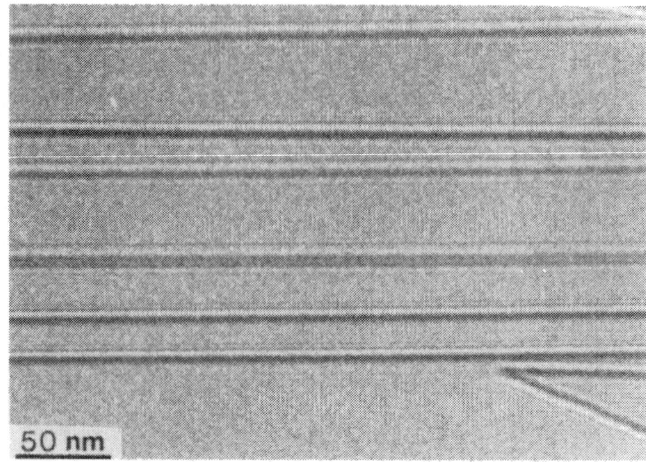

Figure 3.14 A REM image of GaAs (110) surface taken using a FEG installed microscope: 100kV, beam divergence of $5x10^{-6}$ rad. Note many Frensel fringes at steps. (Courtesy Mr. N. Osakabe).

condition and is seen dark when the flat areas are at the exact Bragg condition (Osakabe et al. 1981a).

The RHEED pattern in (b) for the [211] azimuthal orientation is similar to that in Fig. 3.3 but the relative intensities of the 7x7 spots are different from those in Fig. 3.3, as is expected.

Figure 3.14 shows a REM image of a GaAs (110) cleaved surface taken by Osakabe et al. (1989) using a 100kV electron microscope with a FEG. The 880 and 620 reflections are excited simultaneously. A small beam divergence of about $5x10^{-6}$ rad is the reason for an increase in the number of Fresnel fringes at the steps. Holographic analysis showed that the phase difference of reflected waves across the steps is $0.4\,\pi$.

Domains and domain boundary contrast

Figure 3.15 shows REM images of the same area of a Si(001) 2x1 surface taken under slightly different Bragg conditions. The incident beam is along the [110] direction. The 2x1 surface is composed of two domains, the 1x2 and 2x1 domains, where doubling of the unit cell is caused by the reconstruction to form dimers. Two domains are separated by single (one-atom high) steps with the height of a quarter of the unit cell edge of the Si lattice. In the present azimuthal orientation the 2x1 domains, whose two-fold direction is perpendicular to the electron beam, give superlattice reflections in the zero-th Laue zone while the 1x2 domains do not. Thus, the Bragg conditions and hence the intensities of the specular beam, for the two domains are different and image contrast between the two domains is expected as seen in (a) and (b). Generally, intensities of the specular beam from domains which give superstructure reflections in the zero-th Laue zone is reduced due to the excitation of these reflections, and these domains are seen to be dark in comparison with domains which do not show reflections in the zero-th Laue zone (Osakabe et al., 1980b; Inoue et al., 1987). A RHEED calculation by Kawamura et al. (1988b) in the case of Si (001)2x1 showed this general tendency for large glancing angles where the half-order reflections are excited appreciably, in good agreement with the observation. Fig. 3.15(a) is this case. However, in (b) the contrast is reversed, which means that the above rule is not always the case. For the unambiguous identification of the domains, dark field images using the superlattice reflections are needed.

Figure 3.16 shows a REM image of a Si(111) surface taken at 830°C, which illustrates an image contrast of phase boundaries between the 7x7 and 1x1 phases and out-of-phase boundaries

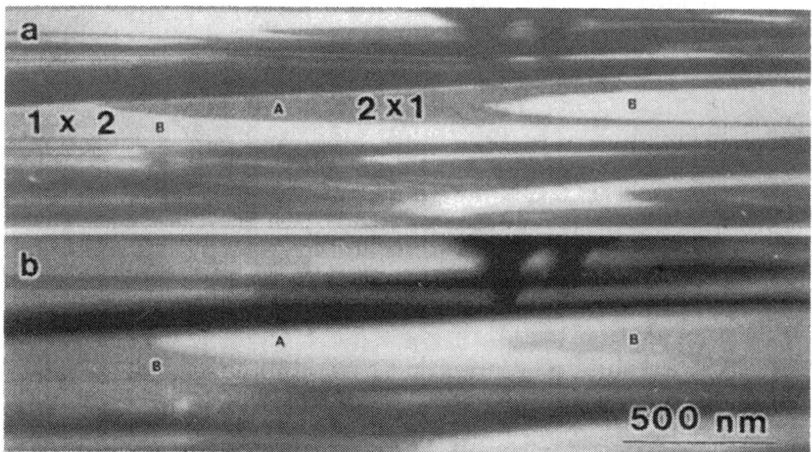

Figure 3.15 REM images of a Si (001)2x1 surface taken at slightly different Bragg conditions. Two domains 1x2 and 2x1 are contrasted in opposite ways in the two micrographs.

(OPBs) in the 7x7 phase. Dark contrast regions are the 7x7 low temperature phase, nucleated on the upper-side terraces along the steps on cooling. Bright regions are the 1x1 high temperature phase. The dark contrast of the 7x7 regions is considered to be due to the excitation of the 7x7 reflections in the zero-th Laue zone as mentioned above (contrast reversal is sometimes noted; Tanishiro et al. 1982). Since the nucleation of the 7x7 regions takes place without any mutual phase relations, OPBs are formed between the nucleated domains and are seen as white lines in Fig. 3.16. A shift of 7x7 lattice fringes is actually noted in high resolution REM images (Tanishiro et al. 1986). OPBs are generally seen as dark lines or bright lines depending on the contrast of the domains where the OPBs are formed (Kahata and Yagi, 1989a). Since OPBs and steps have line tensions, the steps are slightly bent at the meeting point with the OPBs in Fig. 3.15, as schematically illustrated in the lower part (Tanishiro et al., 1983. See also Fig. 3.27 (Kahata and Yagi, 1989a)).

Figure 3.17 shows a REM image of a Si(111)5x2-Au surface, where orientational domains bright (B), dark (D) and intermediate (I) contrast regions) and an OPB in an (I) domain are seen (Yagi et al., 1980).

Figure 3.18 shows REM images of an Au-deposited Pt surface, taken under slightly different glancing angle as indicated. Coverage of Au is about a half monolayer. The REM image (a) was taken under Bragg conditions such that the Pt surface before the deposition has a maximum reflected intensity (the brightest in the image). In (a) the Au covered regions are dark. Atomic steps on the Pt surface are wavy boundaries on the right hand side of the dark region and their sense is such that the right hand side is higher, indicating that monolayer Au is formed along the lower side of the steps. The image (b) was taken under a slightly smaller glancing angle condition and contrast reversal is noted. An intuitive explanation is that the atomic radius of Au is larger than that of Pt and the spacing between the Au layer and underlying Pt layer is slightly larger than that between Pt layers, resulting in a slightly smaller "Bragg" angle for the Au covered regions (Yagi et al., 1988). Dynamical calculations by Ohse et al. (1988) showed that a difference of scattering factors between Au and Pt is not the cause of the contrast and that the slightly larger spacing between Au and Pt layers is responsible for it.

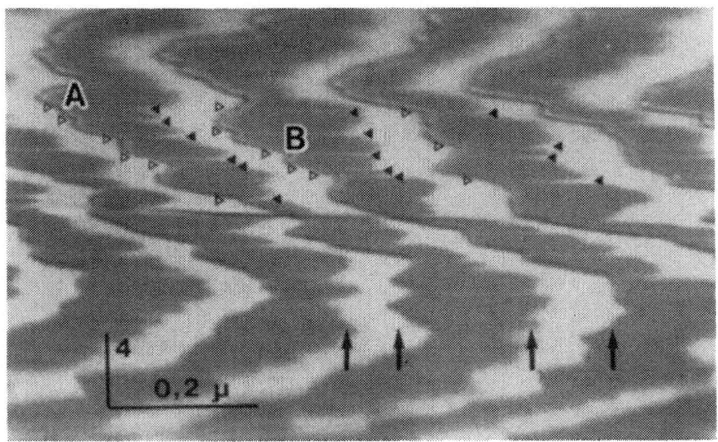

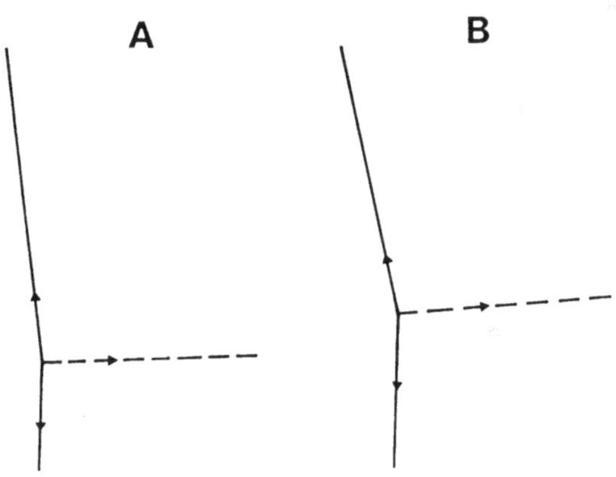

Figure 3.16 A REM micrograph of a Si(111) surface taken at 830°C, showing a phase transition process from the 1x1 high temperature phase (bright regions) to the 7x7 low temperature phase (dark regions). Bright lines in the dark regions are out of phase boundaries. In the lower part steps (solid lines) and out of phase boundaries (broken lines) at A and B in (a) are shown after a correction of the foreshortening, which shows a balance of the line tensions.

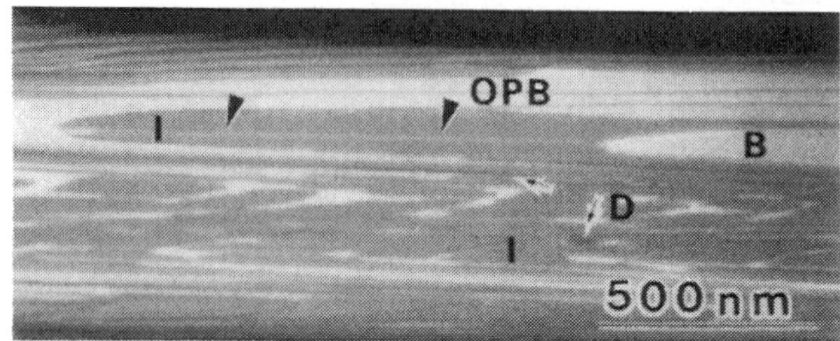

Figure 3.17 A REM image of a Si(111) 5x2-Au surface. Domains of bright (B), dark (D) and intermediate (I) contrast regions and out of phase boundary (OPB) are seen.

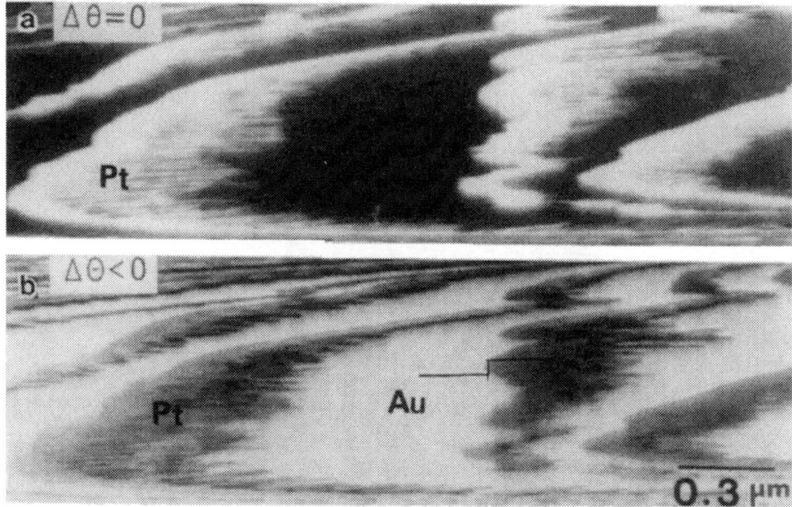

Figure 3.18 REM images of a Pt(111) surface on which Au of a half monolayer was deposited. (a) was taken at exact Bragg condition for the Pt surface, and regions not covered by Au are bright. (b) was taken at a slightly smaller glancing angle than (a). Note the contrast reversal from (a) to (b).

Lattice fringes in REM

Figure 3.19 shows a high resolution REM image of Si(111)7x7 surface taken along the [110] direction by Koike et al. (1989) with use of a UHV 1MV electron microscope (H1250S, Honjo et al., 1980), which was also used to resolve surface atomic arrangements of reconstructed Au (001), (110) and (111) surfaces (Hasegawa et al. 1987). Seven electron beams including the specular beam were included in the objective aperture. Wide vertical fringes correspond to the unit cell spacing (2.3nm) of the 7x7 structure. These fringes had been taken with use of 100kV (Tanishiro et al. 1985) and 200kV (Takayanagi et al., 1987) microscopes. However, in Fig. 3.19 finer fringes corresponding to one-third of the main fringe spacing are clearly seen in a focus range of about 2μm. Usually when the number of imaging beams is increased, half-period fringes appear first, but this is not so in the present case, reflecting that the third and the fourth order 7x7 reflections are relatively strong (characteristic of the DAS structure). This seems to suggest that the observed fine fringes reflect the 7x7 structure to some extent, though generally this is not expected as discussed in the previous section. A kinematical image simulation showed that dark 7x7 fringes correspond to dimer rows in the DAS model and the one-third fringes correspond to adatom rows. The fact that the steps always coincide with dark fringes of the 7x7 structure (see Fig. 1 of Takayanagi et al.'s paper (1987), see also Fig. 3.20) also suggests that the fringes reflect surface structure to some extent. This is because STM studies show that terrace edges generally coincide with dimer rows of the 7x7 structure (Becker et al., 1985).

Figure 3.20 shows another example of lattice fringes for the five times periodicity of the Si(111)5x2-Au structure. A corresponding RHEED pattern formed by 1MeV electrons is also shown. Arrow heads indicate the position where the vertical steps coincide with dark fringes.

Observations of Surface Dynamic Processes

One of the most important applications of REM is to observe surface dynamic processes using the capabilities to characterize surface structures as mentioned above. Most of the surface dynamic processes are inhomogeneous except cases which take place at

Figure 3.19 A high resolution REM image of a SI (111) 7x7 surface taken by a UHV 1000kV microscope ([110]) incidence). 7x7 lattice fringes (2.3nm) and fine fringes with a spacing of one third of the fundamental fringes are noted.

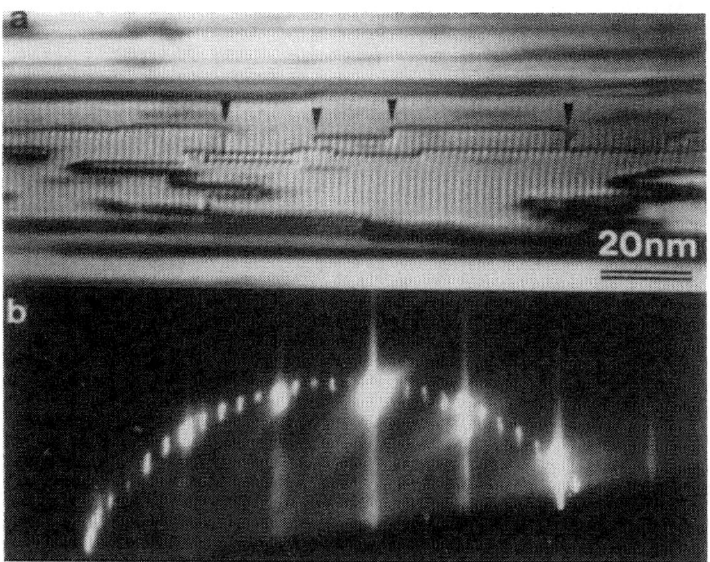

Figure 3.20 A high resolution REM image of a Si(111)5x2 surface (a), and a corresponding RHEED pattern (b), taken by the UHV 1000kV microscope. Note that the dark fringes of the 5 times superlattice coincide with the steps at positions indicated by arrow heads.

relatively low temperature where the relevant "diffusion" distance is small. Inhomogeneous processes closely related to surface defects such as steps and phase boundaries are interesting subjects. The processes include sublimation, surface structure phase transition, adsorption of metals or gas atoms or molecules, oxidations, sputtering and annealing, surface electromigration of adsorbed foreign metal atoms, current induced surface structure changes. The purpose of the present chapter is not to show detailed results. So in the following some examples are briefly illustrated to show how the above mentioned capabilities are useful for surface science problems.

Sublimation

Sublimation is seen as motions of single steps on the surfaces at high temperture. From the temperature dependence of such motions, the activation energy of sublimation is obtained. In the case of Si(111) it is about 3.5eV (Yagi, 1982).

Phase transition

Reconstructed surface structures and adsorbate structures transform to disordered or other ordered structures when the surfaces are heated. Nucleation sites of the new phases can be easily noticed. Fig. 3.16 is an old but typical example.

Adsorption processes

Fig.. 3.21 shows the case of Au deposited on a Pt (111) surface. The surface steps down from right to left. After a deposition of Au of about 0.2 monolayer, Au nucleates at the lower side of the terraces as dark regions (see also Fig. 3.17). The fact that dark Au regions are wide along the steps whose lower side terraces are wide, as indicated by large and small vertical arrows, means that adsorbed Au atoms on a terrace do not contribute to the growth on the lower side terrace (Schwoebel effect (Schwoebel; 1969)). Therefore, the dynamics of adatoms can be known by the detail observations of the growth features.

Another point to mention is that it is possible to observe the processes in reference to the lattice fringes of the surface adsorbate structure. Identification of the oxidized area during oxidation (Shimizu et al., 1985) and observations of conversion from the 7x7 fringes to 5x2 fringes during growth of Si(111)5x2-Au on Si(111)7x7 (Tanishiro and Takayanagi, 1989) are typical examples.

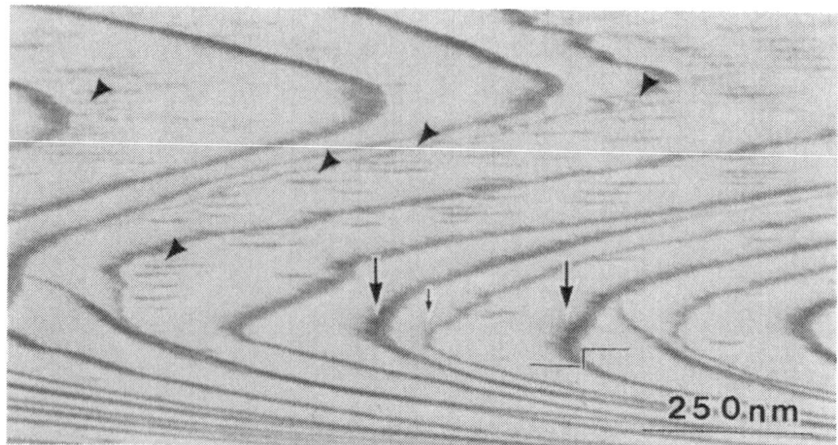

Figure 3.21 A REM image of a Pt(111) surface on which 0.2 monolayer of Au was deposited. Monatom high Au nuclei (dark regions) were formed on lower side terraces (left hand side) along steps. Widths of Au covered regions depend on widths of terraces as shown by big and small arrows. Note also that the monolayer nuclei are also formed on the left had side of the wide terraces (arrow heads).

Oxidation

Oxidation is one of the typical surface dynamic processes. Figures 3.22 shows a series of REM images of Si(001) surface taken during exposure to oxygen of 10^{-8} torr level at 700° C. Under these conditions, oxygen gas reacts with Si atoms to form SiO and sublimation of SiO takes place leaving vacancies on the surface. Some of the vacancies, formed close to surface steps, migrate to the steps and are captured there. Others may coalesce into hollows of unit depth at the central areas of wide terraces (Shimizu et al., 1985, 1987). In Fig. 3.22 (b) a hollow with bright contrast is seen to be formed in a dark 2x1 domain terrace. The hollow is seemingly elliptic with major axis in the horizontal direction, but its real shape after a correction of the foreshortening is elliptic with major axis in the vertical direction. The anisotropic shape reflects the anisotropic energy of surface steps (Kahata and Yagi, 1989b).

When the oxygen pressure is as high as 10^{-7} torr, SiO_2 is formed and the 7x7 structure transforms to oxidized surface.

Ion sputtering and annealing

Ion sputtering and annealing have been generally used for surface cleaning. REM studies of these processes have been carried out through changes of surface atomic steps and defect formation and elimination (Ogawa et al., 1988; Claverie et al. 1986).

Surface electromigration

As is the case for bulk and thin films, adsorbate metal atoms on clean Si surfaces were found to move on an adsorbate metal layer or layers in the direction depending on the current in the substrate Si. This was called surface electromigration by Yasunaga et al. (1986, 1988). Figure 3.23 reproduces the processes as studied by REM (Yamanaka et al., 1989). Au was deposited between masked areas of Si(111) surface. The micrograph (a) shows an edge of the Au-deposited area which is on the right hand side of the micrograph. The current to the left was applied to the specimen and micrographs (b) and (c) were taken afterwards. It is noted that the Au adsorbate moves to the right. The fact that isolated Au adsorbates indicated by arrow heads in (b) disappear in (c) means that Au atoms migrate not only on adsorbate layer as was suggested previously but also on clean Si(111) surfaces. At the right-hand side edge of the Au deposited area a movement of the edge to the right was noted. A reversal of the current direction caused the reversal of the migration direction of Au atoms. Indium atoms move in the opposite way (parallel to the current direction).

Current effect on surface structures of Si

The phenomenon of surface electromigration seems to suggest that current may affect surface atoms or adatoms on clean Si surfaces resulting in current-dependent surface structure. Figure 3.24 shows one example, where changes of major domains on a Si(001)2x1 surface, depending on the current direction, are illustrated (Kahata and Yagi, 1989a). In the upper right corner of each micrograph the current direction is shown by an arrow. The surface steps down from left to right. Two-domain structure as in Fig. 3.15 is

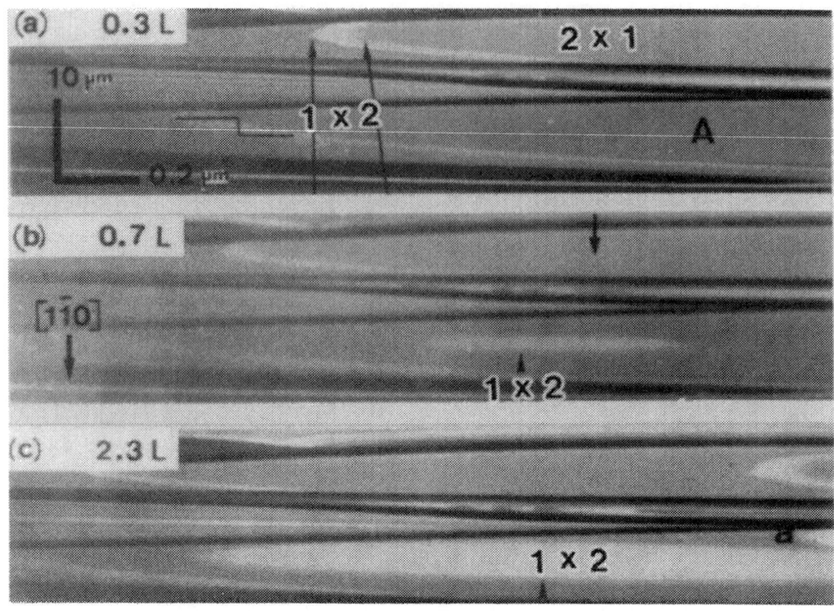

Figure 3.22 A sequence of REM images of a Si(001)2x1 surface which show formation of a hollow in (b) and its growth in (c). The shape is elliptic with the major axis in the vertical direction (after correction of foreshortening) so as to reduce a total step energy.

seen. In (a) the major domain is 1x2, but in (b), after the reversal of the current direction, it is 2x1. REM images of (c) to (f) shows the process of a conversion of a major domain from 2x1 to 1x2 after the next reversal of the current direction: expansions of narrow domains by the movement of surrounding steps are seen. The observation revealed that the major domain depends on whether the current is in the step -up or step-down direction.

Characterization of cleaved surfaces

Cleaved surfaces can be easily used as specimens for REM observations. Cleaved (001) MgO (Osakabe et al., 1981a; Yamamoto and Spence, 1983), (110)GaAs (and other zinc blend type compound semiconductors) (Iijima and Hsu, 1982, Yamamoto and Spence, 1983, Hsu et al., 1984), (001)PbS (Iijima and Hsu, 1982), graphite (Hsu and Cowley, 1985) and Al_2O_3 (Yao et al., 1989) have so far been studied. The step structure on the cleaved surface may be used

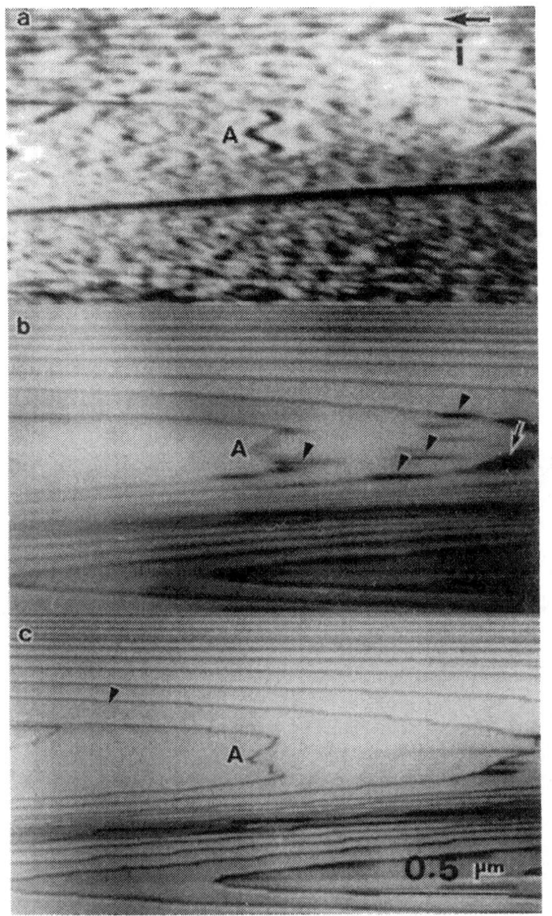

Figure 3.23 A sequence of REM images producing surface electromigration of Au deposited on Si(111). The current is from right to left as indicated and Au adsorbate moved to right.

to know the process of the cleavage. The fact that image contrast of REM is also determined by subsurface structures allows us to detect local difference in stacking in the subsurface (Hsu and Cowley, 1985).

Another application of the technique is to know the internal structure by observing its cross section with a cleavage plane.

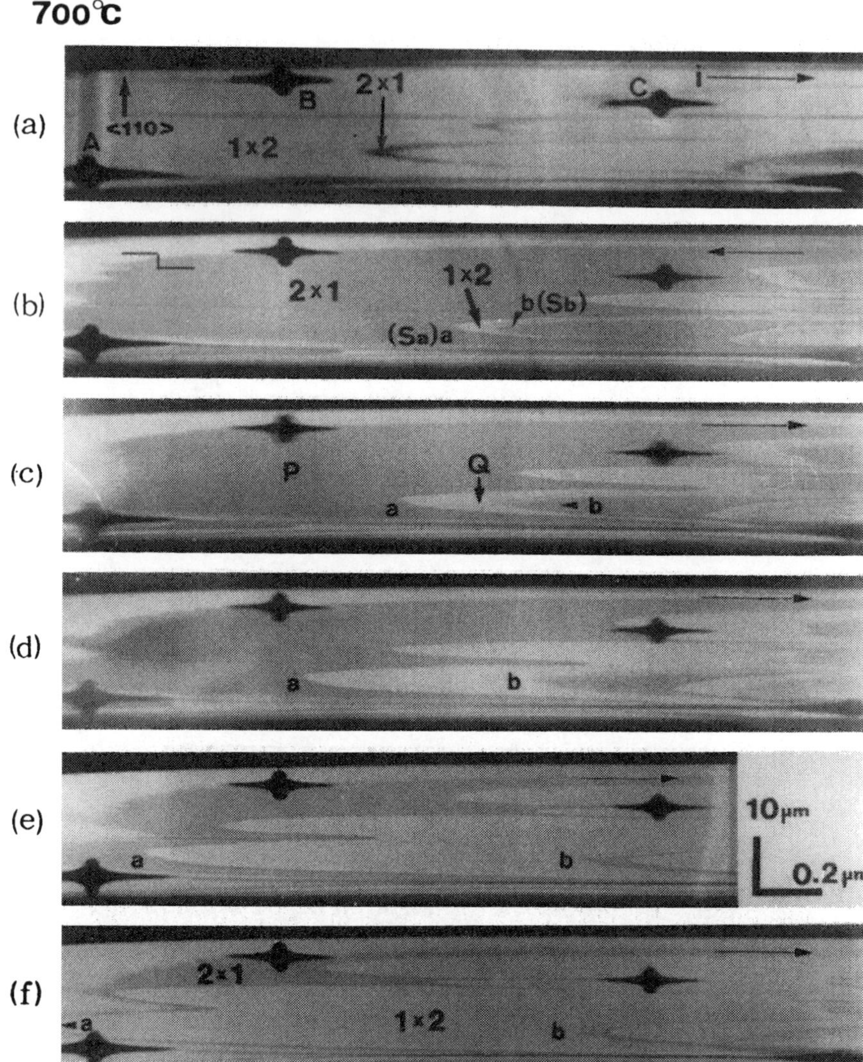

Figure 3.24 A sequence of REM image which shows reversal of major domains between 1x2 and 2x1 depending on the specimen current direction. (b) to (f) shows the reversal process.

A typical example is shown in Fig. 3.25 which illustrates the superlattice structure of $Al_{0.3}Ga_{0.7}As$ (12nm)/GaAs(8nm) (Yamamoto and Muto, 1984; Yamamoto, 1986, Hsu, 1985). Irregularities of the interfaces are seen (arrow heads). The micrograph clearly shows how undulations of the substrate (a right hand side bright region) cause undulations of the overlayer interface and how the undulations decrease with increase of the number of superlattice layers. Careful analysis of the bright and dark contrast and its focus dependence show that the contrast is partly due to difference in the structure factors of the two regions and partly due todifference in surface height caused by surface relaxation of each tetragonally-distorted superlattice layer (Yamamoto, 1986).

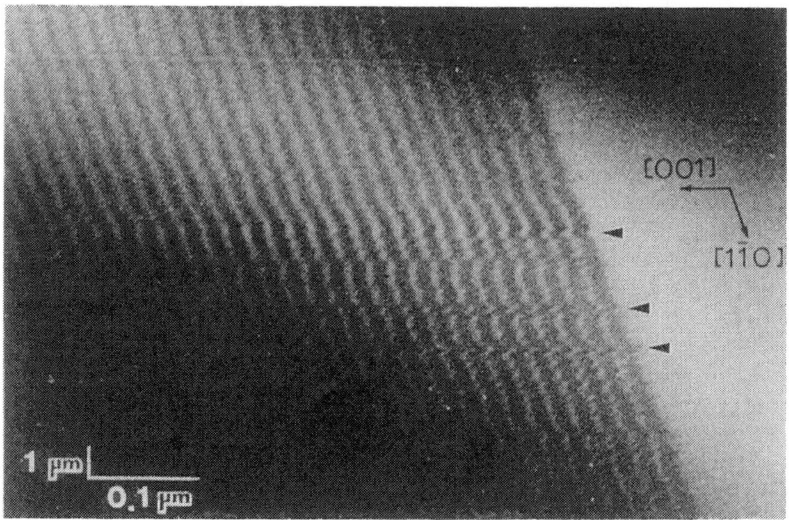

Figure 3.25 A REM image of a (110) cleaved surface of $Al_{0.3}Ga_{0.7}As$(12nm)/GaAs(8nm) supperlattice. Undulations of interfaces in a wide area are clearly seen. Note that the undulations are seen in an exaggerated way due to foreshortening.

3.4 Conclusions

RHEED and REM studies of surfaces are surveyed. RHEED can be easily combined to UHV surface analyzing instruments and it can be used to characterize surface structures and to monitor surface dynamic processes such as MBE processes. REM, on the other hand, needs special instruments and widespread use of it is not expected. SREM needs attachments for formation of micro- or nano-beam and its scanning. For high resolution with nano-beams the working distance is small and it cannot easily be combined to other instruments.

REM and SREM studies have not been as widely used as in TEM and there are many basic problems to be solved. They are image contrast analyses of surface structures such as domains and domain boundaries and other surface defects and of high resolution surface lattice fringes. However, as shown in the present chapter real space observations give direct information of surface structure and processes and give sound bases for real space understanding of RHEED patterns and their changes.

For the development of REM there are several future problems in addition to the developments of the image contrast mentioned above. One is, as is always the case, high resolution. Developments of in-situ techniques to observe various surface processes and combinations of surface chemical analyzing techniques such as EELS, AES and X-ray analysis are needed.

There are several surface imaging methods: scanning tunneling microscopy (STM) (see for example a recent proceedings of international conference of STM (Ichinokawa (1990)), SEM (Venables, 1982), low energy REM (Bauer and Telieps., 1988), photoemission electron microscopy (PEEM) (Bethge et al., 1982; Mundschau et al., 1988), field ion or emission microscopy (FIM or FEM) (Tsong, 1969). In them, SEM (in UHV conditions) and PEEM give images related to locally emitted electron intensities and give no or little surface crystallographic information. STM gives surface microtopography at the atomic resolution level and details of atomic structures of reconstructed structures and surface defects such as steps on semiconductor surfaces. Scanning tunnelling spectroscopy (STS) used with STM gives electronic structures of not only surface atoms but also subsurface atoms in the case of semiconductors. However, the latter kind of analysis is only for limited cases and generally subsurface structures cannot be studied. RHEED, on the other hand, can be used for studies of subsurface structures. Therefore, REM and RHEED are complemental to other techniques and combinations of REM, SREM and RHEED with other surface imaging techniques are important future problems and some of the them are now in progress.

REFERENCES

Albrecht, M. and Meyer-Ehmsen, G. (1988) (See Larsen and Dobson, 1988, p. 211).

Anderson, S.K. and Howie, A. (1975) Surf. Sci. 50, 197.

Anstis, G.R. (1989) Private Communication.

Baines, M., Howie, A. and Andersen, S.K. (1975) Surf. Sci. 53, 546.

Banzhof, H., Hermann, K.M. and Lichte, H. (1988) Proc. 9th Eur. Cong. Electron Microscopy (York), vol. 1, 263.

Bauer, E. and Telieps, W. (1988) (See Larsen and Dobson, 1988, p.381).

Becker, R.S., Golvovchenko, J.A., McRae, E.G., Swartzentruber, R.S. (1985) Phys. Rev. Lett. 55, 2028.

Bennett, P.A. and Johnson, A.P. (1988) (See Larsen and Dobson, 1988, p. 371).

Bethge, H., Gerth, D. and Matern, D. (1982) Proc. Int. Cong. Electron Microsoc. (Hamburg) Vol. 1, 69.

Bleloch, A.L., Howie, A., Milne, R.H. and Walls, M.G. (1989). Ultramicroscopy 29, 175.

Bolger, M. and Larsen, P.K. (1986) Rev. Sci. Inst. 57, 1362.

Cates, J. (1933) Trans. Faraday Soc. 29, 823.

Chang, C.C. (1977) Appl. Phys. Lett. 31, 304.

Clark, S. and Vvedensky, D.D. (1987a) Surf. Sci. 189/190. 1033.

Clark, S. and Vvedensky, D.D. (1987b) Phys. Rev. Lett. 51, 340.

Claverie, A., Faure, J., Vieu, C., Beauvillain, J. and Jouffrey, B. (1986). Proc. 11th Int. Cong. Electron Micorsc. (Kyoto) vol. 2, 1357.

Colella, R. (1972). Acta Crystallogr. A28, 11.

Colella, R. and Menadue, J.F. (1972) Acta Crystallogr, A28, 16.

Cowley, J.M. (1981) Diffraction Physics. North Holland, Amsterdam.

Cowley, J.M. (1982a) Ultramicroscopy 9, 231.

Cowley, J.M. (1982b) Surf. Sci. 114, 587.

Cowley, J.M. (1988a) in Surface and Interface Characterization by Electron Optical Methods, A. Howie and U. Valdre, Eds, (Plenum Press, New York and London) pp. 127-58.

Cowley, J.M. (1988b) (see Larsen and Dobson p. 26).

Cowley, J.M. and Kang, Z.-C. (1983) Ultramicroscopy 11, 131.

Cowley, J.M. and Neumann, K.D. (1984) Surf. Sci. 145, 301.

Cowley, J.M. and Peng, L.M. (1985) Ultramicroscopy 16, 59.

Cowley, J.M., Albain, J.L., Hembree, G.G., Nielsen, P.E.H., Koch, F.A., Landry, J.D. and Shuman, H. (1975) Rev. Sci. Instrum. 46. 826.

Dobson, P.J., Joyce, B.A. and Neave, J.H. (1987) J. Cryst. Growth 81, 1.

Doi, T. and Ichikawa, M. (1989) J. Cryst. Growth 95, 468.

Eades, A.E. and Shannon, M.D. (1988) (See Larsen and Dobson, 1988. p. 237).

Germer, L.H. and Hartman, C.H. (1960) Rev. Sci. Instrum. 31, 784.
Halliday, J.S. and Newman, R.C. (1960) Br. J. Appl. Phys. 11, 158.
Hasegawa, T. Ikarashi, N., Kobayashi, K., Takayanagi, K. and Yagi, K. (1987) Proc. 2nd Int. Conf. Structure of Surfaces (The Structure of Surfaces II) p. 42, Springer, Berlin.
Hasegawa, S., Ino, S., Yamamoto, Y. and Daimon, H. (1985) Jpn. J. Appl. Phys. 24, L387.
Heinemann, K. and Poppa, H. (1986) J. Vac. Sci. Tech. A4, 127.
Henzler, M. (1988) (See Larsen and Dobson, 1988. p. 193).
Hirsch, P.B., Howie, A., Nicholson, R.B., Pashley, D.W. and Whelan, J.M. (1965) Electron Microscopy of Thin Crystals. Butterworth, London.
Honjo, G., Yagi, K. Takayanagi, K., Kobayashi, K., Nagakura, S., Katagiri, S., Kubosoe, M. and Matui, I. (1980) Electron Microscopy vol. 4, 22.
Horio, Y. and Ichimiya, A. (1983) Surf. Sci. 133, 393.
Horio, Y. and Ichimiya, A. (1985) Surf. Sci. 164, 589.
Horio, Y. and Ichimiya, A. (1989) Surf. Sci. 219, 128.
Howie, A. (1981) Inst. Phys. Conf. Ser. 61, 419.
Howie, A. (1983) Ultramicroscopy 11, 141.
Howie, A. and Basinski, Z.S. (1968) Phil. Mag. 17, 1038.
Howie, A. and Milne, R.H. (1985) Ultramicroscopy, 18, 427.
Hsu, T. (1983) Ultramicroscopy 11, 167.
Hsu, Y. (1985) J. Vac. Sci. Tech. B3, 1035.
Hsu. T. and Cowley, J.M. (1983) Ultramicroscopy 11, 239.
Hsu, T. and Cowley, J.M. (1985) In: "The Structure of Surfaces", (Van Hove, M.A. and Tong, S.Y. eds.) p. 55, Springer-Verlag, Berlin.
Hsu, T. and Peng, L.-M. (1987) Ultramicroscopy 22, 217.
Hsu, T. and Lehmpfuhl, G. (1989) Ultramicroscopy 27, 359.
Hsu, T., Iijima, S. and Cowley, J.M. (1984) Surf. Sci. 137, 551.
Ichikawa, M. and Doi, T. (1987) Appl. Phys. Lett. 50, 1141.
Ichikawa, M. and Doi, T. (1988) (see Larsen and Dobson (1988) p. 343.
Ichikawa, M., Doi, T., Ichihashi, M. and Hayakawa, K. (1984) Jpn. J. Appl. Phys. 23, 913.
Ichimiya, A. (1983) Jpn. J. Appl. Phys. 22, 176.
Ichimiya, A. (1987a) Surf. Sci. 187, 194.
Ichimiya, A. (1987b) Surf. Sci. 192, L983.
Ichimiya, A. and Takeuchi, Y. (1983) Surf. Sci. 128, 343.
Ichimiya, A. and Mizuno, S. (1987) Surf. Sci. 191, L765.
Ichimiya, A. , Kambe, K. and Lehmpfuhl, G. (1980) J. Phys. Soc. Jpn. 49, 684.
Ichimiya, A. , Kohmoto, S., Fujii, T. and Horio, Y. (1989) Appl. Surf. Sci., 41/42, 82.
Ichinokawa, T. (1990) J. Vac. Sci. Tech. A8 p. 153-720.

Iijima, S. and Hsu, T. (1982) Proc. Int. Cong. Electron Microscopy (Hamburg) vol. 2, 293.
Inoue, N., Tanishiro, Y. and Yagi, K. (1987) Jpn. J. Appl. Phys. 26, L293.
Ino, S. (1977) Jpn. J. Appl. Phys. 16, 891.
Ino, S. (1980a) Jpn. J. Appl. Phys. 19, L61.
Ino, S. (1980b) Jpn. J. Appl. Phys. 19, 1277.
Ino, S. (1988) (see Larson and Dobson, 1988, p. 1).
Ino, S., Ichikawa, T. and Okada, S. (1980) Jpn. J. Appl. Phys. 19, 1451.
Joyce, B.A., Neave, J.H., Zhang, J. and Dobson, P.J. (1988) (See Larsen and Dobson, 1988, p. 397).
Kahata, H. and Yagi, K. (1989a) Jpn. J. Appl. Phys. 28, L858.
Kahata, H. and Yagi, K. (1989b) Surf. Sci. 220, 131.
Kambe, K. (1988) Ultramicroscopy 25, 259.
Kawamura, T. (1988) (See Larsen and Dobson, 1988 p. 501).
Kawamura, T. and Maksym, P.A. (1985) Surf. Sci. 161, 12.
Kawamura, T., Maksym, P.A. and Iijima, T. (1984) Surf. Sci. 148, L671.
Kawamura, T., Sakamoto, T. and Ohta, K. (1986) Surf. Sci. 171, L409.
Kawamura, T., Natori, T., Sakamoto, T. and Maksym. P.A. (1987) Surf. Sci. 181, L171.
Kawamura, T., Ichimiya, A. and Maksym, P.A. (1988a) Jap. J. Appl. Phys. 27, L1098.
Kawamura, T., Sakamoto, T., Sakamoto, K., Hashiguchi, G. and Takahashi, N. (1988b) in "Structure of Surfaces II" eds. J.F. van der Veen and M.A. Van Hove, Springer-Verlag, Berlin, p. 298.
Kikuchi, S. (1928) Proc. Jpn. Acad. Sci. 4, 201.
Kikuchi, S. and Nakagawa, S. (1933) Sci. Pap. Inst. Phys. Chem. Res. Tokyo 21, 256.
Knibb, M.G. and Maksym, P.A. (1988) (See Larsen and Dobson, 1988, p. 43).
Koike, H., Kobayashi, K., Ozawa, S. and Yagi, K. (1989) Jpn. J. Appl. Phys. 28, 861.
Krivanek, O.L., Tanishiro, Y., Takayanagi, K. and Yagi, K. (1983) Ultramicroscoy 11, 215.
Kubosoe, M., Tomita, M., Matsui, I., Isakozawa, S. and Kamimura, S. (1989) MRS Symposium Proceedings 139, p. 259
Larsen, P.K. and Dobson, P.J. (1988) Reflection High-Energy Electron Diffraction and Reflection Electron Imaging of Surfaces (Plenum Press, New York and London).
Lehmpfuhl, G. and Dowell, W.C.T. (1986) Acta Crystallogr. A42, 569.
Lehmpfuhl, G. and Uchida, Y. (1986) Proc. 44th Meeting of Electron Microscope Society of America 376.
Lehmpfuhl, G. and Uchida, Y. (1988) Ultramicroscopy 26, 177.
Maksym, P.A. and Beeby, J.L. (1981) Surf. Sci. 110, 423.

Maksym, P.A. and Beeby, J.L. (1984) Surf. Sci. 140, 77.
Marten, H. (1988) (See Larsen and Dobson, 1988, p. 109).
Marten, H. and Meyer-Ehmsen, G. (1985) Surf. Sci. 151, 570.
Materlik, M.J., Zegenhagen, J. and Uelhoff, W. (1985) Phys. Rev. B32 5502.
McRae, E.G. (1971) Surface Sci. 25, 491.
Metois, J.J., Nitsche, S. and Heyraud, J.C. (1989) Ultramicroscopy 27, 349.
Menadue, J.F. (1972) Acta Crystallogr. A 28, 1.
Menter, J.W. (1952-3) J. Inst. Metal. 81, 163.
Miyake, S. (1962) J. Phys. Soc. Jpn. 17, 1642.
Miyake, S. and Hayakawa, K. (1970) Acta Crystallogr. A26, 160.
Miyake, S., Hayakawa, K. and Miida (1968) Acta Crystallogr. A24, 182.
Miyake, S., Kohra, K. and Takagi, M. (1954) Acta Crystallogr. 7, 393.
Mizuno, S. and Ichimiya, A. (1988) Appl. Surf. Sci. 33/34, 38.
Moon, A.R. (1972) Z. Naturforsch 27, 390.
Mori, N. Oikawa, T., Katoh, T., Miyahara, J. and Harada, Y. (1988) Ultramicroscopy 25, 195.
Mundschau, M., Bauer, E., Sweich, W. (1988) Surf. Sci. 203, 412.
Nakahara, H. and Ichimiya, A. (1989) J. Cryst. Growth 95, 472.
Neave, J.H., Joyce, B.A., Dobson, P.J. and Norton, N. (1983) Appl. Phys. A31, 1.
Newkirk, J.B. and Mallett, G.R. (1967) Advances in X-ray analysis. Prenum Press, New York.
Nielsen, P.E.H. and Cowley, J.M. (1976) Surf. Sci. 54, 340.
Ogawa, S., Tanishiro, Y, Takayanagi, K. and Yagi, K. (1987) J. Vac. Sci. Tech. A5, 1735.
Ogawa, S., Tanishiro, Y. and Yagi, K. (1988) Nucl. Inst. Meth. Phys. Sci. B33, 474.
Ohse, H., Yagi, K. and Kawamura, T. (1988) Read at Annual Meeting of Phys. Soc. Jpn. April.
Osakabe, N., Yagi, K. and Honjo, G. (1980a) Jpn. J. Appl. Phys. 19, L309.
Osakabe, N., Tanishiro, Y., Yagi, K. and Honjo, G. (1980b) Surf. Sci. 97, 393.
Osakabe, N., Tanishiro, Y., Yagi, K. and Honjo, G. (1981a) Surf. Sci. 102, 424.
Osakabe, N., Tanishiro, Y., Yagi, K. and Honjo, Gl. (1981b) Surf. Sci. 109, 353.
Osakabe, N., Matsuda, T., Endo, J. and Tonomura, A. (1988) Jpn. J. Appl. Phys. 27, L1772.
Osakabe, N., Endo, J., Matsuda, T., Tonomura, A. and Fukuhara, A. (1989) Phys. Rev. Lett. 62, 2969.
Peng, L.M. and Cowley, J.M. (1986) Acta. Crystallogr. A42, 545.
Peng, L.M. and Cowley, J.M. (1989) Ultramicroscopy 29, 168.

Putike, P.R., Cohen, P.K. and Batra, S. (1988) (See Larsen and Dobson, 1988, p. 427).
Ruska, E. (1933) Z. Phys. 83, 492.
Sakamoto, T., Funabashi, H., Ohta, K., Nakagawa, T., Kawai, N.J. Kojima, T. and Bando, Y. (1985) Superlattice and Microstructures 1, 347.
Sakamoto, T., Kawamura, T. and Hashiguchi, G. (1986) Appl. Phys. Lett. 48, 1642.
Schwoebel, R.L. (1969) J. Appl. Phys. 40, 614.
Shannon, M.D., Eades, J.A., Maichle, M.E. and Turner, P.S. (1985) Ultramicroscopy 16, 175.
Shimizu, N. and Muto, S. (1987) Appl. Phys. Lett. 51, 743.
Shimizu, N., Tanishiro, Y., Kobayashi, K., Takayanagi, K. and Yagi, K. (1985) Ultramicroscopy 17, 453.
Shimizu, N., Tanishiro, Y., Takayanagi, K. and Yagi, K. (1987) Surf. Sci. 191, 28.
Shuman, H. (1977) Ultramicroscopy 2, 361.
Smith, A.E. (1988) (See Larsen and Dobson, 1988, p. 151).
Smith, A.E. and Lynch, D.F. (1988) Acta Crystallogr. A44, 780.
Spence, J.H.C. and Kim, Y. (1988) (See Larson and Dobson, 1988, p. 117).
Swann, P.R., Jones, J.S., Krivanek, O.L., Smith, D.J., Venables J.A. and Cowley, J.M. (1987) Proc. 45th Annual Meeting of EMSA (San Francisco Pres, San Frnacisco) p. 136.
Takayanagi, K., Tanishiro, Y., Kobayashi, K., Akiyama, K. and Yagi, K. (1987) Jpn. J. Appl. Phys. 26, L957.
Takayanagi, K., Yagi, K., Kobayashi, K. and Honjo, G. (1978) J. Phys. E11, 441.
Takayanagi, K., Tanishiro, Y., Takahashi, M. and Takahashi, S. (1985) Surf. Sci. 164, 367.
Takayanagi, K., Tanishiro, Y., Kobayashi, K., Yamamoto, N., Yagi, K., Ohi, K., Kondo, Y, Hirano, H., Ishibashi, Y., Kobayashi, H. and Harada, Y. (1986) Proc. 11th Int. Cong. Electron Micros. (Kyoto) 2, 1337.
Tanishiro, Y. and Takayanagi, K. (1989) Ultramicroscopy 31, 20.
Tanishiro, Y., Takayanagi, K., Kobayashi, K. and Yagi, K. (1982) Proc. 10th Int. Cong. Electron Microsc. (Hamburg) vol. 2, 299.
Tanishiro, Y., Takayanagi, K. and Yagi, K. (1983) Ultramicroscopy 11, 95.
Tanishiro, Y., Takayanagi, K. and Yagi, K. (1986) J. Microsc. 142, 211.
Tong, S.Y., Zhao, T.C. and Poon, H.C. (1988) (See Larsen and Dobson, 1988, p. 63).
Tsong, T.T. (1969) Field Ion Microscopy, American Elsevier, New York.
Uchida, Y. and Lehmpfuhl, G. (1987) Ultramicroscopy 23, 53.

Uchida, Y., Lehmpfuhl, G. and Jaeger, J. (1984) Ultramicroscopy 15, 119.
Uyeda, R. (1942) Proc. Phys. Math. Soc. Jpn. 24, 809.
Uyeda, R. and Miyake, S. (1957). Acta Crystallogr. 10, 53.
Venables, J.A. (1982) in Chemistry and Physics of Solid Surfaces. Vol. 4, ed. R. Vanselow and R. Howe, 123. Springer Verlag, Berlin.
Wang, Z.L. (1988) Ultramicroscopy 24, 371.
Wang, Z.L. and Cowley, J.M. (1988a) J. Microsc. Spectros. Electron 13, 189.
Wang, Z.L. and Cowley, J.M. (1988b) Surf. Sci. 193, 501.
Wang, Z.L. and Egerton, R.F. (1988) Surf. Sci. 205, 25.
Wang, Z.L., Lu, P. and Cowley, J.M. (1987) Ultramicroscopy 23, 205.
Wang, Z.L., Lu, P. and Cowley, J.M. (1989) Ultramicroscopy 27, 101.
Watanabe, M. (1957) J. Phys. Soc. Jpn. 12, 874.
Wilson, R.J. and Petroff, P.M. (1983) Rev. Sci. Instr. 54, 1534.
Woodruff, D.P. (1975) Surf. Sci. 53, 538.
Yagi, K. (1980) Proc. 38th Meeting of Electron Microscope Society of America 290.
Yagi, K. (1982) In Scanning Electron Microscopy (O. Johari, ed.) Vol. 4, p. 1421, SEM, Inc., Chicago.
Yagi, K. (1987) J. Appl. Cryst. 20, 147.
Yagi, K. (1989a) In: "High Resolutin Electron Microscopy" (P. Buseck, J.M.Cowley and L. Eyring ed.), Chap. 13, Oxford University Press, Oxford.
Yagi, K. (1989b) Advances in Optical and Electron Microscopy, vol. 12, p. 47.
Yagi, K. , Osakabe, N., Tanishiro, Y., and Honjo, G. (1980) Proc. 4th Int. Conf. Solid Surfaces (Cannes), I, 1007.
Yagi, K., Takayanagi, K., and Honjo, G. (1982). In "Crystals: Growth, Properties and Applications", Vol. 7(H.C. Freyhardt, ed.) p. 47, Springer, Berlin.
Yagi, K., Takayanagi, K., Kobayashi, K. and Nagakura, S. (1983) Proc. 7th Int. Conf. High Voltage Electron Microsc. (Berkeley) 11.
Yagi, K., Ogawa, S. and Tanishiro, Y. (1988) (see Larsen and Dobson, 1988, p. 285).
Yamaguti, T. (1935) Proc. Phys.-Math. Soc. Jpn. 17, 58.
Yamamoto, N. (1986) Proc. 11th Cong. Electron Microsc. (Kyoto) vol. 2, 1481.
Yamamoto, N. and Spence, J.C.H. (1983) Thin Solid Films 104, 43.
Yamamoto, N. and Muto, S. (1984) Jpn. J. Appl. Phys. 23, L804.
Yamanaka, A., Yagi, K. and Yasunaga, S. (1989) Ultramicroscopy 29, 161.
Yao, N., Wang, Z.L. and Cowley, J.M. (1989) Surf. Sci. 208, 533

Yasunaga, H., Kubo, Y. and Okuyama, N. (1986) Jpn. J. Appl. Phys. 25, L400.
Yasunaga, H., Sakamura, S., Asaka, T., Kanayama, S., Okuyana, N. and Natori, A. (1988) Jpn. J. Appl. Phys. 27, L1603.

4

Electron Diffraction Effects due to Modulated Structures

S. Amelinckx and D. Van Dyck

4.1. INTRODUCTION

The diffraction pattern of a modulated structure is often characterized by the appearance of weak reflections derived from and often close to the reflections of the basic structure. These so-called satellites often form linear sequences or 2D or 3D arrays ssociated with the basic reflections. When arrays of satellites belonging to different basic spots meet, they do not necessarily match, i.e. they do not belong to a common reciprocal lattice. In the simplest case of linear arrays a "spacing" anomaly or an "orientation" anomaly or a combination of both may occur (Figure 4.1). In this case we call the diffraction pattern "incommensurate". In some systems the positions of the satellites may change continuously upon varying one parameter such as temperature.

The first attempts for X-ray structure refinement including these satellites made use of a common reciprocal lattice, corresponding with an artificially large unit cell in real space on the basis of which the refinement can be carried out. Such a treatment, however, is rather artificial and cumbersome. In this chapter we will study the characteristics of electron diffraction in modulated structures from a more physical and systematic point of view.

4.2.1. Basic concepts

4.2. TYPES OF MODULATED STRUCTURES

A modulated structure usually originates from a basic structure which contains entities that can be altered, located on a perfectly

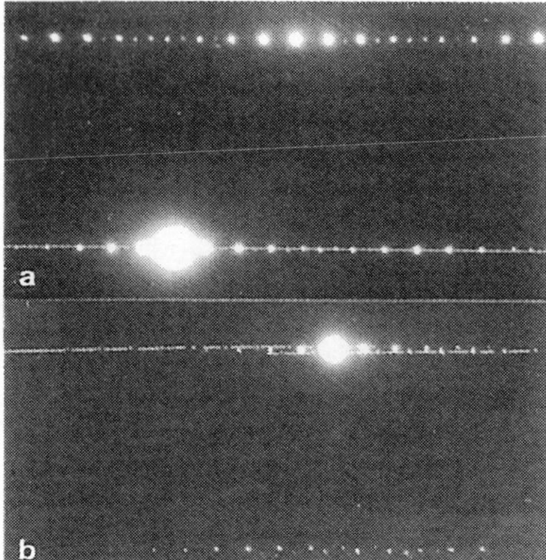

Fig. 4.1 Examples of incommensurate diffraction patterns in $MnSi_{2-x}$ (Nowotny phases)

(a) Spacing anomaly.
(b) Orientation anomaly.

periodic lattice. The modulation then consists of the modification of these entities by one or more driving forces which tend to impose their own periodicities which need not be the same as those of the basic lattice. If this periodicity is not rationally related in spacing and/or orientation to the basic periodicity, the modulation is called incommensurate. The different types of modulated structures can be classified either in terms of the type of entity that is modified or in terms of competing driving forces. According to the relative strength of the interaction between competing forces, the modulation can range from continuous to discrete [§ 4.3].

4.2.2. Modifying entities

In deformation (displacive) modulated structures, the altered entity is the position of one or more atoms; in occupation modulated structures it is the occupation of one or more particular sites of the basic structure; in spin modulated structures it is the spin of one or more atoms: in charge density wave modulations it is the distribution of the electric charge and the accompanying atom displacements and in interface modulated structures it is the presence of planar interfaces which may be either translation interfaces (antiphase boundaries, out of phase boundaries, stacking faults, crystallographic shear planes), twin planes, inversion boundaries or a combination of these.

Interface modulated structures are to be considered as a particular case since the introduction of an interface usually changes the basic structure as a whole which has a particular effect on the diffraction pattern (see 4.5.6). The superstructure now consists of juxtaposed "modules" of a simpler basic structure.

A class of structures that can still be considered as a kind of modulated structures are the mixed layer compounds which consist of a 1D, 2D or 3D stacking of slabs and building blocks of two or more different basic structures. Quasi-crystals can, to some extent, also be interpreted as modulated structures of this type in which building blocks such as the Mackay polyhedra are arranged in a quasi-periodic way.

Some modulated structures may be classified in more than one category, or can be considered as a combination of several types. For instance, non-conservative Long-Period Anti-Phase Boundary alloys (LPAPB) can be considered as composition modulated or interface modulated structures. Interface modulated structures can be described as mixed layer compounds in which the interface region is considered as one type of layer and the region between interfaces as another type of layer. Discommensuration walls are translation interfaces occurring in deformation modulated commensurate superstructures (also called "locked-in" structures) (see 4.6.2) [1].

4.2.3. Driving forces

Modulated structures can be classified according to the driving force that causes the modulation.

In magnetic systems the driving force can result from different contributions to the magnetic energy.

In deformation modulated structures, the driving force may be the electronic energy. This energy can be lowered by creating a periodicity (modulation) which generates a band gap along flat parts of the Fermi surface. This effect is called Fermi nesting. The formation of LPAPB structures in some alloys has also been attributed to Fermi surface effects [2]. However, recently successful theories have been introduced which invoke competing interactions between nearest and next nearest neighbour atoms (e.g. ANNNI-model) [3].

Another modulating phenomenon is the mismatch between lattices (misfit structures). When two structures with different sublattices coexist in one material, both structures may influence one another so as to create a mutual modulation with a periodicity which is close to a common multiple of both basic periodicities. If these are incommensurate, the modulation is also incommensurate. Two-dimensional examples are the planar interfaces between structures as in the case of adsorbate layers and epitaxial layers, or interfaces between domains of the same structure but with different orientations (as e.g. in some grain boundaries). Other examples are intercalation compounds in which layers between sandwiches of the host structure are filled with atoms which tend to form a different periodicity, or channel structures in which channels of the host lattice are filled with strings of atoms with a different periodicity. In the chimney ladder structures or Nowotny phases [4] the channels in the tetragonal transition metal sublattice are filled with helices of Silicon or Germanium.

Composition may be another driving force for the formation of modulated structures and mixed layer compounds. Also in interface modulated structures it can play a role provided the interfaces are non-conservative. Periodic shear structures and polysynthetically and "chemically" twinned structures belong to this category. In interface modulated structures, the elastic interaction energy between the interfaces may also be a driving factor.

Some modulated structures are generated by the crystal growth mechanism. For instance some polytypes of SiC or ZnS, which are in fact translation interface modulated structures, can show very large periodicities which are formed during the spiral growth of the crystal. A special case are the artificially grown epitaxial layer structures such as GaAs-AlAs which can be considered as mixed layer compounds. Here the modulation, i.e. the sequence of the layers, can be controlled layer by layer during growth by molecular beam epitaxy.

4.3. CONTINUOUS VERSUS DISCRETE MODULATION

4.3.1. Model

Modulated structures can also be classified according to the strength of the interaction between the phenomena causing the two competing periodicities. In order to make this clear we will use a very simple model which nevertheless contains the essential features of most modulated structures. Consider a linear array of equidistant sites with period l onto which we want to position a number of identical atoms in such a way that the distribution is as uniform (homogeneous) as possible. Calling Δ the average distance between the atoms, the density, i.e. the number of atoms divided by the number of sites in the same interval, equals $\rho = l/\Delta$. If ρ is irrational the chain of atoms is incommensurate with the array of sites. If ρ is rational it can always be written as the ratio of two relative primes, i.e. $\rho = P/Q$. The atom chain is now commensurate with the array of sites, i.e. periodically an atom will coincide with a site with a period $\Lambda = P\Lambda = Ql$ which is the smallest common multiple of Λ and l. We will now consider two limiting cases: weak interaction and strong interaction.

4.3.2. Continuous modulation

If the interaction between the "substrate" of sites and the atoms in the surface layer is negligible the two periods will subsist simultaneously and the diffraction pattern will be the superposition of the diffraction patterns due to the two periodic structures. We shall term such compounds "doubly periodic" (Figure 4.2a).

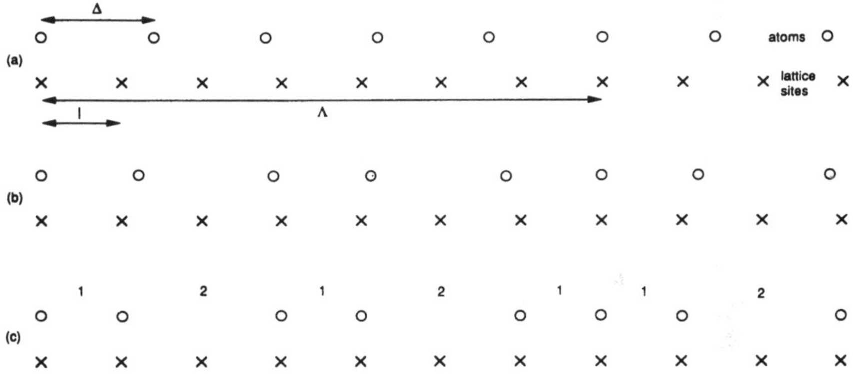

Fig. 4.2 Schematic illustration of different models corresponding with different strengths of the interaction between atoms and lattice sites..
(a) no interaction
(b) weak interaction
(c) strong interactioon

If the interaction is no longer negligible but still weak, the site lattice, which we consider to be rigid for simplicity, will influence the atom positions and will displace the latter towards the positions of minimum energy which we assume to coincide with the site positions. In the commensurate case the displacement pattern has a superperiod Λ In the incommensurate case the modulation quasiperiod will be somewhat different from Λ, i.e. it will be a multiple neither of Λ nor of l; strictly speaking the displacement pattern does not repeat (Figure 4.2b).

4.3.3. Discrete modulation: uniform sequence

In the limit of strong interaction the atoms will occupy the closest lattice positions. This will lead to a uniform distribution of two different interatomic spacings. In the commensurate case the sequence repeats after P atoms (or Q sites). Calling M = $1/\rho$ and N the integer for which N < M < N+1, the spacing between adjacent atoms is either Nl or (N+1)l. In the case shown in Figure 4.2c, M = 1.4 so that the spacings are respectively 1 and 2 (in units l). The sequence is labelled as (1121211212) which corresponds with an alternation of clusters of 2 and 3 adjacent atoms. The total period contains 5 atoms and 7 sites. If M (or ρ is irrational, the sequence consists of an infinitely long repeat period. It is clear that the minority spacing (in this case 2) always occurs isolated.

Uniform sequences can also be generated without an underlying lattice. In general, an infinite sequence of two different spacings, say S and L, is called uniform if the sequence is such that at any point the average spacing is as close as possible to a certain value Δ, say S < Δ < L. S, L and Δ may be mutually irrational. The nodes of the sequence form a so-called one-dimensional quasi-lattice. The idea can be generalized to more dimensions.

If S, L and Δ are given, the whole sequence can be determined unambiguously up to infinity. Hence, although not being periodic, the sequence maintains a long range order, i.e. a correlation up to infinity. As a consequence, the diffraction pattern will show very sharp satellites at positions which are multiples of the inverse average spacing (see § 4.4.2.). Hence the geometries of the diffraction patterns of discretely and continuously modulated structures are eventually the same; they only differ in the relative intensity of the satellites.

Different methods exist to derive the stacking sequence from the values of S, L and Δ. We will discuss two different important methods in more detail.

i) 1D method

Calling m the fraction of L spacings and 1-m the fraction of S spacings, the average spacing is:

$$\Delta = (1-m)S + m\,L \qquad (4.1)$$

from which we deduce:

$$m = (S-\Delta)/(S-L) \qquad (4.2)$$

and thus

$$1 - m = (\Delta - L)/(S - L) \tag{4.3}$$

The total length of a sequence of n spacings is ideally $n\Delta$. The ideal number of L spacings is nm and of S spacings n(1-m). Since both numbers are not necessarily integers we should take their integral parts. As a consequence the end point x_n of the sequence differs from the ideal end point $n\Delta$ by a distance

$$\begin{aligned} n\Delta - x_n &= L\{nm\} + S\{n(1-m)\} \\ &= (L-S)\{nm\} \end{aligned} \tag{4.4}$$

where { } denotes the fractional part, so that finally

$$x_n = n\Delta - (L-S)\{nm\} \tag{4.5}$$

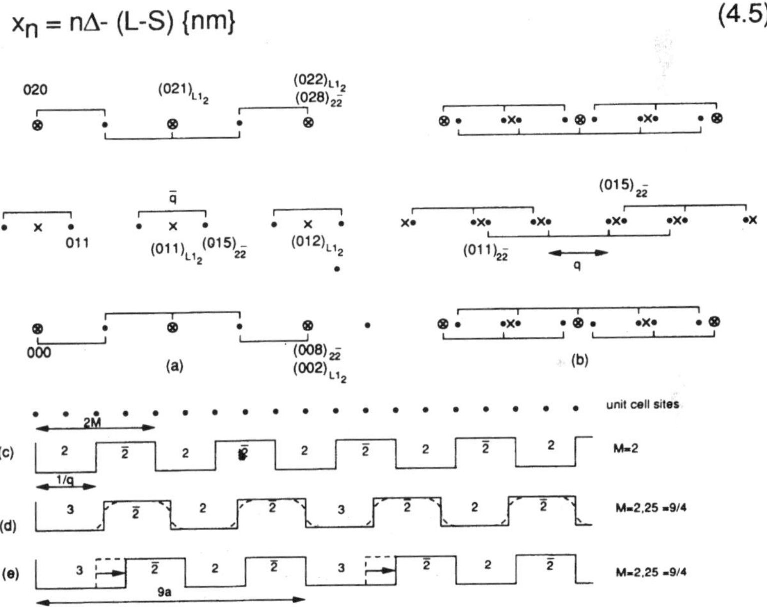

Fig. 4.3.
(a) Schematic representation of the diffraction pattern of the $2\bar{2}$ long period superstructure (M=2) of the $L1_2$ structure. The crosses represent the positions of the $L1_2$ diffraction spots.
(b) Diffraction of the M = 2,25 phase of $Au_{3+}Zn$: the crosses now represent the positions of the $2\bar{2}$ diffraction spots.
(c) Fujiwara construction leading to the $2\bar{2}$ sequence (M = 2).
(d) Fujiwara construction corresponding with M = 9/4. The dotted lines represent the occupation probability at higher temperatures.
(e)Construction simulating the periodic phase slip at <3> bands.

For symmetry reasons S and L may be interchanged and correspondingly m with 1-m without affecting the results.

If L = (N+1)a and S = Na with Na < Δ< (N+1)a the simple graphical square wave algorithm, first used by Fujiwara [5], allows to deduce the stacking sequence corresponding with a given Δ ≡ Ma value. This is represented in Figure 4.3c for the simple case M = 2. The period of the square wave is 2Ma with Ma = 1/q, being the average separation of anti-phase boundaries. The positions of successive layers of $L1_2$ unit cells are represented by dots in Figure 4.3c. Points falling in throughs of the square wave correspond with unit cells which have suffered a displacement **R** = 1/2 $[110]_{FCC}$ with respect to unit cells represented by dots falling outside the throughs. Points coinciding with the left endpoints of throughs are considered to be in the throughs. Undisplaced unit cells are indicated in the stacking symbol by numbers topped with a minus sign. The construction can only be applied to sequences of anti-phase boundaries, i.e. translation interfaces with a displacement vector equal to half a lattice vector. For the simple case of Figure 4.3c one finds obviously the sequence $2\bar{2}$; the corresponding structure is represented in Figure 4.4, together with a high resolution image of the structure of Au_3Zn along the [010] zone.

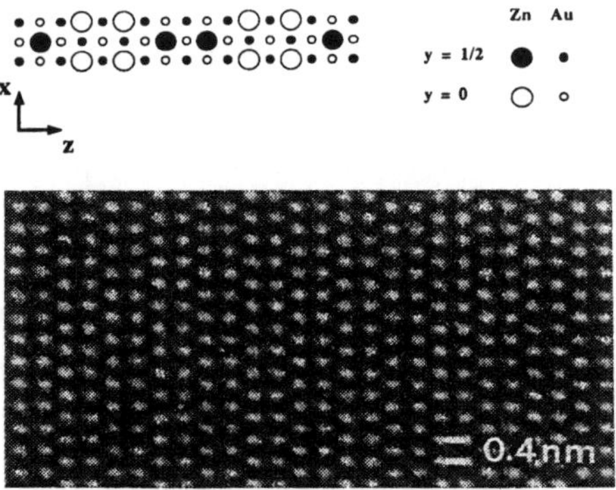

Fig. 4.4 Schematic structure of the $2\bar{2}$ long period structure compared with the high resolution image of $Au_{3+}Zn$ in which the basic columns are imaged as bright dots.

ii) 2D method

Another well-known geometrical method to deduce uniform sequences is the "cut and projection" method (cp method), also called the strip method [6] which proceeds by embedding 1D space into 2D space. Consider a two-dimensional orthogonal lattice (Figure 4.5) based on unit vectors with respective lengths a and b. For simplicity we will assume the lattice to be orthogonal but the method can easily be generalized without changing the final results. Every 1D-sequence of a and b spacings can uniquely be represented in 2D-space as a stepped zigzag line.

Consider now a sequence of n segments containing a mixture of nm b spacings and n(1-m) a spacings. The slope of the line connecting the origin with the end point of the sequence is then

$$\operatorname{tg} \Theta = mb/(1-m)a \tag{4.6}$$

When the sequence has to be uniform, the average spacing at any point has to be as close as possible to the average value d = m b + (1-m)a. Hence the stepped line has to be as close as possible, at any point, to a straight line E passing through the origin, with a slope given by

$$\operatorname{tg} \Theta = mb/(1-m)a = (a-d)b/(d-b)a \tag{4.7}$$

It is easy to see that all the nodes of the sequence derived in this way belong to a strip obtained by sliding one unit cell along and parallel to E (Figure 4.5). Hence the method is sometimes called "strip" method. This strip represents a half closed interval; only points coinciding with one of the limiting lines being acceptable.

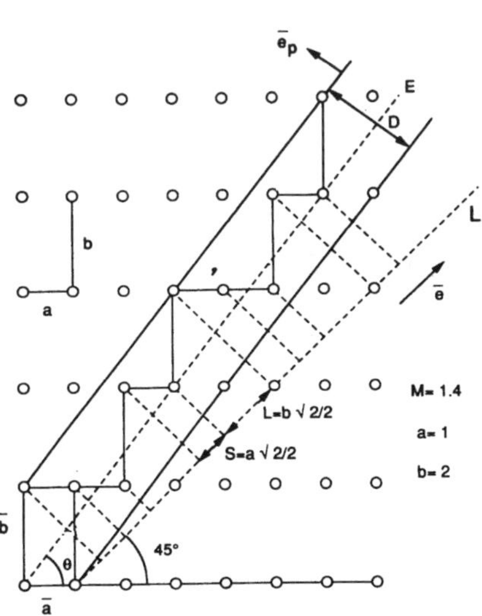

Fig. 4.5 Cut and projection method used to drive uniform sequences of long (L) and short (S) segments (b/a = 2; M =1.4): S S L S L ... .

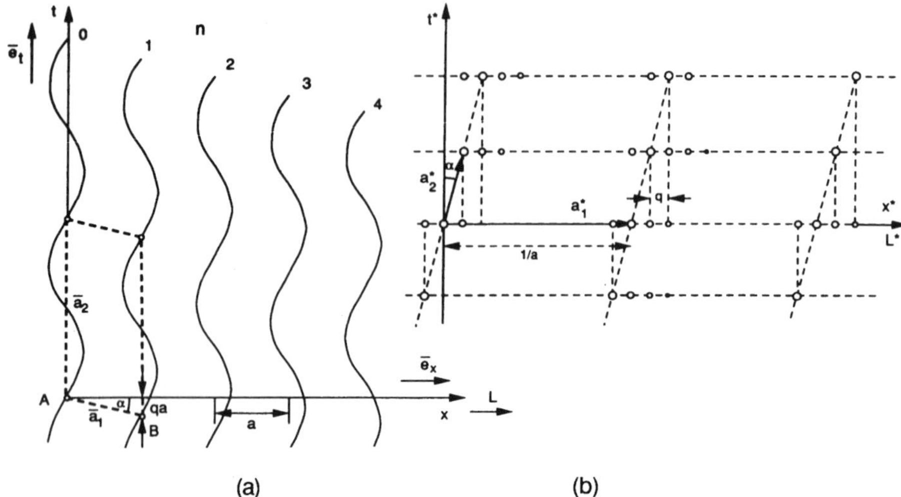

(a) (b)

Fig. 4.6 Two-dimensional representation of a one-dimensional displacively modulated structure.
(a) Derivation of the site locations.
(b) Derivation of the spot positions in the diffraction pattern

The sequence of L = b√2/2 and S = a√2/2 spacings can now be obtained by projecting the zigzag line within the strip (including the top line) onto a line L with a slope of 45°. This particular angle ensures that the ratio of the lengths of the L and S units remains the same as that of b and a through the projection.

Both methods presented here to derive the uniform stacking sequence will be used to calculate the diffraction pattern of uniformly modulated crystals (§ 4.4.2.).

4.3.4. Embedding in a higher dimensional space

The "cut and projection" method is a simple example of "embedding" a non-periodic structure (1D) in a higher dimensional space (2D) in order to make it periodic. This has advantages since the description of periodic structures is simpler than that of non-periodic ones, even at the expense of having to work in a hyperspace with more than three dimensions. In particular Fourier methods and symmetry considerations are much easier in periodic structures. We shall not discuss this procedure in its general form but rather demonstrate its use on the simplest non-trivial example: the 1D deformation modulated non-commensurate crystal. The scattering units are assumed to be located at positions:

$$x_n = na + A \sin(2\pi nqa + \phi) \qquad (4.8)$$

i.e. the structure with lattice parameter a is modulated by means of a longitudinal wave. We parametrize this expression by introducing a continuous dimensionless variable t, which is essentially analogous to a "phase" of the sine wave. This allows to represent the longitudinal wave as a transverse wave in an (x,t) reference system.

$$x_n = na + A \sin[2\pi(qna+t)+\phi] \qquad (4.9)$$

For each value of the integer n this expression represents a sine wave in the (x,t) plane (Figure 4.6a). The intersection of this set of curves with the line $t = 0$ then produces a sequence of points which represent the positions of the scattering units in physical space, i.e. in the modulated structure. Each sine wave with index n+1 is phase shifted along t with respect to the previous one n over a distance $\Delta t = qa$. As a result, points with equal "elongation" or phase such as A and B are connected by a vector with components a and -aq along the reference axis x and t, i.e. by $\mathbf{a}_1 = a\,\mathbf{e}_x - aq\,\mathbf{e}_t$. Thus AB encloses an angle α with the x-axis such that $q = \mathrm{tg}\,\alpha$. The set of curves corresponding with different n-values (Figure 4.6a), but for a given q-value, forms a periodic arrangement in two dimensions with a unit mesh based on the vectors:

$$\mathbf{a}_1 = a\,\mathbf{e}_x - aq\,\mathbf{e}_t\ ;\ \mathbf{a}_2 = \mathbf{e}_t \qquad (4.10)$$

The geometrical meaning of this is as follows. A lattice translation of the basic structure, i.e. changing $n \to n+1$ is <u>not</u> a symmetry translation for the modulated structure since it changes the phase of the modulation by $2\pi qa$ and hence changes the distance between successive scattering units. Such a translation cannot be a symmetry operation for the modulated structure. This phase shift can be compensated by applying simultaneously a translation along t over the same amount, but in the opposite sense, i.e. over -qa. The resulting translation $a\,\mathbf{e}_x - aq\,\mathbf{e}_t$ then leaves the modulation invariant, i.e. superposes two adjacent waves in Figure 4.6a.

The periodic character of the two-dimensional representation also follows directly from the expression for x_n. Applying the translation $\mathbf{a}_1$, i.e. making the substitution $na \to (n+1)a$ and $t \to t-aq$ leaves the argument $2\pi(nqa+t) + \phi$ unchanged, whereas applying the second translation $\mathbf{a}_2$, i.e. making the substitution

$$n \to n;\ t \to t+1$$

increases this argument by 2π and thus also leaves the configuration unchanged. We thus have represented a one-dimensional incommensurately modulated structure as the intersection of a periodic two-dimensional arrangement by a straight line t = 0.

4.4. DIFFRACTION: GENERAL EXPRESSIONS

In the kinematical approximation, the electron diffraction amplitude is given by the Fourier transform of the electrostatic potential of the scattering system. Two different approaches can now be considered, each having different advantages for generalization to modulated structures. Although the derivation of the following expressions is quite straightforward and even common knowledge, we like to rederive them here, since they will serve as the basis for the rest of the work.

4.4.1. Scattering unit approach

In this approach we consider the system as consisting of units i with electrostatic potential $V_i(\mathbf{r})$ located at positions $\mathbf{r}_i$. These units can possibly be individual atoms. We assume that the total potential is the superposition of the potentials of the units:

$$V(\mathbf{r}) = \sum V_i(\mathbf{r}-\mathbf{r}_i) \qquad (4.11)$$

Fourier transforming (4.11) yields for the diffraction amplitude:

$$A(\mathbf{h}) = \int V(\mathbf{r}) \exp(-2\pi i \mathbf{h}.\mathbf{r})\, d\mathbf{r} \qquad (4.12)$$

or

$$A(\mathbf{h}) = \sum_i f_i(\mathbf{h}) \exp(-2\pi i \mathbf{h}.\mathbf{r}_i) \qquad (4.13)$$

with

$$f_i(\mathbf{h}) = \int_{\text{unit } i} V_i(\mathbf{r}) \exp(-2\pi i \mathbf{h}.\mathbf{r})\, d\mathbf{r} \qquad (4.14)$$

The interpretation of (4.13) is straightforward. Each unit contributes to the scattering in the direction $\mathbf{h}$ with a scattering amplitude, or scattering factor $f_i(\mathbf{h})$. The phase factor $\exp(-2\pi i \mathbf{h}.\mathbf{r}_i)$ accounts for the phase shift due to the position $\mathbf{r}_i$ of the unit.

Several units can be grouped into a larger unit; the scattering amplitude of the larger unit (from (4.13)) is then given by

$$F_i(\mathbf{h}) = \sum_j f_j(\mathbf{h}) \exp(-2\pi i \mathbf{h}.\mathbf{R}_j) \tag{4.15}$$

where the summation extends over all subunits, located at relative positions $\mathbf{R}_j$, within the unit i.

If all units are identical, i.e. $f_i(\mathbf{h}) = f(\mathbf{h})$, (4.13) becomes

$$A(\mathbf{h}) = f(\mathbf{h}) \sum_i \exp(-2\pi i \mathbf{h}.\mathbf{r}_i) \tag{4.16}$$

In a perfect crystal, the unit cell is taken as scattering unit. The summation then extends over the position vectors $\mathbf{r}_j$ of the unit cells, i.e. the lattice vectors of the crystal so that the diffraction amplitude A(**h**) yields delta-like maxima at the nodes **g** of the reciprocal lattice. The amplitude is proportional to f(**g**) and the intensity to $|f(\mathbf{g})|^2$. In quasi-crystals the scattering units may for instance be McKay polyhedra [7].

When a system contains several types of units, each type undergoing a different modulation, it is possible to subdivide the system into subsystems each containing only one type of unit. Each subsystem can then be described independently by an expression such as (4.16). The diffraction amplitude is then the superposition of the corresponding diffraction amplitudes (4.16). This might apply when different atoms in the unit cell undergo a different modulation.

In § 4.5 Eq. (4.16) will be used as a starting point for introducing modulation by assuming the unmodulated system to consist of identical scattering units. Two ways are conceivable to introduce modulation in (4.16). Either the scattering units f(**h**) are modified throughout the system while keeping the positions $\mathbf{R}_j$ unchanged or by altering the positions as in a frozen-in vibrational mode while keeping the scattering units rigid. A combination of both is also possible. Most of the modulated structures can be treated in either of both ways, although usually one is more appropriate than the other.

4.4.2. Continuum approach

In the continuum approach we start from a perfect crystal in which the electrostatic potential has the translation periodicity and can thus be written as a Fourier series:

$$V(\mathbf{r}) = \sum_{\mathbf{g}} V_g \exp(2\pi i \mathbf{g}.\mathbf{r}) \tag{4.17}$$

where the summation extends over the reciprocal lattice vectors **g** of the perfect crystal. We will consistently denote the set of reciprocal lattice vectors of the unmodulated crystal by {**g**}.

The Fourier coefficients, closely related to the structure amplitudes, are obtained by Fourier transforming the potential of one unit cell yielding

$$V_{\mathbf{g}} = \frac{1}{V} \int_{\text{unit i}} V(\mathbf{r}) \exp(-2\pi i \mathbf{g}.\mathbf{r}) \, dr \qquad (4.18)$$

$$= \frac{1}{V} \sum_j f_j(\mathbf{g}) \exp(-2\pi i \mathbf{g}.\mathbf{R}_j) \qquad (4.19)$$

with V the volume of the unit cell and the summation extending over the atoms of the unit cell with positions $\mathbf{R}_j$ and scattering factors $f_j(\mathbf{g})$. From (4.15) and (4.19) it is clear that

$$V_{\mathbf{g}} = F(\mathbf{g})/V \qquad (4.20)$$

The diffraction amplitude per unit cell can now be obtained by Fourier transforming (4.17) using (4.12) yielding

$$A(\mathbf{h}) = \sum_{\mathbf{g}} V_{\mathbf{g}} \, \delta(\mathbf{g}-\mathbf{h}) \qquad (4.21)$$

with δ the Dirac functional.

It is thus clear, in agreement with (4.16), that the scattering is confined to the reciprocal nodes with the amplitude proportional to $V_{\mathbf{g}}$ or $F(\mathbf{g})$ and the intensity proportional to $|V_{\mathbf{g}}|^2$ or $|F(\mathbf{g})|^2$.

In summary: the Fourier transform of the structure represented by the lattice potential, is the reciprocal lattice of delta functions, weighted by the Fourier coefficients of the lattice potential, i.e. by the structure factors.

In the following sections we will use either (4.13) or (4.17) as the starting point for introducing modulation; (4.13) is more appropriate for describing discrete modulation whereas (4.17) is more appropriate for continuous modulation. It should be noticed that these expressions remain valid for X-ray diffraction provided the electrostatic potential V(**r**) is replaced by the electron density ρ(**r**).

Furthermore, it is generally believed that the kinematical approximation for electron diffraction is only valid for very thin objects. This is true if one considers individual atoms as scattering units. However, rigorously speaking, the electron diffraction amplitude is the Fourier transform of the wave function at the exit face of the crystal. Due to the forward scattering of high energy electrons ($\geq$100 keV) the wave function at the exit face still shows a local character even for the "thicker" objects used in transmission electron microscopy. This is the basis for the column approximation which has been successfully used for the study of defects [8]. More recently it has been shown (e.g. ref. [9]) that, if a crystal is viewed along a zone axis, i.e., along the atom columns, the wave function at the exit face of a column is still typical for that column irrespective of its neighbouring columns provided the intercolumn distance is not too small. The reciprocal lattice is then the corresponding two-dimensional zone. Hence the expressions (4.13) or (4.17) remain valid for electron diffraction in thicker objects provided the potential is replaced by the wave function at the exit face so that the scattering factor of the units (4.14) is replaced by a "dynamical" structure factor.

4.5. DIFFRACTION BY MODULATED STRUCTURES

4.5.1. Modulation of the scattering units

We start from (4.16) where it is assumed that the scattering units vary as a function of the position $\mathbf{r}_i$ while keeping their positions unchanged as in composition modulation. If this variation is periodic in $\mathbf{r}_i$, we can expand f($\mathbf{h}$) in a Fourier series

$$f(\mathbf{h},\mathbf{r}_i) = \sum_{\mathbf{G}} a_{\mathbf{G}}(\mathbf{h}) \exp(2\pi i \mathbf{G}.\mathbf{r}_i) \qquad (4.22)$$

where {**G**} represents the set of reciprocal lattice vectors corresponding with the periodicity of the modulation wave. This periodicity may be one-, two- or three-dimensional and is usually larger than that of the unmodulated structure. Henceforth we will consistently denote the reciprocal lattice vectors of the modulation by {**G**}. It has to be noted that the modulation function (4.22) only takes values at the positions $\mathbf{r}_i$ (i.e. discrete modulation). Substitution of (4.22) into (4.16) now yields for the diffraction amplitude

$$A(\mathbf{h}) = \sum_{\mathbf{G}} a_{\mathbf{G}}(\mathbf{h}) \sum_{i} \exp[-2\pi i(\mathbf{h}-\mathbf{G}).\mathbf{r}_i)] \qquad (4.23)$$

In case the unmodulated system is a perfect crystal, the second factor is summed over the lattice vectors of the perfect crystal yielding delta-like maxima at the nodes **g** of the reciprocal lattice hence.

$$\mathbf{h} - \mathbf{G} = \mathbf{g} \text{ or } \mathbf{h} = \mathbf{G} + \mathbf{g} \tag{4.24}$$

i.e. each Bragg reflection **g** of the unmodulated crystal gives rise to an array of satellites with vectors **G**. All satellites can be characterised by two reciprocal lattice vectors **g**,**G**, respectively corresponding with the basic crystal and with the modulation. They can thus be labelled by 4, 5 or 6 indices depending on whether the modulation is one-, two- or three-dimensional. The amplitude of the satellite at **G** is then given by $a_G(\mathbf{h})$ and the intensity by $|a_G(\mathbf{h})|^2$. The detailed calculation of $a_G(\mathbf{h})$ requires a detailed model for the modulation wave (4.22).

In principle nearly all types of modulated structures originating from a perfect crystal can be described in this way. However the success of this approach depends largely on the number of Fourier coefficients required to describe (4.22) adequately and on the ease to calculate these coefficients. In case of a sinusoidal modulation of f(**h**), as for instance in the initial stage of spinodal decomposition, only two coefficients are needed in the expansion (4.22) so that in this case the method is very appropriate. Each basic reflection acquires two satellites, one on each side, so-called "side-bands".

Continuous deformation modulation can also be described by a real-space approach (§ 4.5.5) which in some cases may be more appropriate.

4.5.2. Rigid displacement of scattering units

In case the scattering unit itself remains unchanged but its position is displaced from $\mathbf{r}_i$ to $\mathbf{r}_i+\boldsymbol{\delta}_i$, the diffraction amplitude (4.16) becomes

$$A(\mathbf{h}) = f(\mathbf{h}) \sum_i \exp -2\pi i \mathbf{h}(\mathbf{r}_i+\boldsymbol{\delta}_i)$$

$$= f(\mathbf{h}) \Sigma_i \exp(-2\pi i \mathbf{h}.\boldsymbol{\delta}_i) \exp(-2\pi i \mathbf{h}.\mathbf{r}_i) \tag{4.25}$$

Since $f(\mathbf{h}) \exp(-2\pi \mathbf{h}.\boldsymbol{\delta}_i)$ can be interpreted as a modified scattering unit, still located at $\mathbf{r}_i$, it is clear that this description can be considered as a special case of the foregoing one where the displacement $\boldsymbol{\delta}_i$ is a periodic function of $\mathbf{r}_i$; one can then expand

$$\exp(-2\pi\mathbf{h}.\boldsymbol{\delta}_i)=\sum_{\mathbf{G}} a_{\mathbf{G}}(\mathbf{h}) \exp(+2\pi i\mathbf{G}.\mathbf{r}_i) \quad (4.26)$$

remembering that this function takes only values at the positions $\mathbf{r}_i$ (i.e. discrete modulation). Substitution into (4.25) leads to similar results as (4.23): however several satellites now appear at the reciprocal lattice positions given by (4.24). In principle, the problem of calculating the diffraction amplitude is reduced to the problem of calculating the Fourier coefficients $a_{\mathbf{G}}(\mathbf{h})$. This can be particularly simple in case the displacements $\boldsymbol{\delta}_i$ are sufficiently small. The exponential in (4.26) can then be expanded up to first order as

$$\exp(-2\pi i\mathbf{h}.\boldsymbol{\delta}_i) = 1 - 2\pi\mathbf{h}.\boldsymbol{\delta}_i$$

This approximation holds if the displacement obeys: $|\delta_i| \ll 1/|\mathbf{h}|$

The displacement itself being periodic, it can be expanded in a Fourier series

$$\delta_i =\sum_{\mathbf{G}} \mathbf{A}_{\mathbf{G}} \exp(2\pi i\mathbf{G}.\mathbf{r}_i) \quad (4.27)$$

where the Fourier coefficients are now vectors. Within this approximation and comparing (4.26) with (4.27) one obtains

$$a_{\mathbf{G}}(\mathbf{h}) = -2\pi i\mathbf{h}.\mathbf{A}_{\mathbf{G}} \quad (4.28)$$

and $a_{\mathbf{G}}(\mathbf{h}) = 1$ for $\mathbf{G} = 0$

From this it is clear that extinction of the satellites occurs when the polarisation vector $\mathbf{A}_{\mathbf{G}}$ is perpendicular to the diffraction vector $\mathbf{h}$. Different harmonics (Fourier terms) may have different polarisation vectors leading to different extinction conditions. Note that the satellites associated with the origin are lost in this approximation.

Calling $I_{\mathbf{h},o}$ the intensity of the basic Bragg reflection and $I_{\mathbf{h},\mathbf{G}}$ the intensity of the satellite **G** associated with this Bragg reflection, one has:

$$I_{\mathbf{h}\mathbf{G}} / I_{\mathbf{h},o} = 4\pi^2|\mathbf{h}.\mathbf{A}_{\mathbf{G}}|^2$$

i.e. the relative intensity of a satellite **G** is proportional to the square of the diffraction vector of the basic reflections: i.e. satellites belonging to higher order basic reflections will be relatively stronger. The absolute value of this intensity is proportional to $|f(\mathbf{h})|^2$. These features are characteristic for deformation modulation with small displacements. (In case the modulation (4.27) is sinusoidal only two Fourier coefficients $\mathbf{A}_G$ persist in the approximation (4.28) and only two satellites are predicted at each Bragg reflection.)

As a concrete example we shall exactly treat the case of a sinusoidal displacement wave with wave vector **q**. The positions of the scattering units can then be represented by

$$\mathbf{r}_i + \mathbf{A} \sin 2\pi\mathbf{q}.\mathbf{r}_i$$

where the $\mathbf{r}_i$ represent the nodes of the undeformed lattice of scattering units with $f(\mathbf{h}) \equiv 1$; the amplitude of the diffracted beam then becomes

$$A(\mathbf{h}) = \sum_i \exp 2\pi i\mathbf{h}.(\mathbf{r}_i + \mathbf{A} \sin 2\pi\mathbf{q}.\mathbf{r}_i)$$

Making use of the Jacobi-Anger relation one finds

$$A(\mathbf{h}) = \sum_i \exp 2\pi i\mathbf{h}.\mathbf{r}_i \sum_{n=-\infty}^{+\infty} \exp(in\, 2\pi\mathbf{q}.\mathbf{r}_i)\, J_n(2\pi\mathbf{h}.\mathbf{A})$$

where J_n represents the Bessel functions of the first kind of order n (n = integer)

$$A(\mathbf{h}) = \sum_{i,n} J_n(2\pi\mathbf{h}.\mathbf{A}) \ \exp[2\pi i(\mathbf{h}+n\mathbf{q}).\mathbf{r}_i]$$

The lattice sum over i is zero unless $\mathbf{h}+n\mathbf{q}$ is a reciprocal lattice vector **g**, hence reflections are formed at positions $\mathbf{h} = \mathbf{g}-n\mathbf{q}$. Introducing m = -n we can write $\mathbf{h} = \mathbf{g}+m\mathbf{q}$ and

$$A(\mathbf{g}+m\mathbf{q}) = J_{-m}[2\pi(\mathbf{g}+m\mathbf{q}).\mathbf{A}] \qquad (4.29)$$

For small values of n the Bessel functions of order n can be approximated by $J_n(x) = x^n/2^n n!$ and thus

$$A(\mathbf{g}+m\mathbf{q}) = [2\pi(\mathbf{g}+m\mathbf{q}).\mathbf{A}]^m/2^m m!$$

If $|\mathbf{q}.\mathbf{A}| \ll |\mathbf{g}.\mathbf{A}|$ which is often the case,the ratio of the intensities of two successive reflections is given by

$$I_{m+1}/I_m = \pi^2 |\mathbf{g}.\mathbf{A}|^2/(m+1)^2 \qquad (4.30)$$

The intensities in a sequence of satellites decrease rapidly with increasing order m of the satellite for a given **g**-vector.

It is well known that the intensity of diffraction spots is temperature dependent through the Debye-Waller factor exp(-2W) with $W = 1/3\, g^2 \langle u^2 \rangle$ (g = diffraction vector of the considered reflection; $\langle u^2 \rangle$ = mean-square displacement of the atoms from their equilibrium position). Incommensurate modulated structures exhibit specific gapless elementary excitations, called "phasons", corresponding with the sliding of modulation waves with respect to the basic structure [10]. Phasons reduce the intensity of satellite reflections as compared with the intensity of satellites associated with a static modulation wave. The phasons contribute a g-independent term to W.

The modulation waves may further interact with pinning points and as a result loose partly their periodic character, which causes additional broadening and weakening of the satellite spots. As a consequence of these phenomena the results of Eqs. (4.29) and (4.30) are only approximate.

It should further be noted that this intensity ratio depends on the form of the modulation wave; only at temperatures close to T_C is the modulation sinusoidal, as was assumed in deriving (4.30). With decreasing temperature the wave becomes more square. Such a gradual change in wave shape with temperature can be interpreted as a continuous transition from deformation modulation at high temperature to anti-phase boundary modulation at low temperature. The positions of the anti-phase boundaries being restricted to well defined lattice planes separating certain atom layers in the unit cell their spacing is in general an integer number of lattice parameters of the basic structure, i.e. the spacing is either constant or it is a mixture of two commensurate spacings forming a uniform sequence. The low temperature "locked-in" structure is therefore very often interface modulated to a good approximation. If the period remains constant with temperature the transition does not affect the geometry of the diffraction pattern, but it does change the relative intensities of the satellites: the higher order satellites acquire relatively larger intensities as the wave becomes square.

4.5.3. Quasi-periodic stacking of scattering units

A special but important system which requires extra attention is the one-dimensional quasi-lattice constructed of segments S and L as described in § 4.3.2, the nodes of which are decorated with identical scattering units.

The more general case of a uniform stacking of two types of units with different widths, such as in mixed layer compounds can also be reduced to this case. Furthermore the results can be generalized to two- or three-dimensional quasicrystals. In that case the scattering units are tiles respectively polyhedra.

The case in which the scattering units are translation interfaces, in particular out-of-phase boundaries, has to be treated differently since the insertion of a translation interface translates part of the crystal with respect to the rest, i.e. the notion of average structure looses its meaning.

We will now restrict ourselves to a one-dimensional quasi-lattice of identical scattering units f(**h**), separated by segments S and L and an average spacing Δ, in the sense as described in § 4.3.2. In this case the diffraction amplitude (4.16) becomes:

$$A(\mathbf{h}) = f(\mathbf{h}) \sum_{n} \exp(2\pi i\mathbf{h}.x_n) \qquad (4.31)$$

we will now derive the diffraction amplitude using two different methods, based on the two different ways in which we have deduced in § 4.3.2 the uniform stacking sequence, i.e. the positions x_n given by (4.5).

<u>i) 1D method</u>

From (4.5) the position of the nth scattering unit is

$$x_n = n\Delta - (L-S)\{nm\} \qquad (4.5)$$

with

$$m = (S-\Delta)/(S-L)$$

so that (4.26) becomes

$$A(h) = f(h) \sum_{n} \exp(2\pi ih.n\Delta) \exp[-2\pi ih.(L-S)\{nm\}] \qquad (4.32)$$

The function {zm} is periodic in z with periodicity 1/m. Hence the function exp $2\pi i\alpha\{zm\}$ with α = h(L-S) is also periodic and can be expanded in a Fourier series, yielding:

$$\exp -2\pi i\alpha\{zm\} = \sum_p \exp(-2\pi ipzm)\exp[i\pi p+\alpha)]\sin[\pi(p+\alpha)]/\pi p+\alpha)] \quad (4.33)$$

we thus finally obtain, using (4.33) with z ≡ n and (4.32) for the diffraction amplitude

$$A(h) = f(h) \sum_p \exp[i\pi(p+\alpha)] \sin[\pi(p+\alpha)]/\pi(p+\alpha) \sum_n \exp[-2\pi in(h\Delta+pm)] \quad (4.34)$$

$$\text{with } \alpha = h(L-S) \quad (4.35)$$

The sum over n produces a sequence of δ-functions, i.e. satellites, at positions given by $h\Delta + pm = l$(integer) (4.36)

i.e. at positions

$$h = (l/\Delta) - p/(\Delta m) \quad (4.37)$$

or using the expression for $m = (S-\Delta)/(S-L)$

$$h = (l/\Delta) - [p(S-\Delta)/\Delta(S-L)] \quad (4.38)$$

The satellites are thus labelled by two indices l and p, one corresponding with a real space periodicity Δ and the other with a spacing Δ/m. The amplitude of the satellites is, apart from a phase factor, given by

$$a(h) = f(h) \sin\pi u/\pi u \quad (4.39)$$

with from (4.34) and (4.35)

$$u = p + h(L-S)$$

or using (4.38)

$$u = p(S/\Delta) + l(L-S)/\Delta \quad (4.40)$$

The geometry of the diffraction pattern described by (4.38) is essentially the same as for the previous cases § 4.5.1 and § 4.5.2. This can be seen as follows: l/Δ (l = integer) forms a one-dimensional reciprocal lattice corresponding with the modulation repeat Δ. Let us denote these one-dimensional reciprocal lattice vectors by **g**; (p/Δ) (S-Δ)/(S-L) (p = integer) represents the one-dimensional reciprocal lattice corresponding with the spacing $\Delta(S-L)/(S-\Delta)$; let us denote these vectors by **G**. Then (4.38) becomes

$$\mathbf{h} = \mathbf{g}+\mathbf{G} \tag{4.41}$$

In case a common sublattice with periodicity l exists for S and L so that $S = N_1\, l$ and $L = N_2\, l$ (N_1, N_2 integers) as in the simplified example of § 4.3.2 or in most uniform sequences occurring in real modulated crystals, it is easy to show that the repeat is equivalent to l so that **g** then corresponds with the Bragg positions of the original crystal and (4.41) can be considered as the 1D analog of (4.24). Extension to higher dimensions is obvious.

It is thus clear that, from the geometry of the diffraction pattern alone, it is impossible to discriminate between the various models used thus far. Only the intensity of the satellites allows to distinguish the models. It may seem surprising that very sharp satellites appear while the stacking of scattering units is not strictly periodic. This is due to the fact that the sequence, albeit not periodic, has long range order (i.e. correlation) up to infinity (§ 4.3.2). The same argument also explains the existence of sharp diffraction spots in incommensurate 2D or 3D quasicrystals.

ii) 2D method

In order to derive the diffraction pattern from the 2D "cut and projection" method, we follow the idea outlined in § 4.3. The uniform sequence is obtained by constructing a two-dimensional orthogonal lattice with unit vectors a and b, cutting a strip by moving the unit cell parallel to the direction given by (4.7) and projecting the nodes within the strip onto a line L with a slope of 45° (Figure 4.7a).
Hence the diffraction pattern can be obtained by Fourier transforming these successive operations. First we draw the reciprocal orthogonal lattice with lattice parameters 1/a and 1/b. Then the selection of a strip with width D can be represented by multiplying the original lattice with a window function which is 1 within the strip and 0 outside the strip. The Fourier transform of this operation is a convolution product of the reciprocal lattice with the Fourier transform of the window function. This means that every node of the reciprocal lattice is replaced by a line perpendicular to the strip and that the amplitude along the line changes with distance U from the node following

$$\sin \pi UD/\pi UD \tag{4.42}$$

Now the final operation of projecting the strip onto the line L corresponds in reciprocal space with an intersection with the line L* (projection theorem) (Figure 4.7b). Scattering then occurs at the intersection points of the line L* with the projection lines perpendicular to the slit direction through the reciprocal lattice nodes. The amplitude of the satellite is given by the structure amplitude associated with the node multiplied with (4.42) in which U is the distance between the node and the line L* measured along the projection line.

From this picture it follows that each satellite spot can be associated with one or several nodes of the 2D-reciprocal lattice. If the **q**-value is irrational the set of projected points on L* is dense but only nodes within a distance of the order 1/D from the line L* will produce an observable spot. If the **q**-value is rational a number of nodes will project in the same point on L*. The density of spot positions is then finite and a commensurate diffraction pattern results. The procedure discussed so far is only applicable to a 1D sequence of unit scatterers. The relative magnitudes of the scattered amplitudes derived by this procedure are nevertheless in qualitative agreement with the experimental results for materials in which the "scattering power" is uniformly distributed within the blocks. Below we extend the method to "decorated" 1D sequences. We will now show that the positions and amplitudes derived in this way are essentially the same as those derived in (4.38), respectively (4.39).

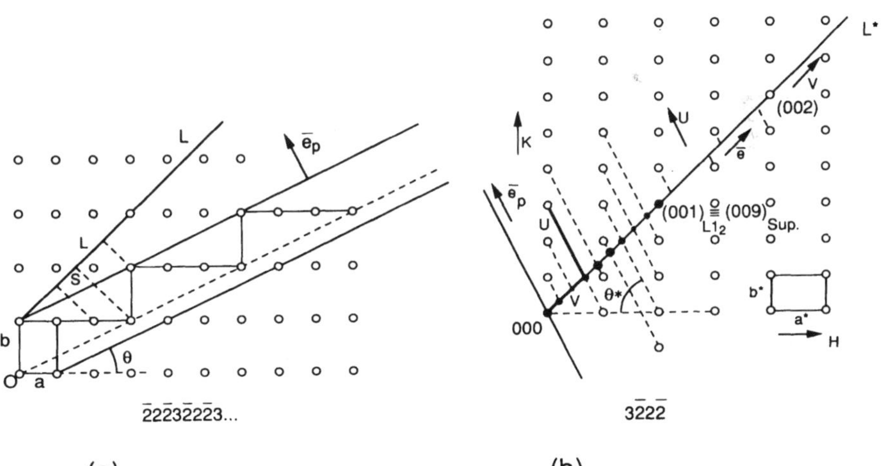

Fig. 4.7. Cut and projection method: derivation of the diffraction pattern corresponding with the sequence $2\bar{2}2\bar{3}$ b/a = 3/2; M = 9/4

(a) Derivation of the sequence.

(b) Derivation of the central row of the diffraction pattern.

The position vector of a reciprocal lattice node with indices H,K (Figure 4.7b) is given by

$$H\mathbf{a}^* + K\mathbf{b}^*$$

We can also take the unit vectors **e**, $\mathbf{e}_p$ respectively parallel to L* and perpendicular to the slit. Calling V and U the coordinates of the node in this coordinate system, i.e. measured along L* and perpendicular to the slit we have

$$H\mathbf{a}^* + K\mathbf{b}^* = V\,\mathbf{e} + U\,\mathbf{e}_p \tag{4.43}$$

By projecting (4.43) onto **a** and **b** and after straightforward but lengthy calculations one obtains

$$V = [(H-K)/(S-L)] + (KS-HL)/\Delta(S-L) \tag{4.44}$$

and

$$U = (KS-HL)/\Delta D \tag{4.45}$$

where D is the strip width. The position V of the satellite in the 1D diffraction pattern is given by (4.44). Its amplitude, from (4.42), is given by

$$\sin \pi UD/\pi UD$$

with U obtained from (4.45). Calling H = l and K = l-p and UD = u, expressions (4.44) and (4.45) become identical with the expressions (4.38) and (4.39) respectively derived with the 1D method.

However, thus far we did not decorate the 2D lattice with scattering units. This can be done as follows: We have to decorate each node of the lattice with a scattering unit which, in projection onto L (the z-axis) should give the required 1D scattering unit of the uniform sequence: The scattering unit in 2D may never be partly in and out of the strip, otherwise the projection would depend on the exact position of the strip, which is unacceptable on physical grounds. This can only be realised by decorating each lattice node with the required 1D scattering unit, taken parallel to the strip. In a sense, the lattice is convoluted with the 1D scattering unit and subsequently limited by the slit. Fourier transforming then corresponds with the product of the whole reciprocal space with the Fourier transform of the scattering unit.

Since the scattering unit is one-dimensional and represented by a segment parallel to L its Fourier transform depends only on the distance u along the line L and is independent on the distance v perpendicular to L. Hence the amplitude (4.42) of the satellites, of which the positions V (4.44) along L* are given by the intersection with the lines through the reciprocal nodes, has to be multiplied with f(**h**), the scattering factor of the scattering unit. This is in complete agreement with the result (4.39) obtained by the 1D method.

This result also follows directly from the fact that the reciprocal lattice has to be weighted with the scattering factor of the scattering unit, associated with the corresponding node.

The method of embedding a 1D deformation modulation structure in 2D space (§ 4.3.4) allows also to deduce graphically the diffraction pattern in a qualitative manner. It is sufficient to translate into reciprocal space the construction used to obtain the sequence of scattering centers. The base vectors of the 2D reciprocal lattice, as defined by $\mathbf{a}_i . \mathbf{a}_j = \delta_{ij}$ (i,j = 1,2) are given by:

$$\mathbf{a}_1^* = (1/a)\mathbf{e}_x^* = g\,\mathbf{e}_x^* \;;\; \mathbf{a}_2^* = q\,\mathbf{e}_x^* + \mathbf{e}_x^*$$

where the unit vectors $\mathbf{e}_x^*$ and $\mathbf{e}_t^*$ are such that

$$\mathbf{e}_x . \mathbf{e}_x^* = \mathbf{e}_t . \mathbf{e}_t^* = 1 \text{ and } \mathbf{e}_x . \mathbf{e}_t^* = \mathbf{e}_x^* . \mathbf{e}_t = 0$$

The operation of intersection with the line L in physical space corresponds with projecting the reciprocal lattice nodes along a direction perpendicular to L* onto L*. The basic spots are situated on L* whereas the satellites are obtained as projections of the other reciprocal lattice nodes onto L*. The intensities of the satellites decrease with increasing distance from the line L* (Figure 4.6b).

4.5.4. Short range ordered stacking of scattering units

Sometimes the scattering units are stacked with a mixture of spacings exhibiting only short range order in which the distance between adjacent units is only given in probabilistic terms. The satellites are then again located approximately at multiples of the inverse average spacing but the intensity of the (higher order) satellites of increasing order decreases drastically with decreasing degree of atomic order whereas their width increases. Often only two broadened satellites are observed, one on each side of the basic reflection of the sublattice.

4.5.5. Continuous deformation: deformation field approach [11]

A deformation modulation can be modelled by means of a continuous vector field in real space, the displacement field **R**(**r**). It expresses the fact that the electrostatic potential (or charge density in the case of X-ray diffraction) at the point **r** in the deformed crystal is the same as that of the point **r**-**R**(**r**) in the undeformed crystal. Contrary to the approach in § 4.5.2 this approximation assumes the atoms or the units to be deformable. The vector field **R**(**r**) can then be written as a continuous function of **r**.

One thus has from (4.17)

$$V_{def}(\mathbf{r}) = V(\mathbf{r}-\mathbf{R}(\mathbf{r})) = \sum_{\mathbf{g}} V_{\mathbf{g}} \exp 2\pi i\mathbf{g}.\mathbf{r} \exp[-2\pi i\mathbf{g}.\mathbf{R}(\mathbf{r})] \qquad (4.46)$$

In case the deformation field **R**(**r**) is periodic the last factor in (4.46) is also periodic and can be expanded in a Fourier series (similar to (4.22) and (4.26)

$$\exp[-2\pi i\mathbf{g}.\mathbf{R}(\mathbf{r})] = \sum_{\mathbf{G}} a_{\mathbf{G},\mathbf{g}} \exp(2\pi i\mathbf{G}.\mathbf{r}) \qquad (4.47)$$

with {**G**} the reciprocal lattice vector corresponding with the periodicity of the deformation modulation. Note that although the lattice potential V(**r**), as well as the deformation field **R**(**r**) are both periodic, in general with different periods though, the final structure obtained by applying this deformation field to the basic structure will in general be non-periodic. Also the average positions of homologous atoms remain the same as in the basic structure after modulation by **R**(**r**). The periodicity of **R**(**r**) may be one-, two- or three-dimensional and the "unit cell" is usually larger than that of the basic crystal. Expression (4.47) is very similar to (4.26) but, whereas in the latter the modulation takes only values in discrete positions $\mathbf{r}_j$ (discrete modulation), it is continuous in this case. Substitution of (4.47) into (4.46) now yields:

$$V_{def}(\mathbf{r}) = \sum_{\mathbf{g},\mathbf{G}} V_{\mathbf{g}} a_{\mathbf{G},\mathbf{g}} \exp[2\pi i(\mathbf{g}+\mathbf{G}).\mathbf{r}] \qquad 4.48)$$

The scattering amplitude is given by the Fourier transform of (4.48) yielding

$$A(\mathbf{h}) = \sum_{\mathbf{g},\mathbf{G}} V_{\mathbf{g}}\, a_{\mathbf{G},\mathbf{g}}\, \delta(\mathbf{g}+\mathbf{G} - \mathbf{h}) \tag{4.49}$$

The diffraction pattern then exhibits peaks at the position $\mathbf{h} = \mathbf{g}+\mathbf{G}$, i.e. the same as derived in the previous cases (§ 4.5.1, § 4.5.2, § 4.5.3). Each Bragg reflection $\mathbf{g}$ of the unmodulated crystal gives rise to an array of satellites defined by vectors $\{\mathbf{G}\}$ of the reciprocal lattice of the modulation. The amplitude of the satellite at $\mathbf{g}+\mathbf{G}$ is in this case given by $V_{\mathbf{g}}\, a_{\mathbf{G},\mathbf{g}}$ and the intensity by $|V_{\mathbf{g}}\, a_{\mathbf{G},\mathbf{g}}|^2$. This result is similar to that of § 4.5.2. Also in this case the determination of the Fourier coefficients $a_{\mathbf{G},\mathbf{g}}$ is much simplified if the deformation is small, i.e. $\mathbf{R}(\mathbf{r}) \ll 1/|\mathbf{g}|$. The exponential in (4.47) can then be expanded up to the first order:

$$\exp[-2\pi i\mathbf{g}.\mathbf{R}(\mathbf{r})] = 1 - 2\pi i\mathbf{g}.\mathbf{R}(\mathbf{r}) \tag{4.50}$$

Expanding the displacement field in the Fourier series

$$\mathbf{R}(\mathbf{r}) = \sum_{\mathbf{G}} \bar{A}_{G} \exp(2\pi i\mathbf{G}.\mathbf{r}) \tag{4.51}$$

where the coefficients $\bar{A}_G$ are now polarization vectors. Substitution of (4.51) in (4.50) and comparing with (4.47) yields

$$\begin{aligned} a_{\mathbf{g},\mathbf{G}} &= -2\pi i\mathbf{g}.\bar{A}_G \\ a_{\mathbf{g},0} &= 1 \qquad \text{for } \mathbf{G} = 0 \end{aligned} \tag{4.52}$$

These results can directly be compared with (4.28) so that also the conclusions remain the same. Extinction occurs for the satellites for which $\mathbf{g}.\bar{A}_G = 0$. The absolute value of the amplitude is proportional to $V_{\mathbf{g}}$, to the scattering amplitude of the Bragg reflection, and to the length $|\mathbf{g}|$ of the reciprocal lattice vector. These results are characteristic for deformation modulation. In case the modulation is sinusoidal, only two satellites per Bragg reflection will be sufficiently intense to be observed.

The use of Parseval's theorem allows to prove that the sum of the intensities of the array of satellites, including that of the residual

Bragg reflection is equal to the intensity of the corresponding Bragg spot of the unmodulated structure. Parseval's theorem states that if the Fourier transform of f(**r**) is F(**h**) we have:

$$\int_{-\infty}^{+\infty} |F(\mathbf{h})|^2 d\mathbf{h} = \int_{-\infty}^{+\infty} |f(\mathbf{r})|^2 d\mathbf{r}$$

We apply the theorem to the function exp(-2πi**g**.**R**), The Fourier transform, using its Fourier expansion (4.47) is given by

$$\sum_{\mathbf{G}} a_{\mathbf{g+G}} \, \delta(\mathbf{G}-\mathbf{h}).$$

For the node **g** it represents the array of satellites {G} around **g**. The left hand side of Parseval's relation thus becomes

$$\sum_{\mathbf{G}} |a_{g+G}|^2$$

Since exp(-2π**g**.**R**) is periodic in physical space with the unit cell of **R**(**r**) the right hand side is given by

$$(1/V)_v \int |\exp(-2\pi i \mathbf{g}.\mathbf{R})|^2 \, dr \equiv 1$$

and does not depend on **g** the modulus of the integrandum being always unity. One thus obtains for a given **g**:

$$\sum_{G} I_{\mathbf{g+G}} = \sum_{G} |V_g a_{g+G}|^2 = |V_g|^2 \sum_{G} |a_{g+G}|^2 = I_{\mathbf{g}} \tag{4.53}$$

4.5.6. Translation interface modulated structures [12]

We assume the basic structure to be modulated i.e. divided in "modules" by a set of planar interfaces normal to **e** and separated by a constant spacing Δ; the displacement vector associated with the interfaces being $\mathbf{R}_O$ (Figure 4.8a). The wavevector of the modulation is then **q** = (1/Δ)**e**. If **u**(**r**) represents the displacement of the atom at **r**, this function has the property

$$\mathbf{u}(\mathbf{r} + N\,\Delta\,\mathbf{e}) = \mathbf{u}(\mathbf{r}) + N\,\mathbf{R}_O \tag{4.54}$$

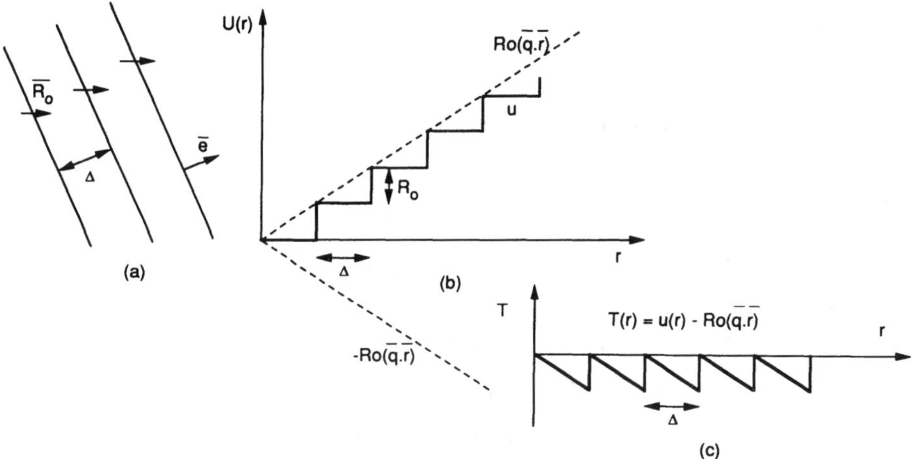

Fig. 4.8. Translation interface modulated structure.
(a) Model and notations.
(b) Displacement field $\mathbf{u}(\mathbf{r})$.
(c) Periodic function $\mathbf{T}(\mathbf{r})$.

(N = integer), which follows from the cumulative nature of the displacements $\mathbf{R}_o$. We can furthermore define a function (Figure 4.8b,c)

$$\mathbf{T}(\mathbf{r}) = \mathbf{u}(\mathbf{r}) - (\mathbf{q}.\mathbf{r})\, \mathbf{R}_o \qquad (4.55)$$

It is easy to show that:

$$\mathbf{T}(\mathbf{r} + \Delta\, \mathbf{e}) = \mathbf{T}(\mathbf{r}) \qquad (4.56)$$

provided (4.54) is satisfied; this shows that $\mathbf{T}(\mathbf{r})$ is periodic with the same period Δ along $\mathbf{e}$ (Figure 4.8c). The function $\exp 2\pi i \mathbf{h}.\mathbf{T}(\mathbf{r})$ being periodic with the same period as $\mathbf{T}(\mathbf{r})$ itself, it can be expanded in a Fourier series with the same $\mathbf{q}$ vector:

$$\exp 2\pi i \mathbf{h}.\mathbf{T}(\mathbf{r}) = \sum_m A_m(\mathbf{h}) \exp (2\pi i m \mathbf{q}.\mathbf{r}) \qquad (4.57)$$

Using (4.13) the amplitude of the diffracted beam can be obtained by summing over the atoms at $\mathbf{r}_n$ with scattering factor f_n

$$A(\mathbf{h}) = \sum_n f_n \exp 2\pi i \mathbf{h}.[\mathbf{r}_n + \mathbf{u}(\mathbf{r}_n)]$$

Using (4.55) and (4.57) leads to

$$A(\mathbf{h}) = \sum_n \sum_m f_n A_m(\mathbf{h}) \exp 2\pi i\{[\mathbf{h} + (\mathbf{h}.\mathbf{R}_o)\mathbf{q} + m\mathbf{q}].\mathbf{r}_n\} \qquad (4.58)$$

This function exhibits peaks at reciprocal space positions given by **H**:

$$\mathbf{H} = \mathbf{h} + (m + \mathbf{h}.\mathbf{R}_o)\mathbf{q} \qquad (4.59)$$

i.e. the superstructure spots **H** associated with **h** and corresponding with successive values of the integer m form equidistant linear arrays of satellites in the direction of **q**. These arrays are shifted with respect to the basic spots **h** over a fraction $\mathbf{h}.\mathbf{R}_o$ of the interspot distance **q**.

In many cases **q** is simply related to the unit cell of the basic structure; the arrays of satellites associated with the different basic reflections belong to a common reciprocal lattice.

Let $\mathbf{h} = \sum \mathbf{h}_j \mathbf{b}_j$, where the $\mathbf{h}_j$ are integers and the $\mathbf{b}_j$ are the reciprocal lattice base vectors, we can then rewrite **H** as a vector with four components

$$\mathbf{H} = \sum \mathbf{h}_j \mathbf{B}_j + m\mathbf{q} \qquad (4.60)$$

where $\mathbf{B}_j = \mathbf{b}_j + (\mathbf{b}_j.\mathbf{R}_o)\mathbf{q}$.

The vectors $\mathbf{B}_j$ together with **q** form the basis of a four-dimensional reciprocal space, which is often used in the interpretation of X-ray diffraction patterns. The generalization of the expression (4.59) to the case of 3D modulation is evident. Each basic spot is now surrounded by a 3D array of satellites, which is in general fractionally shifted.

The diffraction effects due to such a translation interface modulated or shear structure can semi-quantitatively be understood on the basis of general diffraction principles. In physical space the shear structure is the convolution product of the superlattice generated by the shearing operation, with a slab of basic structure. The diffraction pattern is then the Fourier transform of this product. In Fourier space this convolution product becomes the ordinary product of the two Fourier transforms i.e. the product of the reciprocal lattice of the shear structure with the Fourier transform of a slab of basic structure. This last transform can be obtained as follows: In physical space a slab of basic structure is the ordinary product of the basic structure with the window-function limiting the slab. The transform of such a slab thus becomes the convolution product of the reciprocal lattice of the basic structure with the shape transform of the window. This transform thus consists of the weighted reciprocal lattice nodes due to the basic structure convoluted with the function sin $\pi u\Delta/\pi u\Delta$ along the direction normal to the shear planes and with a δ–function along the direction parallel to the shear planes and with the basic node as the origin (Figure 4.9b). The amplitude to be associated with a superlattice reflection is thus determined by the value of the Fourier transform at the considered reciprocal space position: it then depends on the amplitude associated with the closest basic reflection and through the shape transform on the distance u from this basic reflection. A graphical representation of these different operations is given in Figure 4.9.

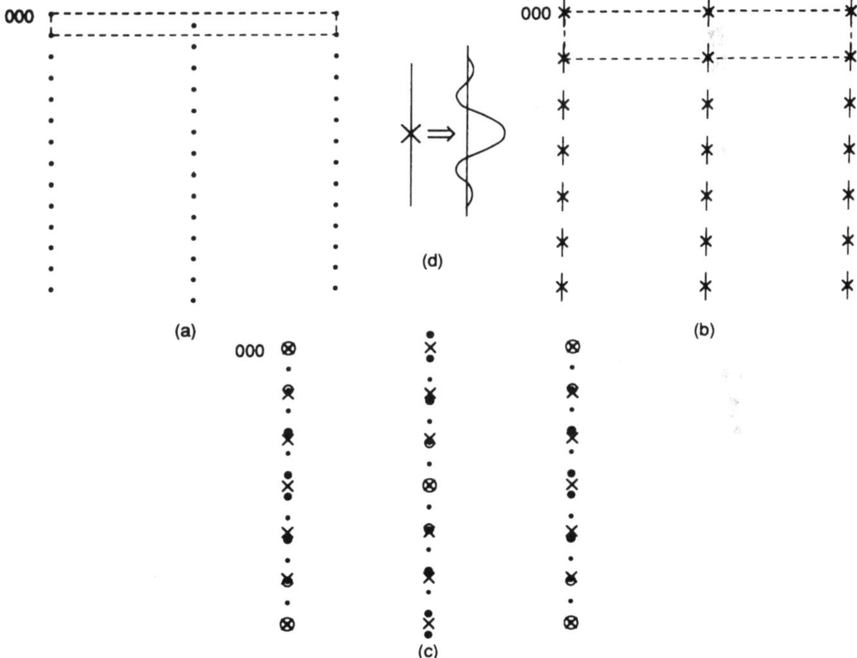

Fig. 4.9.
(a) Reciprocal lattice sites of shear structure.
(b) The crosses indicate the reciprocal lattice sites due to the basic structure. With each nodepoint an amplitude distribution given by the Fourier transform of a slab (shown in d) has to be associated.
(c) Diffraction pattern of the shear structure. The size of the dots reflects the relative intensities of the diffraction spots.

4.5.7. Translation interfaces considered as scattering units [13]

The insertion of a translation interface in a crystal displaces one part of the crystal with respect to the adjacent one over the displacement vector $\mathbf{R}_0$. Hence a translation interface cannot be considered as an isolated scattering unit and necessitates a somewhat more sophisticated approach.

Let us consider a crystal containing a sequence of parallel translation interfaces, all with the same displacement vector $\mathbf{R}_0$. We call **e** the unit vector perpendicular to the interface and x_n the coordinate of the n-th interface along **e**. We can now use the deformation field approach of § 4.5.5. In between the n-th and n+1-th interface the deformation field is given by

$$\mathbf{R}(\mathbf{r}) = n\,\mathbf{R}_0 + \delta(\mathbf{r}) \qquad (4.61)$$

where $\delta(\mathbf{r})$ accounts for relaxation deformation in the vicinity of the interface [13]. The diffraction amplitude is now obtained by Fourier transforming (4.46). This calculation is rather tedious especially when the stacking of the interfaces is not periodic. We will limit ourselves here to summarize the results and refer to [13] for the details of the calculation.

In the diffraction pattern, each Bragg reflection of the perfect crystal acquires either a streak of diffuse intensity, or an array of satellites, parallel to **e**. The diffraction amplitude along the streak, as a function of the distance **h** to the Bragg reflection **g** is given by

$$A_g(\mathbf{h}) = V_{\mathbf{g}}\, f(\mathbf{g},\mathbf{h})\, p(\mathbf{g},\mathbf{h}) \qquad (4.62)$$

where $V_{\mathbf{g}}$ is the structure amplitude of the perfect crystal and $f(\mathbf{g},\mathbf{h})$ is a generalized scattering factor for the translation interface including relaxation effects. The general expression can be found in [13]. In case of an "ideal" translation interface with displacement vector $\mathbf{R}_0$ and without relaxation, this scattering factor becomes

$$f(\mathbf{g},h) = \frac{1}{2\pi i h}\,[1 - \exp(2\pi i \mathbf{g}.\mathbf{R}_0)] \qquad (4.63)$$

From this it is clear that the streaks or satellites disappear for Bragg reflections for which $\mathbf{g}.\mathbf{R}_0$ = integer

The second factor is

$$p(\mathbf{g},h) = \sum_n \exp(2\pi i n \mathbf{g}.\mathbf{R}_0) \exp(-2\pi i h.x_n) \tag{4.64}$$

where the summation extends over all interfaces. This is a geometrical factor which contains the information about the stacking sequence of the interfaces and which describes the scattering amplitude along the streak **e**. $n\mathbf{g}.\mathbf{R}_0$ accounts for the cumulative displacement of the n-th domain. For a periodic stacking of interfaces with repeat distance Δ one has $x_n = n\Delta$ so that

$$p(\mathbf{g},h) = \sum_n \exp[2\pi i n(\mathbf{g}.\mathbf{R}_0 - h\Delta)] \tag{4.65}$$

It is clear that satellites occur at the positions for which $\mathbf{g}.\mathbf{R}_0 - h\Delta = m$ or

$$\mathbf{h} = \frac{1}{\Delta} [m - \mathbf{g}.\mathbf{R}_0]\mathbf{e} \qquad \text{with } m = \text{integer} \tag{4.66}$$

which is identical with the result derived in § 4.5.6.

Each Bragg reflection **g** of the basic crystal is replaced by a one-dimensional array of satellites, parallel to **e** and with spacing $1/\Delta$. The satellite array thus corresponds with the reciprocal lattice of the modulation, similar to the previous types of modulation. However, contrary to the previous cases, the array of satellites is shifted with respect to the position of the original Bragg reflection **g** over a distance.$-\mathbf{g}.\mathbf{R}_0/\Delta$ caused by the cumulative displacement. Also the original Bragg reflection has disappeared. The amplitude of the satellites is given by (4.62). The results (4.63) and (4.66) have been derived originally for describing the diffraction effects of periodic crystallographic shear planes [14]. The results have since been successfully used to determine the displacement vector $\mathbf{R}_0$ of shear planes and out-of-phase boundaries, by measuring the shift $\mathbf{g}.\mathbf{R}_0$ of the satellites for various **g**. In case of a uniform sequence of translation interfaces, with distances S and L, the values of x_n in (4.64) are given by (4.5). Hence (4.64) becomes

$$p(\mathbf{g},h) = \sum_n \exp[-2\pi i n(h\Delta\mathbf{g}.\mathbf{R}_0)] \exp 2\pi i[h(L-S)\{nm\}] \tag{4.67}$$

comparing (4.64) with (4.32) it is immediately clear that the results of § 4.5.3 remain valid provided h is replaced by $h-\mathbf{g}.\mathbf{R}_0/\Delta$. From (4.38) it then follows that the satellite positions are given by

$$h = (l/\Delta) - [p(S-\Delta)]/[\Delta(S-L)] + (\mathbf{g}.\mathbf{R}_0)/\Delta \qquad (4.68)$$

i.e. also for uniform sequences the results remain the same but the satellites are "fractionally" displaced over $-\mathbf{g}.\mathbf{R}_0/\Delta$ The same holds for the short range ordered stacking of translation interfaces (§ 4.5.4).

4.5.8. Summary and conclusions

In all cases, the periodic modulation of a perfect crystal causes each Bragg reflection to be replaced by an array of satellites. The array corresponds with the reciprocal lattice of the modulation, which may be one-, two- or three-dimensional. According to Parsefal's theorem the sum of the intensities of the satellites added to that of the remaining Braggpeak, is equal to that of the original Bragg reflection so that, in a sense, the intensity is redistributed over the satellites.

Two major different cases can obviously be distinguished. If the origin of the satellite sequence coincides with the position of the Bragg reflection, the average position of a given atom, or the average scattering unit, remains the same as that in the unmodulated structure. This is usually the case in most types of modulated crystals.

The position of the satellites is then given by

$$\mathbf{h} = \mathbf{g}+\mathbf{G} \qquad (4.24)$$

with $\mathbf{g}$ the reciprocal vector of the original lattice and $\mathbf{G}$ the reciprocal vector of the modulation.

If the origin of the satellite array is displaced with respect to the position of the original Bragg reflection, the shift is due to a cumulative displacement caused by periodic planar translation interfaces with repeat distance Δ. This is the case for translation interface modulation. Here from (4.66) the position of the satellites is given by

$$\mathbf{h} = \mathbf{g}+\mathbf{G}-(\mathbf{g}.\mathbf{R}_0)\mathbf{e}/\Delta \qquad (4.69)$$

with $\mathbf{e}$ the normal to the interfaces. From the geometry of the diffraction pattern alone, it is then relatively easy to deduce the periodicity of the modulation and, in case of translation interface modulation to deduce the displacement vector $\mathbf{R}_0$ of the interfaces as well as their orientation.

If disorder is present in the modulation periodicity, the satellites broaden so that their intensity is smeared out over a larger area which causes the higher order satellites to disappear. The correlation length of the order can be deduced from the inverse of the satellite width. However, from the geometry of the diffraction pattern alone, one cannot discriminate further between the different types of modulation, even not between continuous and discrete modulation (uniform sequence).

Differences between the various types of modulation are revealed in the relative intensity of the satellites whereas the symmetry of the modulation field is reflected in the extinctions of the satellites. In fact, the diffraction amplitude of a satellite contains the information about the corresponding Fourier component of the modulation wave. In order to discriminate between different models, one has to calculate the intensity of the satellites using (4.22), (4.27), (4.47), (4.51), depending on the model used, and then compare these with the experimental intensities. In some cases, qualitative conclusions can be drawn without detailed calculations. For instance, if the intensity of the satellites, relative to the intensity of the Bragg reflection increases with $|\mathbf{g}|$, it is indicative for a deformation modulation. Since the satellites are absent for those reflections for which the reciprocal vector is perpendicular to the polarization vector of the wave, this polarization vector can be deduced from the extinction conditions [11]. Since, from (4.24) and (4.41) each satellite site can be reached by a sum of reciprocal lattice vectors of the original lattice and of the modulation, each satellite can be labelled by four, five or six indices, depending on whether the modulation is one-, two- or three-dimensional. Hence it is possible to describe the modulated structure in a four-, five- or six-dimensional space, and describe the symmetry operations using a four-, five- or six-dimensional space group. For more details we refer to [15].

4.6. SPECIAL CASES

4.6.1. Simultaneous modulation

In case a crystal contains different sublattices of atoms, each subject to a different modulation, it is clear, as has been stated in § 4.4.1, that each of the sublattices can be treated independently and the diffraction amplitude is the superposition of the corresponding amplitudes. In case the satellites do not overlap, the intensity in the diffraction pattern is also the superposition of the respective diffraction patterns. In order to deduce information about the modulations, one then has to disentangle the corresponding diffraction pattern and to analyze each of them separately. Overlapping satellites can complicate this analysis.

A crystal may also be modulated simultaneously by two independent modulations, acting on the same atoms. It is then usually possible to describe the total modulation as a succession of the two modulations, i.e. the first modulation creates a superstructure upon which then acts the second modulation. The diffraction pattern can then be described by a two-step process. The first modulation creates satellites which can be considered as new Bragg reflections for the second modulation. For instance, two-dimensional satellite arrays of a two-dimensional deformation modulated structure can be considered as being generated by two successive 1D deformation modulations. The first modulation creates a one-dimensional satellite array and in the second modulation each satellite is in its turn replaced by a one-dimensional array in the other direction so as to give rise to a two-dimensional array. If both modulations are of the same type (e.g. deformation) the sequence of the modulations is commutative. In case one deformation is a translation interface modulation, the two-dimensional satellite array is shifted in the corresponding direction, i.e. perpendicular to the interfaces. If both modulations are translation interface modulations the total displacement of the satellite array is the vector sum of both respective vector displacements. Also in this case the result is independent of the order in which the operations are performed.

4.6.2. Discommensurations

A number of deformation modulated structures undergo a phase transformation which can be schematized by the following sequence

high temperature average structure	--->	intermediate temperature incommensurately modulated structure	--->	low temperature commensurately modulated structure

The lower temperature phase is then called the locked-in phase. At the higher temperature the modulation is close to harmonic and it is convenient to describe the modulation, starting from the average structure, using the approach adopted in § 4.5. However, starting from the low temperature end it may be more convenient to describe the modulation as a two-step process. First a commensurate deformation modulation is introduced, giving rise to the locked-in superstructure. Secondly, in this superstructure, translation interfaces are introduced to give rise to the final modulated structure. These interfaces are the so-called discommensurations or discommensuration walls [1]. The discommensuration causes a local "phase slip" in the locked-in structure and is sometimes called a phase soliton. If the sequence of discommensurations is uniform, the resulting modulated structure is incommensurate. In this way the system finds a compromise between the elastic energy, which is lowest in the locked-in phase and the electronic energy which is lowered when the modulation wavevector obeys the Fermi nesting conditions (§ 4.1).

At low temperature, the discommensurations may become sharp and acquire the character of planar translation interfaces. Upon increasing temperature the discommensurations broaden and the phase slip becomes more gradual, so that the modulation changes from discrete to continuous. At high temperature the modulation is completely continuous. The transition is often reversible. From this it is clear that at high temperature, the classical approach, which has been used mostly thus far, and in which the modulation is described starting from the average crystal, may be more appropriate. If the deformation wave is close to harmonic, only a few Fourier coefficients are sufficient. However, at low temperature where the discommensurations are sharp, this approach requires an unnecessarily large number of Fourier components whereas the two-step description is a more adequate starting point. This method may

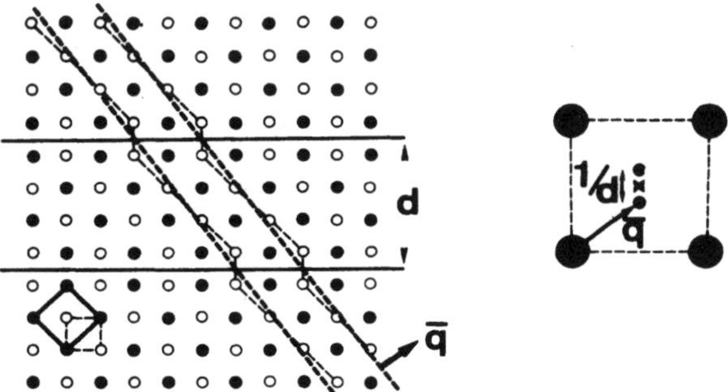

Fig. 4.10. 2D-model illustrating the relation between two descriptions of the same structure. The dotted lines represent the average orientation of the lattice planes in a displacement modulation interpretation. The full lines refer to an interface modulated interpretation of the same diffraction pattern.

be used even in case the locked-in structure is not observed as a stable phase. This will be illustrated by means of a very simple two-dimensional model.

The basic structure has a primitive square unit cell containing only two atoms. The basic spots as well as the satellites are shown in Figure 4.10b. If the superstructure reflections are described as satellites of the basic spots, using a modulation wave vector **q** (Figure 4.10a), two aspects are overlooked. 1) The satellites are located close to the centre of the square, which can be considered as the reciprocal node of a hypothetical structure (locked-in or ordered structure). This hypothetical superstructure contains two different atoms; the unit cell is shown in Figure 4.10a, the line through the centre of the square joining two satellites has a simple crystallographic direction. It is more appropriate to consider the satellites as resulting from the hypothetical structure, in which translation interfaces (discommensurations) are introduced perpendicular to the satellite row, with a spacing which is the inverse of the satellite spacing. The method has been successfully applied to the study of the "discommensurate" structures of $NbTe_4$ [16] and CuV_2S_4 [17]. Thus far the method has only been used in electron diffraction. However, X-ray structure determinations based on the two step description, using (4.64) and (4.66) including relaxation at the discommensurations, may possibly be more useful for the determination of the detailed structure of the discommensurations.

4.6.3. Mixed layer compounds

Consider a one-dimensional stacking of two types of blocks A and B with widths L and S and respective scattering factors F_L and F_S. In practice the blocks may be two-dimensional slabs as in most mixed layer compounds, or even twin related lamellae as in polysynthetically-twinned structures. We assume the stacking to be uniform, i.e. such that the average width is at any point as close as possible to a fixed value, say Δ. Let us again call m the fraction of L blocks and (1-m) the fraction of S blocks, one has, as in § 4.3.3

$$\Delta = m\,L + (1-m)\,S$$

The difference with the treatment in § 4.3.3 is that we now have two types of stacking units instead of one as well as two different spacings L and S. However the problem can be reduced to that of § 4.3.3 in the following way. As is clear from Figure 4.11, the uniform stacking Figure (4.11a) can be decomposed into the superposition of a uniform stacking of identical scattering units F_L with spacings L and S and average spacing Δ (Figure 4.11b (or d)) and a uniform stacking of units F_L-F_s with spacing L and L+S and an average repeat Δ' (Figure 4.11c or 4.11e).

$$\Delta' = \Delta/m \qquad (4.69)$$

Each of these uniform sequences can be treated using the approach introduced in § 4.4 and the final diffraction amplitude is the superposition of the two respective diffraction amplitudes (§ 4.5.3). It should be noted that, when the blocks are sufficiently large as usually is the case in mixed layer compounds or polysynthetically twinned structures, the scattering factors F_L and F_S lead to maxima in reciprocal space at the reciprocal positions of the pure L, respectively S structures.

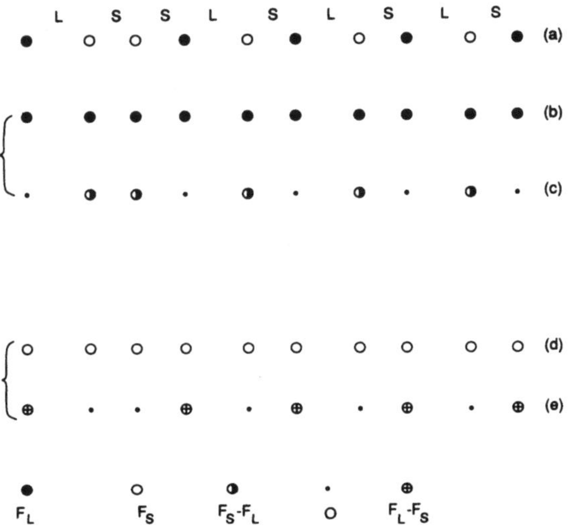

Fig. 4.11. Schematic representation of a 1D mixed layer compound consisting of two different scattering units with different spacings.
(a) Actual structure.
(b) + (c) Decomposition of (a) in two sequences each consisting of identical scattering units.
(d) + (e) Alternative decomposition of (a).

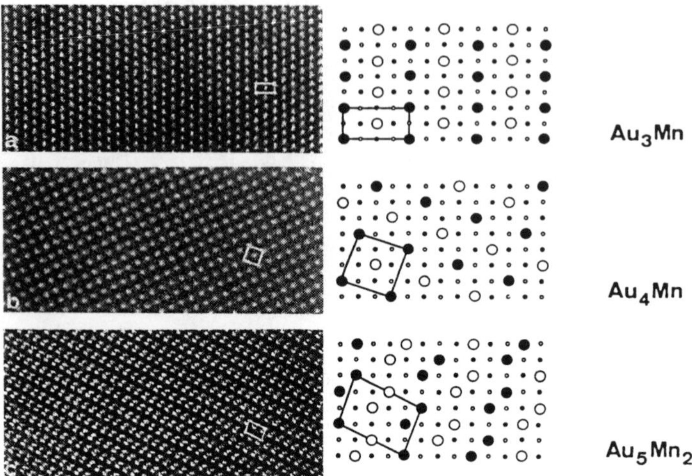

Fig. 4.12. Simple ordered structures in the Au-Mn system.
(a) Au_3Mn (b) Au_4Mn (c) Au_5Mn_2
The left figures are high resolution images revealing the minority atom columns as bright dots.

4.7. CASE STUDIES

4.7.1. Examples of interface modulated structures in alloys

i) The Au-Mn system [18,19]

A few examples illustrating the different aspects of the theory will now be analysed. The gold rich part of the phase diagram of the gold-manganese system is particularly proliferic in phases (Figure 4.12). We shall discuss two composition driven derivatives of the Au_4Mn phase, which has a superstructure based on a face centered cubic lattice, represented in Figure 4.12b. A monodomain

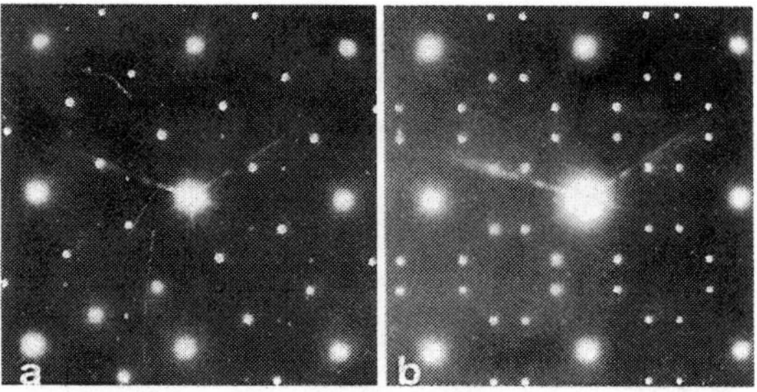

Fig. 4.13 Diffraction patterns of the Au_4Mn structure along the [001] zone.
(a) Single variant.
(b) Two variants with a common c-axis.

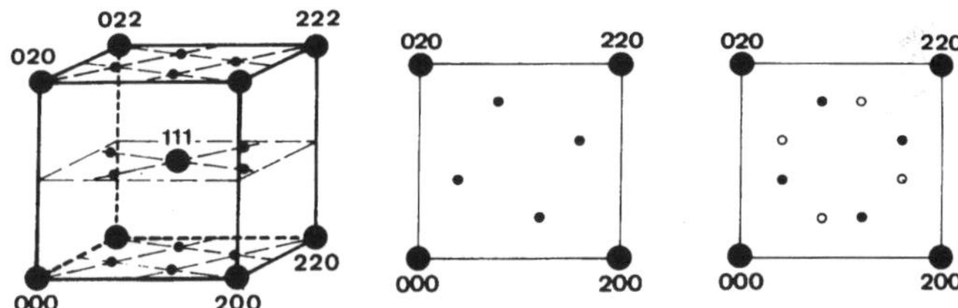

Fig. 4.14. Reciprocal lattice of the Au_4Mn structure.
(a)Spatial view: one variant. (left)
(b)(001)* Section: one variant. (middle)
(c)(001)* Section: two variants. (right)

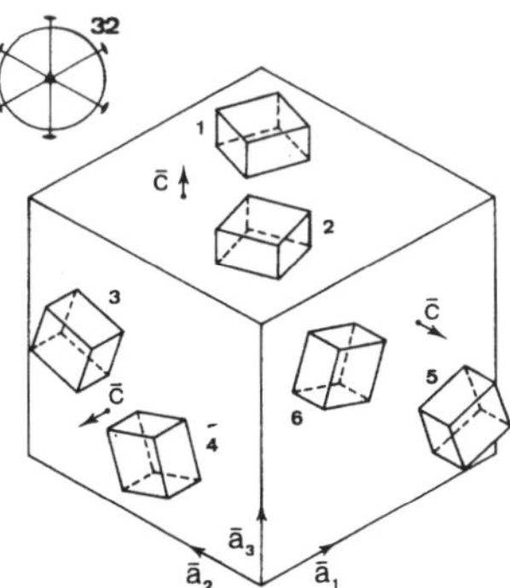

Fig. 4.15. Six orientation variants of the Au_4Mn structure represented in their relation to the basic FCC structure. The variants are related by the symmetry operations of the point group 32.

Fig. 4.16. High resolution image of 1D long period superstructure of Au_4Mn.

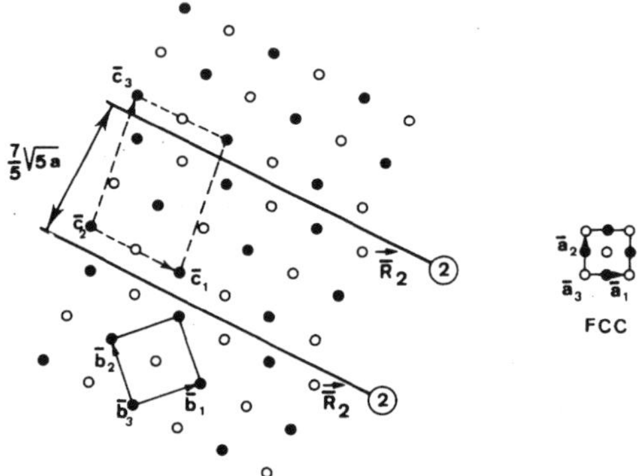

Fig. 4.17. Model for the 1D long period superstructure of Figure 4.16. Only the manganese columns are shown. The monoclinic unit cell is outlined.

diffraction pattern is shown in Figure 4.13 which can be compared with the reciprocal lattice of Figure 4.14. The corresponding high resolution image is shown in Figure 4.12b; it exhibits the positions of the columns of minority Mn-atoms as bright dots. With respect to the face centered cubic sublattice this superstructure can occur in six orientation variants (Figure 4.15). As a result selected area diffraction patterns are mostly composite patterns resulting from the overlap of several monodomain patterns; an example is shown in Figure 4.13b where two orientation variants with a common fourfold axis are present.

In alloys containing a slight excess of manganese a one-dimensional monoclinic long period superstructure is formed as shown in the high resolution image of Figure 4.16. This phase, with theoretical composition $Au_{22}Mn_6$, has a structure derived from the Au_4Mn structure by introducing a periodic array of non-conservative out-of-phase boundaries, represented schematically in Figure 4.17. The composite diffraction pattern of Figure 4.18a is produced by an area of the specimen containing the one-dimensional long period structure as well as the basic Au_4Mn structure. It is therefore possible to observe directly the "fractional" shifts of the sequences of superstructure spots. In Figure 4.18b the superstructure spots are indicated by full dots and the basic spots by open dots. The fractional shifts **g**.**R** are multiples of 1/5. We now show that the geometry of this diffraction pattern contains all the information necessary to determine the superstructure. The direction of the dot sequence is perpendicular to the set of parallel interfaces, the spacing of the full dots being the inverse of the average interface separation. The fractional shifts give the projections on different **g**-vectors of the displacement vector. Let **R**[u,v,w] be the displacement vector we then must have:

$$[110].[u,v,w] = u + v = 1/5 \text{ (mod 1)}$$
$$[200].[u,v,w] = 2u = 2/5 \text{ (mod 1)}$$

leading to u = 1/5 and v = 0. The w component cannot be determined from this single section of reciprocal space but the consideration that the shortest displacement vector in face centered based alloys is of the type 1/2 $[110]_{FCC}$ leads to the conclusion w = 1/2, i.e. **R** = 1/10 [205].

In general it is not possible to produce a composite diffraction pattern containing the pattern of the basic structure as well as that of the superstructure. However, even in such cases, it is still possible to locate the spot positions of the basic structure provided its lattice is known; the fractional shifts can then still be obtained.

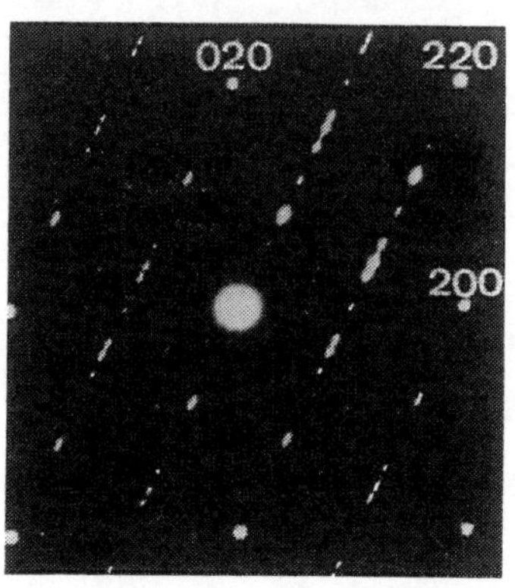

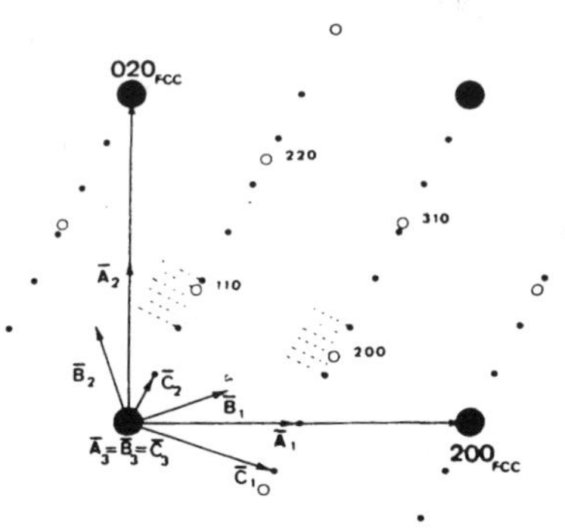

Fig. 4.18. Diffraction pattern of the 1D superstructure of figure 4.17.
(a) [001] zone diffraction pattern. (left)
(b) Schematic representation of the diffraction pattern. (right)

For somewhat larger manganese concentrations, the gold-manganese alloys exhibit a tetragonal two-dimensional interface modulated superstructure of which Figure 4.19 shows a resolution image, the bright dots representing again manganese columns. Figure 4.20 is a model of the structure as deduced from the combined use of the image and of the diffraction pattern, shown in Figure 4.21. The selected area contained also an area of the basic Au_4Mn structure producing the spots surrounded by white circles in Figure 4.22. Note that the most intense superstructure spots are those closest to basic spots. The fractional shifts in the two mutually perpendicular directions can thus be deduced directly from the diffraction pattern. The two families of interfaces can be analyzed independently. The interfaces are perpendicular to the linear sequences of spots, their spacing being the inverse of the spot spacing. The component of the fractional shift along the considered spot row must be used in deducing the displacement vector of the family of interfaces perpendicular to the considered spot row. The observed and calculated fractional shifts for the two families of

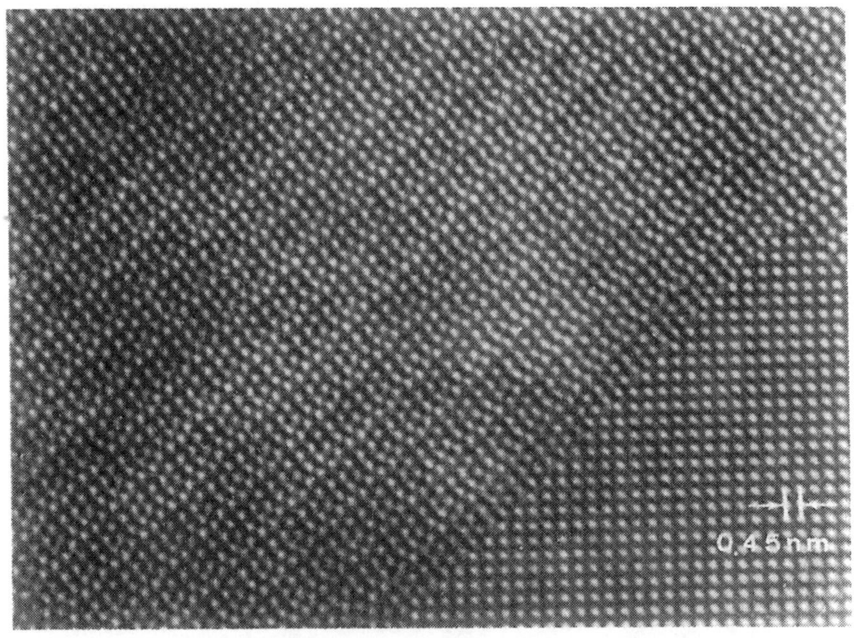

Fig. 4.19. High resolution image of tetragonal 2D long period superstructure of Au_4Mn. An area of basic structure is visible as well.

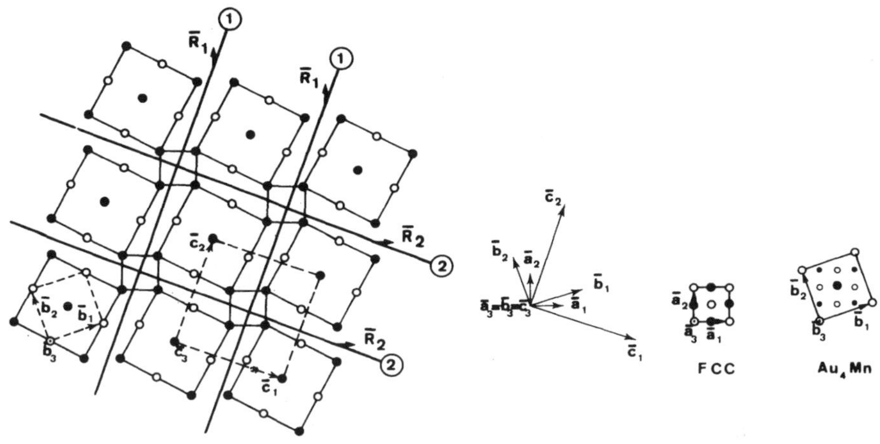

Fig. 4.20. Model of the structure imaged in Figure 4.19.

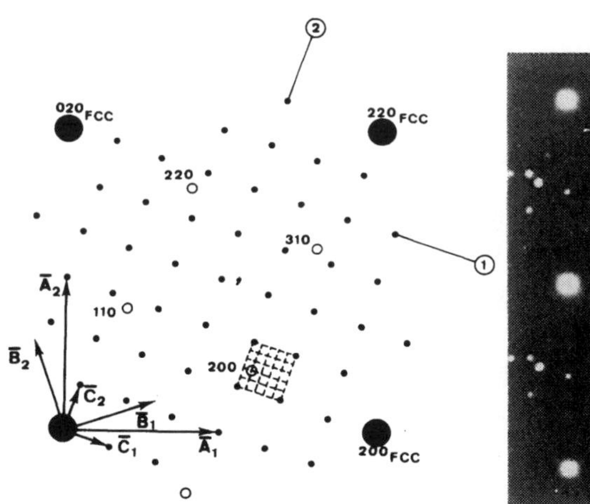

Fig. 4.21. Schematic diffraction pattern of the 2D tetragonal long period superstructure of figure 4.20. The "basic" Au_4Mn spots are indicated by open dots.
Note the fractional shifts in two directions.

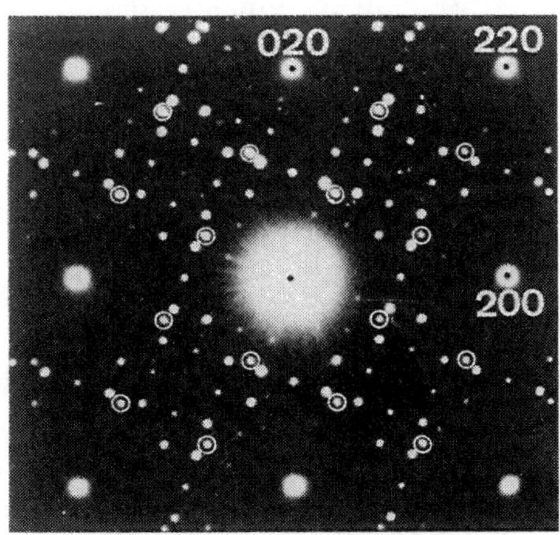

Fig.4 22. Experimental diffraction pattern of Au_4Mn which can be compared with figure 4.21. The basic spots are surrounded by small circles.

interfaces are compared in Table I. They allow to conclude that the structure is as represented in Figure 4.20, i.e. the displacement vectors are respectively $\mathbf{R}_1$ = 1/10 [135] and $\mathbf{R}_2$ = 1/10 [3 $\bar{1}$5]

Table 1. Two-dimensional superstructure.

g	g . R values			
	observed (mod 1)		calculated R_1 = 1/10 [135]	R_2 = 1/10 [3$\bar{1}$5]
	array 1	array 2	array 1	array 2
(1) 200	1/5 or 4/5	2/5 or 3/5	1/5	3/5
(2) 110	2/5 or 3/5	4/5 or 1/5	2/5	1/5
(3) 220	4/5 or 1/5	3/5 or 2/5	4/5	2/5
(4) 310	3/5 or 2/5	1/5 or 4/5	3/5	4/5

ii) The alloy system $Au_{3+}Zn$ [20]

This alloy exhibits a series of long period structures derived from the $L1_2$ ordered structure, which is itself based on an FCC lattice. The long period results from the presence of periodic conservative anti-phase boundaries with a displacement vector of the type 1/2 $\langle 110 \rangle_{FCC}$, the interfaces being perpendicular to one of the cube directions, e.g. the $[001]_{FCC}$. The anti-phase boundaries are conservative.

The simplest among these superstructures produces the diffraction pattern of Figure 4.3a where the crosses represent the positions of the reflections due to the basic $L1_2$ structure. The **q**-vector, as indicated in Figure 4.3a, is 1/2 [001]*, it is rational and the superstructure is thus commensurate. The long lattice parameter of the superstructure is 2M = 4 (in units a). There are extinctions for h + l = odd; the unit cell is centered.

The diffraction pattern, drawn schematically in Figure 4.3a can be interpreted by assimilating a linear array of satellite spots along

[001]* and with a spacing q = 1/M (in units a*) with each $L1_2$ spot. The fractional shifts **g**.**R** are 1/2 for reflections with k = odd such as 011, 012 (referred to $L1_2$) and which are absent in the superstructure; they are zero for reflections with k = even such as 022, 021, ..., which are present. This is evident from Figure 4.3a.

Conversely, assuming that the structure is derived from the $L1_2$ structure by the periodic arrangement of similar out-of-phase boundaries, the geometry of the diffraction pattern contains all the information necessary to deduce the superstructure. The interfaces must be perpendicular to the rows of satellites; they are thus parallel with (001). Their spacing is M = 1/q, where q is known from the diffraction pattern. The displacement vector **R** must be such that all computed fractional shifts agree with the observed ones, i.e. **g**.**R** = 0 for 022, 021, ..., i.e. for k = even, and 1/2 for 011, 012, ..., i.e. for k = odd. In general **g**.**R** = 0 for 1/2(h+k) = even and 1/2 for 1/2(h+k) = odd. From two sections of reciprocal space we can deduce that only **R** = 1/2 [110] (and its crystallographically equivalent vectors) leads for all fractional shifts to the observed values.

An alloy with a slightly different composition produces diffraction patterns such as the ones shown in Figure 4.3b and Figure 4.23c and which look incommensurate. One of these is represented

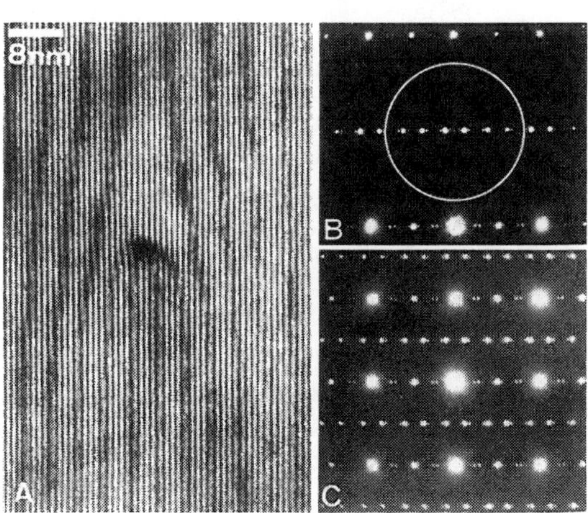

Fig. 4.23.
(a) Discommensuration node in a long period structure $2\bar{2}3\bar{3}$
(b) Diffraction pattern of long period superstructure with M $\neq$ 2 along [110] zone.
(c) Along the [100] zone. The latter pattern can be compared with Figure 4.3b.

schematically in Figure 4.3b. The most intense spots are situated in the vicinity of the spots due to the $2\bar{2}$ structure, suggesting that the superstructure responsible for these diffraction patterns is closely related to the $2\bar{2}$ structure. The average q-vector deduced from the diffraction pattern, for instance by measuring the separation of the two satellites on either side of the $(011)L1_2$ spot position is now q = 4/9 (in units a*). The Fujiwara construction for $M^2 = 2.25 = 9/4$ is shown in Figure 4.3d; it leads to the periodic stacking sequence $3\bar{2}2\bar{2}3\bar{2}2\bar{2}$ ... which is a "uniform" sequence in the sense defined in § 4.3.3. All satellite positions in the diffraction pattern are consistent with a q-vector equal to 4/9. This interpretation implies that the satellites of successive orders associated with a given Bragg reflection of the $L1_2$ structure are those joined by square brackets in Figure 4.3b.

The long period structures of the type 2^k3, $2^k3\ 2^{k+1}3$, ... can also be considered as sequences containing two block sizes 2 and 3. The best approximation to a sequence with a given q-vector and consisting of two block sizes such that $Na < 1/q < (N+1)a$ can be obtained with the "cut and projection" method (see § 4.3.3) as represented in Figure 4.7 for q = 4/9. The shape of the rectangular meshes of the two-dimensional lattice is determined by the ratio of the two block sizes; it is 2X3. The slope of the strip for q = 4/9 is given by the relation (4.6); $tg\alpha = 1/2$ and the resulting sequence is clearly $2\bar{2}2\bar{3}$ (Figure 4.7a).

The central row $00l$ of the kinematical diffraction pattern can now by obtained by constructing the reciprocal lattice and intersecting this with a line L* with a slope of 45°, followed by the projection of the reciprocal lattice nodes onto the line L* along the direction perpendicular to the strip, i.e. such that $tg\ \alpha^* = 2$. The spot positions are given by the intersection points with L*. The closer a reciprocal lattice node is to the line L* the larger is the intensity of the corresponding reflections (see § 4.5.3). The dependence on u being described by $\sin \pi uD/\pi Du$ (Figure 4.7b). The construction thus not only yields the geometry of the diffraction pattern but also the relative intensities according to the kinematical theory. In practice only nodes close to the line L* produce visible spots. Only the row $00l$ is obtained from the line L* through the origin. The reciprocal lattice nodes situated on the line L* are associated with the basic reflections corresponding with the spacing of the largest common subunit of the two blocks. In this particular case of 2a and 3a blocks this is the unit cell with size a ($L1_2$).

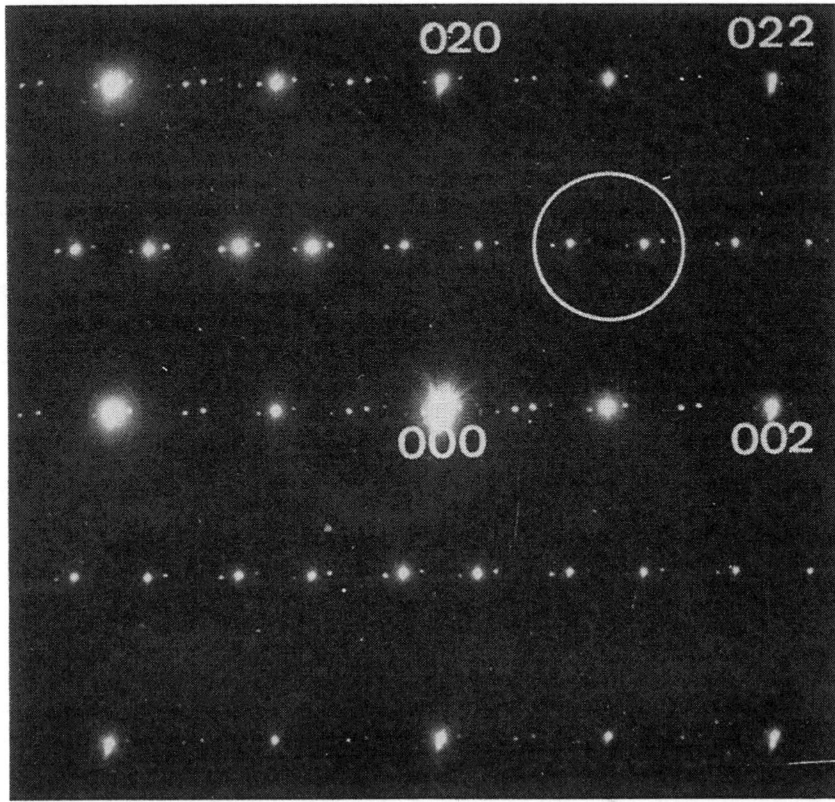

Fig. 4.24. Diffraction pattern of a region of $2\bar{2}2\bar{3}$ phase exhibiting an orientation anomaly due to systematic ledging of the <3> bands.

Other spot rows of the diffraction pattern, and in particular the corresponding fractional shifts cannot be obtained in this way since the phase shifts introduced by the anti-phase boundaries are ignored in this 1D approach. Rather than interpreting the diffraction pattern in terms of the average blocksize M or the average q-vector q = 1/M referred to the $L1_2$ structure, one can note that for q = 4/9 the long period superstructure is commensurate with a true period 9a. It can thus also be described as a periodic out-of-phase boundary superstructure of the basic $2\bar{2}$ structure with a separation between interfaces of 9a and with a displacement vector 1/4 $[001]_{2\bar{2}}$ due to the insertion of extra layers of $L1_2$ unit cells in the <3> bands, which now play the role of discommensuration walls (Figure 4.3e). Each diffraction spot of the $2\bar{2}$ structure will then acquire a satellite sequence with a spacing 1/9a* and with a fractional shift given by 1/4 *l* when referred to the $2\bar{2}$ structure. This clearly leads to the same diffraction pattern (Figure 4.3b).

Nevertheless it is possible to deduce which of the two descriptions is the more adequate one by considering the diffraction pattern of Figure 4.24 which exhibits an orientation anomaly. This phenomenon is a consequence of the presence of systematically ledged <3> bands, leading to an average orientation of the <3> bands enclosing an angle with that of the <2> bands, which remain rigorously in the (001) plane. This could be proved by means of high resolution images. The satellite sequences in Figure 4.24 are perpendicular to the average orientation of the 3 bands, whereas the arrays of superstructure spot positions due to the $2\bar{2}$ basic structure remain strictly oriented along [001]*.

The latter description is thus the most adequate one. One can furthermore conclude that the $2^{k}3$ structures are in fact discommensuration modulated superstructures of the $2\bar{2}$ structure. This can be modelled by means of a Fujiwara-type square wave function as shown in Figure 4.3e where the "phase-slip" associated with the discommensurations is shown to lead to <3> bands.

The presence of discommensuration nodes is obvious from Figure 4.25a which shows the boundary between the top part which has the $2\bar{2}$ structure and the bottom part exhibiting the $2\bar{2}2\bar{3}$ structure. Figure 4.25b shows a high resolution image of the $2\bar{2}2\bar{3}$ structure.

After annealing at temperatures approaching the order-disorder transition temperature the diffraction pattern shows less superstructure satellite spots of measurable intensity. It can be shown that is due to the fact that the atom columns boardering the anti-phase boundaries limiting the <3> blocks acquire a mixed chemical composition. This can be expressed by means of a Fujiwara-type construction by assuming a rounded shape for the steps as indicated in Figure 4.3d by means of a dotted line. The Fujiwara rounded step function is then interpreted as describing an occupation probability of the two FCC sublattices.

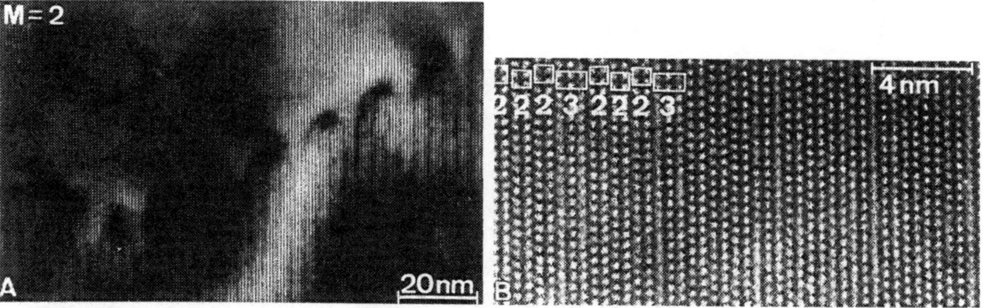

Fig. 4.25.

(a) Lattice fringe image of the transition region between the $2\bar{2}$ region and the $2\bar{2}2\bar{3}$ region. Along the boundary discommensurations (i.e. <3> bands) form fourfold nodes.positions due to the $2\bar{2}$ basic structure remain strictly oriented along [001]*.

(b) High resolution image of the $2\bar{2}2\bar{3}$ structure.

4.7.2. Mixed layer compounds

i) The $(Bi_2Te_3)_n$ GeTe system [21]

The use of the Fujiwara algorithm to derive domain sequences belonging to a given q-value is limited to cases where the domain sizes are of the form Na and (N+1)a, where N is an integer. However this condition is not always satisfied and for instance in the above mentioned family of compounds domain sequences with mixtures of domain widths of 5 and 7 units as well as of 7 and 9 units occur, depending on the composition. The succession of close packed layers in the five layer lamellae is Te-Bi-Te-Bi-Te whereas in the seven layer lamellae it is Te-Bi-Te-Bi-Te-Bi-Te, Germanium occupying Bi-sites, the bonding between successive lamellae is of the Van der Waals type. Periodic sequences such as 57, 557 and 5557 as well as quasi-periodic sequences were observed by means of high resolution electron microscopy and electron diffraction. We shall discuss only one specific example using the "cut and projection" method to predict the semi-quantitative features of the diffraction pattern.

The diffraction pattern is reproduced in Figure 4.26 and represented schematically in Figure 4.27; the q-vector is 3/17 of g_{0006} in the hexagonal description of the BiTe (NaCl-like) structure or 3/17 of g_{00015} in terms of the lattice of the rhombohedral compound

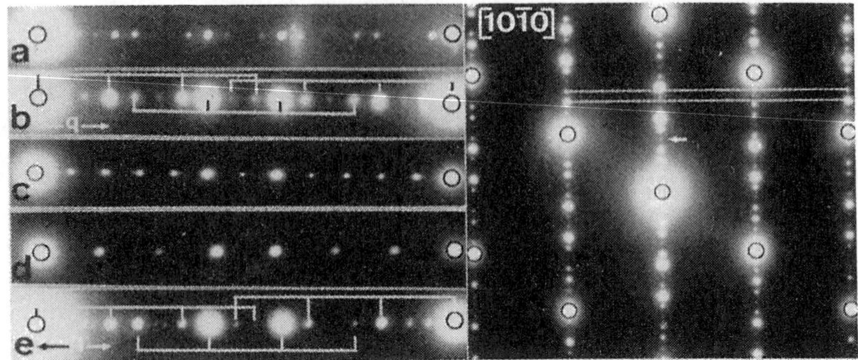

Fig. 4.26. Central rows of the diffraction patterns of members of the series of compounds $(Bi_2Te_3)_n$ GeTe; the q-vector is indicated.

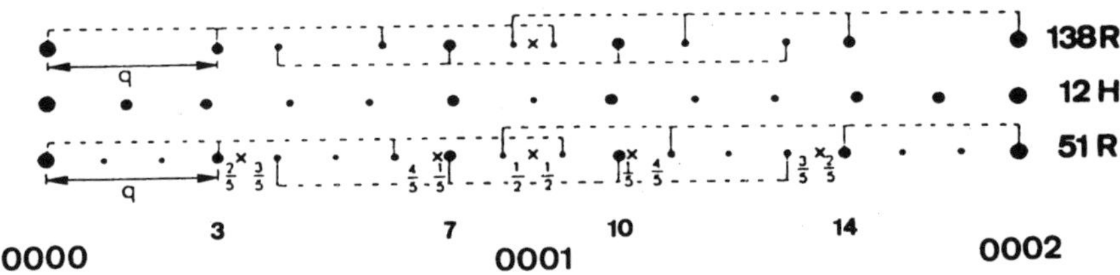

Fig. 4.27. Schematic representations of the diffraction patterns of Figure 4.26.

Bi_2Te_3. This q-value is the inverse of the average lamellae thickness of the 557 sequence (51R). The "cut and projection" construction for q = 3/17, a = 5, b = 7 is represented in Figure 4.28. In (a) the sequence belonging to the q-vector, is derived whereas in (b) the 000l diffraction pattern is deduced. The most intense spots in the 000l row correspond with l = 9, 21, 30, 42; the spots of second "magnitude" are those with $l = 3, 12, 18, 33, 39, 48$; only spots with $l =$ 3-fold are present in this hexagonal description. The qualitative correspondence with the experimental pattern of Figure 4.26 is quite striking. High resolution images confirmed the sequence 557 (51R).

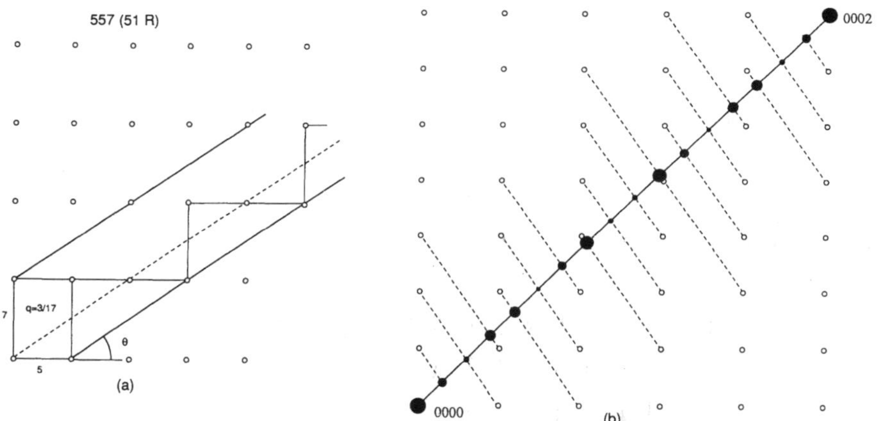

Fig. 4.28.
(a) Cut and projection construction for q = 3/17, a = 5, b = 7, leading to the stacking sequence 557.
(b) Derivation of $000l$ row of the diffraction pattern.

ii) Mixed layer compounds in the system $(GeTe)_n As_2Te_3$ [22]

We illustrate the characteristic features by means of the diffraction. patterns due to compounds of the series $(GeTe)_n As_2Te_3$ (n = integer). The structure of the member n = 5 of this series is represented in Figure 4.29 as viewed along close packed rows of the close packed layers. The diffraction patterns of Figure 4.30

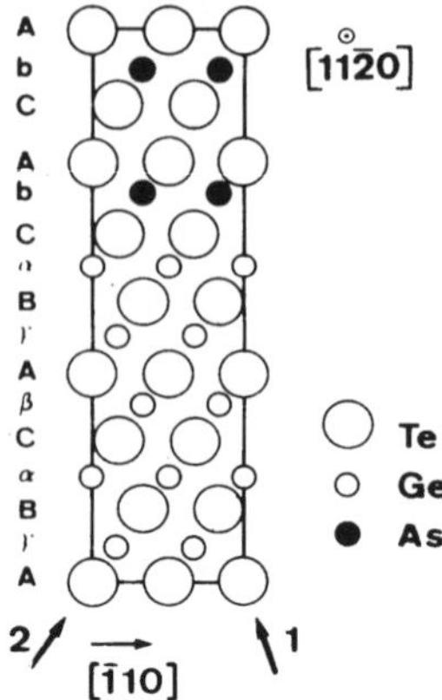

Fig. 4.29. The compound $(GeTe)_5\ As_2Te_3$as viewed along the close packed rows of atoms.

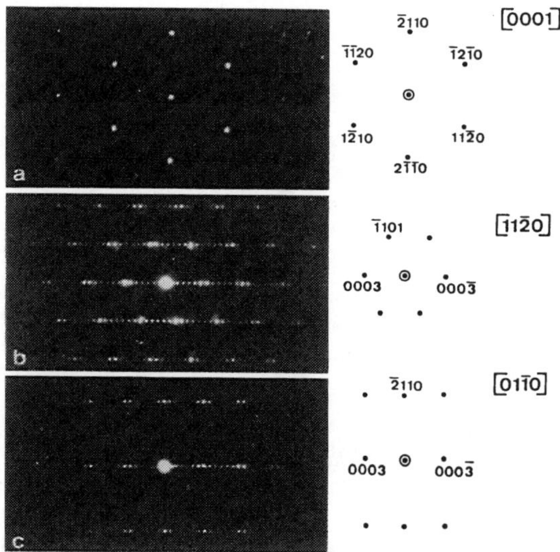

Fig. 4.30. Diffraction patterns along different zones of the compound.
(a) [0001]
(b) [11$\bar{2}$0]
(c) [1$\bar{1}$00]

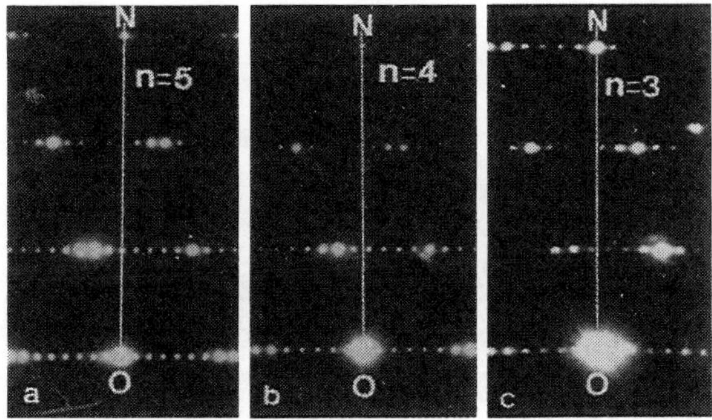

Fig. 4.31. Diffraction patterns along [1120] illustrating the presence of rhombohedral and hexagonal phases.

demonstrate that in the zone [1120] normal to the layer planes no superstructure spots are visible. The [110] zone pattern reveals the stacking. The most intense superstructure spots are clustered around spots due to the basic NaCl-like structure of GeTe, which is the majority component. The layer of As_2Te_3 introduces a shift between successive NaCl-like lamellae and as a result the sequences of superstructure spots are fractionally shifted with respect to the positions of the NaCl spots. This shift also gives rise to either hexagonal ($n = 3m + 2$ e.g. 5) or rhombohedral ($n = 3m + 1$, $3m$ e.g. $n = 3, 4$) superstructures (Figure 4.31) depending on the integer n, i.e. on the thickness of the GeTe lamellae.

4.7.3. Doubly periodic structures

In a number of materials it is possible to distinguish two (or more) interpenetrating sublattices occupied by atoms of different chemical species and forming different crystal structures with different unit cells. The combined structure is then in general incommensurate. Examples are the so-called chimney ladder structures or Nowotny phases such as $MnSi_{2-x}$ [23] and the MTS_3 compounds [24].

The first class of compounds consists of a tetragonal manganese framework within which the silicon atoms are arranged in helices. The c-period of the silicon helices is much longer and not simply related to that of the manganese sublattice and incommensurate diffraction patterns are produced (Figure 4.1).

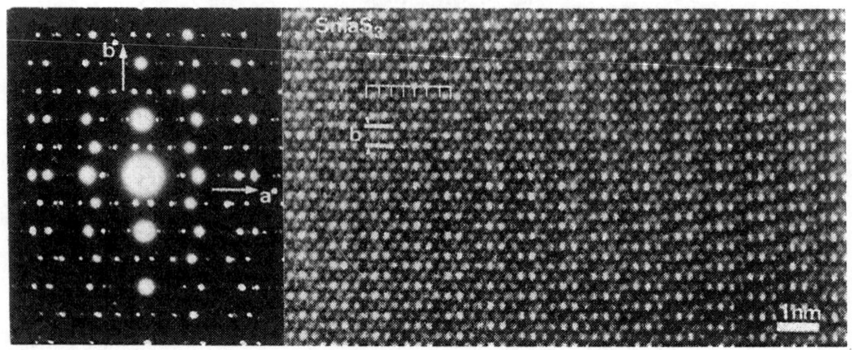

Fig. 4.32. Diffraction pattern along [001] and corresponding high resolution image $SnTaS_3$

In the MTS_3 compounds (T = Ta, Nb, ...; M = Pb, Sn, ...) two different types of layers occur: simple TS_2 layers (having the NbS_2 structure) with hexagonal symmetry are interleaved with double MS (NaCl-like) layers with square symmetry. There is perfect fit along one direction (b) in the layer plane, but a large misfit along the perpendicular direction (a) within the layer plane. Along the c-direction the structure is periodic but along the misfit direction a in the layer plane the structure is modulated.

In both cases the two sublattices interact which induces the modulation of one sublatttice by the other; the TS_2 part is modulated with a period imposed by the MS part and vice versa.

In the chimney ladder structures the modulation is one-dimensional along the c-direction; the spots due to the manganese sublattice acquire linear sequences of satellites which reveal the modulation due to the silicon helices.

In the MTS_2 compounds the modulation occurs along the layer planes. The spots due to the TS_2 sublattice are surrounded by a square array of spots resulting from the modulation by the MS structure and the spots produced by the MS sublattice acquire

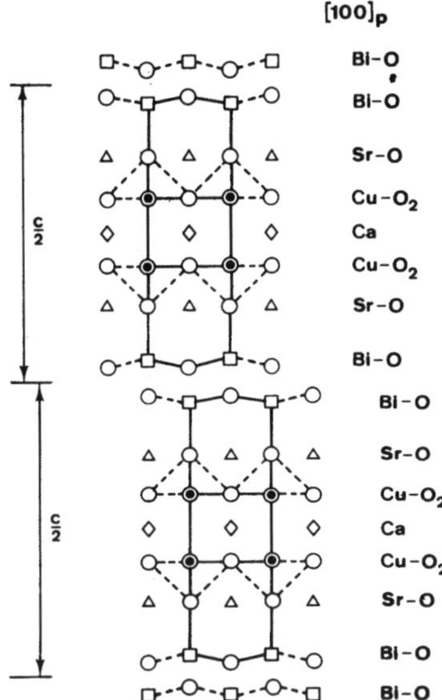

Fig. 4.33. Average structure of $Bi_2Sr_2Ca\text{-}Cu_2O_8$.

hexagonal arrays of satellites due to the modulation by the TS_2 layers (Figure 4.32).

In a general way, in the case of mutual modulation of A and B sublattices the diffraction vectors of the composite structure are given by

$$\{\mathbf{G}\} = \{\mathbf{g}_A\} + \{\mathbf{g}_B\}$$

where $\{\mathbf{g}_A\}$ and $\{\mathbf{g}_B\}$ are the sets of diffraction vectors of the A and B structures respectively.

4.7.4. Displacively modulated structures

i) Layered high T_C superconducting material [25]

The recently discovered high T_C superconducting materials with compositions $Bi_2Sr_2Ca_nCu_{n+1}O_{2n+6}$ (n = integer) offer striking examples of incommensurate displacively modulated structures. Figure 4.33 shows schematically the average structure of one of these complex materials (for n = 1), which can in fact also be considered as mixed layer compounds. The tetragonal average structure consists of perovskite-like lamellae containing the sequence of (001) layers:SrO - CuO_2 - (Ca - $CuO_2)_n$ - SrO interleaved by sodiumchloride-like lamellae BiO - BiO. The thickness of the perovskite-like lamellae depends on n. Also the calcium free compound (n = 0) has a structure based on the same principle.

From tilting experiments the reciprocal lattice, schematically represented in Figure 4.34 was constructed. In particular the section $(110)_p{}^*$ (i.e. referred to the basic perovskite lattice) is of interest; it consists of linear arrays of satellite spots associated with all basic spot rows and which exhibit no fractional shifts. The patterns are different for the n = 0 and n = 1, 2, ... compounds as shown in Figure 4.35. The array of superstructure spots is incommensurate with the basic perovskite lattice. High resolution images, such as Figure 4.36, have shown that the modulation is mainly associated with the BiO - BiO lamellae, which have the most pronounced wavy shape.

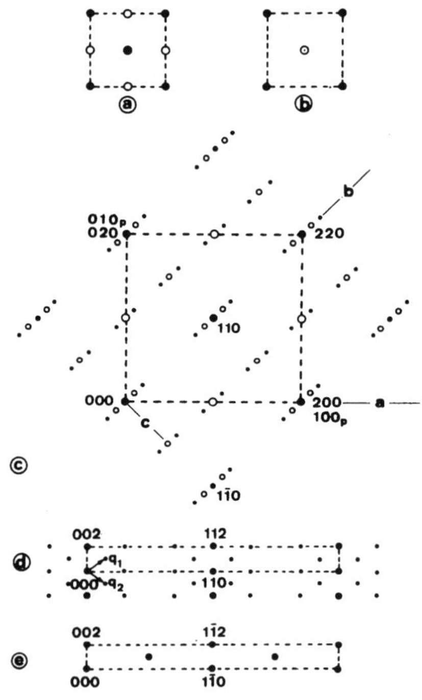

Fig. 4.34. Reciprocal lattice of $Bi_2Sr_2Ca\text{-}Cu_2O_8$.

The pattern of displacements is approximately two-dimensional; it can be represented schematically as in Figure 4.37 for the compounds with n = 0 and n = 1,2, ... In the first case (n=0) the modulation reduces the tetragonal symmetry of the basic structure to

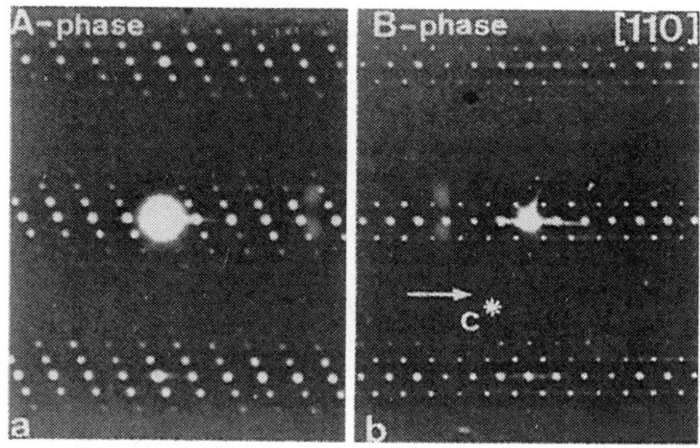

Fig.4.35. Reciprocal lattice section (110)* referred to the perovskite lattice.
(a) n = 0
(b) n = 1,2,...

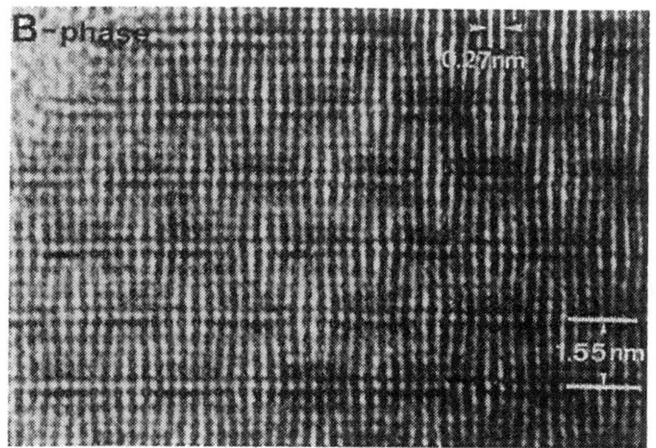

Fig.4.36. High resolution image of $Bi_2Sr_2CaCu_2O_8$ corresponding with figure 4.34

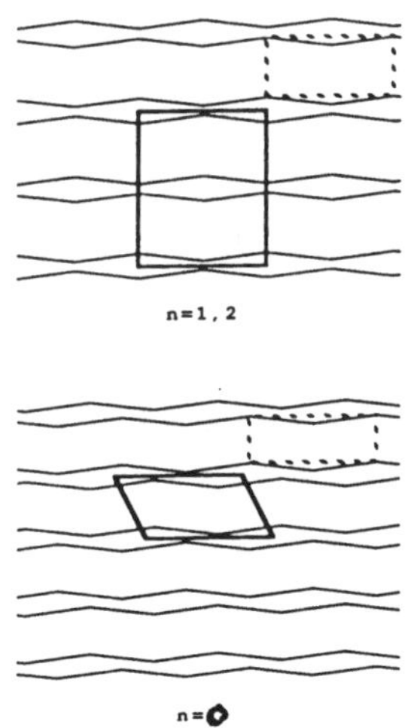

Fig. 4.37 Schematic displacement Pattern in $Bi_2Sr_2Cu_2O_7$(n=o) and Bi_2Sr_2Ca-Cu_2O_8 (n=1)

monoclinic, whereas in the second case (n = 1, 2, ...) the symmetry is reduced to orthorhombic. The origin of the modulations is still a matter of discussion, but there seems to be fair agreement that the primary reason for the modulation is the mismatch between the lattices of the perovskite lamellae and the BiO lamellae, which may induce the uptake of excess oxygen in the BiO layers. The structure is thus also composition modulated.

ii) Quartz [26]

An incommensurate displacively modulated structure occurs in quartz (SiO_2) in a narrow (≈1,5°C) temperature range close to the α–>β phase transition (573°C). The low temperature α-phase of quartz belongs to the point group 32 whereas the high temperature β-form has point symmetry 622. On cooling through the β–>α transition a single crystal of the β-phase breaks up reversibly in two structural variants α_1 and α_2 related by the symmetry operation lost in the transition, i.e. by a rotation over 180° about the threefold axis. These structural variants are known in mineralogy as Dauphinê twins. At temperatures close to the transition temperature a single crystal is broken up in a patchwork of triangular prismatic

domains belonging in an alternating fashion to α_1 and a_2. The interfaces limiting the domains are parallel with the threefold (sixfold) axis; the size of the domains decreases with increasing temperature up to the α–β-transition. The modulation is a 3-q state described by three q-vectors with equal magnitude and enclosing angles of 120°. This "star" configuration of q-vectors is rotated over a small angle $\pm\varepsilon$ with respect to the twofold axis of the quartz structure. This gives rise to two orientation variants +ε and -ε of the modulated structure. ε decreases with increasing temperature and becomes zero at T_c. The modulated structure was first discovered by means of the direct imaging of the domain pattern using the diffraction contrast mode. Such images made at different temperatures close to the transition temperature are shown in Figure 4.38. More recently the structure was also studied by means of X-ray, neutron and electron diffraction. In electron diffraction the satellites associated with all basic spots can be observed; their distance to the basic spot changes continuously with temperature; the spot-splitting due to +ε and -ε can clearly be revealed.

Acknowledgments

We gratefully acknowledge the use of the photographs from joint papers with G. Van Tendeloo, J. Van Landuyt, D. Broddin, S. Kuypers, and H. Zandbergen.

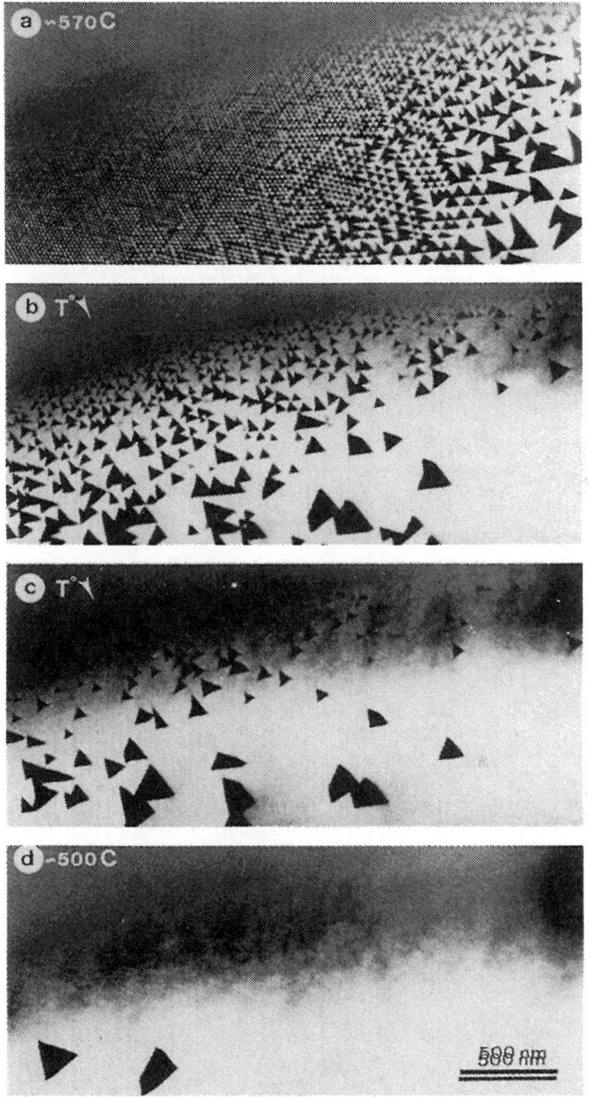

Fig.4.38. Dauphinê twin pattern in quartz as a function of temperature. The modulated structure is visible in (a).

REFERENCES

[1] McMillan, W.L., 1976, Phys. Rev. B14, 1436.

[2] Sato, H., and Toth, R.S., 1961, Phys. Rev. 124, 1833.
Sato, H., and Toth, R.S., 1965, in "Alloying Behaviour and Effects in Concentrated Solid Solutions", T.B. Massalski (ed.), Gordon and Breach, London, p. 295.

[3] Elliott, R.J., 1961, Phys. Rev. 124, 346.
Fisher, M.E., and Selke, W., 1980, Phys. Rev. Lett. 44, 1827.
Selke, W., 1984, in "Modulated Structure Materials", T. Tsakalakos (ed.), Nijhoff, Dordrecht, p. 23-42.
Selke, W., and Fisher, M.E., 1979, Phys. Rev. B20, 257; 1980, Z. Phys. B40, 71.
Yeomans, J., 1987, in "Solid State Physics. Volume 41", H. Ehrenreich (ed.), Academic Press, New York.

[4] Nowotny, H., 1970, in "Chemistry of Extended Defects in Non-Metallic Solids", L. Eyring and O'Keefe (eds.), North-Holland, Amsterdam, p. 223.

[5] Fujiwara, K., 1957, J. Phys. Soc. Japan 12, 7.

[6] Elser, V., 1985, Phys. Rev. B32, 4892.
1986, Acta Cryst. A42, 36.
Levine, D., and Steinhardt, P.J., 1984, Phys. Rev. Lett. 53, 2477 1986, Phys. Rev. B34, 596.
Duneau, M., and Katz, A., 1985, Phys. Rev. Lett. 54, 2688.

[7] Mackay, A.L., and Klinowsky, J., 1986, in "Symmetry", Hargittai (ed.), Pergamon Press, p. 803-824.
Mackay, A.L., 1987, in "Quasicrystals", Materials Science Forum 22-24 (Trans Tech Publications), K.H. Kuo (ed.), p. 11-22.

[8] Hirsch, P.B., Howie, A., and Whelan, M.J., 1960, Phil. Trans. A252, 499 and references therein.

[9] Van Dyck, D., Danckaert, J., Coene, W., Selderslaghs, E., Broddin, D., Van Landuyt, J., and Amelinckx, S., 1989, in "Computer Simulation of Electron Microscope Diffraction and Images", W. Krakow and M. O'Keefe (eds.), The Minerals, Metals and Materials Society, p. 107-134.

[10] Overhauser, A.W., Phys. Rev. B3, 1373.

[11] Van Landuyt, J., Van Tendeloo, G., and Amelinckx, S., 1974, Phys. Stat. Sol. (a) 26, K9.

[12] Amelinckx, S., Van Tendeloo, G., Van Dyck, D., and Van Landuyt, J., 1989, Phase Transitions, Vol. 16/17, p. 3-40.

[13] Van Dyck, D., Broddin, D., Mahy, J., and Amelinckx, S., 1987, Phys. Stat. Sol. (a) 103, 357.

[14] Van Landuyt, J., De Ridder, P., Gevers, R., and Amelinckx, S., 1970, Mat. Res. Bull. 5, 353-362.

[15] Yamamoto, A., 1980, Phys. Rev. B22, 373.1982, Acta Cryst. A37, 838.1985, Acta Cryst. A38, 87.1982, Acta Cryst. B38, 1446, 1451.
De Wolff, P.M., 1974, Acta Cryst. A30, 777.1984, Acta Cryst. A40, 34.
De Wolff, P.M., Janssen, T., and Janner, A., 1981, Acta Cryst. A37, 625.
Bird, D.M., and Withers, R.L., 1986, J. Phys. C19, 3497; 3507.
Pérez-Mato, T.M., Madariaga, G., and Tillo, M.J., 1986, J. Phys. C19, 2613.
Pérez-Mato, T.M., Madariaga, G., and Tillo, M.J., 1984, Phys. Rev. B30, 1534.

[16] Mahy, J., Van Landuyt, J., Amelinckx, S., Uchida, Y., Bronsema, K.D. and Van Smaalen, S., 1985, Phys. Rev. Lett. 55, 1181.
Mahy, J., Van Landuyt, J., Amelinckx, S., Bronsema, K.D. and Van Smaalen, S., 1986, J. Phys. C: Solid State Physics C19, 5049.

[17] Mahy, J., Colaitis, D., Van Dyck, D., and Amelinckx, S., 1987, Journal of Sol. St. Chem. 68, 320-329.

[18] Van Tendeloo, G., and Amelinckx, S., 1981, Phys. Stat. Sol. (a) 65, 73.

[19] Van Tendeloo, G., and Amelinckx, S., 1981, Phys. Stat. Sol. (a) 65, 431.

[20] Broddin, D., Van Tendeloo, G., and Amelinckx, S., 1990, J. Phys. (Condensed Matter) 2, 3459.

[21] Frangis, D., Kuypers, S., Manolikas, C., Van Tendeloo, G., Van Landuyt, J., and Amelinckx, S., 1990, Journal of Sol. St. Chem. 84, 314.

[22] Kuypers, S., Van Tendeloo, G., Van Landuyt, J., Amelinckx, S., Han Wan Shu, Jaulmes, G., Flahaut, J., and Laruelle, P., 1988, Journal of Sol. St. Chem. 73, 192.

[23] Ye, H.-Q., and Amelinckx, S., 1986, Journal of Sol. St. Chem. 61, 8.

[24] Kuypers, S., Van Landuyt, J., and Amelinckx, S., 1990, Journal of Sol. St. Chem. 86, 212.

[25] Van Tendeloo, G., Zandbergen, H.W., and Amelinckx, S., 1988, Solid State Commun. 66, 927.
Zandbergen, H.W., Groen, P., Van Tendeloo, G., Van Landuyt, J., and Amelinckx, S., 1988, Solid State Commun. 66, 397; and many references therein
Matsui, Y., et al., 1988, Jpn. J. Appl. Phys. 27, L372.
Maeda, H., et al., 1988, Physica C153-155, 602.

[26] Dolino, G., Bachheimer, J.P., Berge, B., Zeyen, C.M.E., Van Tendeloo, G., Van Landuyt, J., and Amelinckx, S., 1984, J. Phys. Paris 45, 901.
Van Landuyt, J., Van Tendeloo, G., Amelinckx, S., and Walker, M.B., 1985, Phys. Rev. B31, 2986.

5
Indentification of Unknowns

C.E. Lyman and M.J. Carr

5.1 PHASE IDENTIFICATION

Electron diffraction is most often employed to determine or set the orientation of a particular single crystal of a known phase and to confirm the identity of that phase. An equally important application of electron diffraction is the identification of an 'unknown' phase. An unknown phase in this sense falls into one of two broad classes: it is either a phase that is new-to-science or a phase that has been described previously but was not known to be in the specimen under consideration. If the unknown is truly new-to-science, then its proper identification will require quantitative information about its composition and a determination of its space group, unit cell dimensions, and atom positions within the unit cell. Methods for obtaining such information are covered in Volume 1, Chapters 6 and 7, of this book. However, if the interplanar spacings and angles of the unknown can be matched to those of a previously documented phase by a search through a database of reference compounds, the unknown may be efficiently identified without a detailed structure analysis.

Phase identification by electron diffraction in a modern analytical electron microscope has an important advantage over competing techniques in that submicrometer regions of an unknown phase may be imaged and analyzed both crystallographically and compositionally. Occasionally 'new' compounds, or at least new solid solution phases, are identified by electron diffraction because the method works with objects so small that they would be overlooked by other methods. Identification may be made when the subject phase is either embedded in a matrix or isolated as a particle on a support film. Unfortunately, electron diffraction has the disadvantage of relatively low precision compared to x-ray diffraction. Interplanar spacings can be measured to only about 1%, at best, and the angles between diffraction vectors can rarely be measured to better than 1°. This low precision is compensated for by the ability to obtain compositional data and by the ability to tilt to different low index zone axis directions in the same crystal to obtain three-dimensional crystallographic information.

The basic method in search/match identification is straightforward. One makes observations about the unknown and compares them to a reference collection of similar data for candidate phases. Candidates which cannot account for the current observations are rejected. Under ideal circumstances, one and only one candidate remains, and it is the unique correct match. In practice, this rarely occurs. Observations made on the unknown might be incomplete, imprecise, or inaccurate. Intensity data, useful for phase identification by x-ray diffraction, is less useful in electron diffraction because of multiple diffraction effects, even to the point where forbidden reflections sometimes appear. Conversely, the reference data for candidates may be flawed for our purpose by being derived from a specimen of greater or lesser purity than the unknown or by being obtained with a different analytical method possessing different sensitivities to certain reflections. Worst of all, the particular unknown we are seeking to identify could be absent entirely from the finite collection of reference compounds being used. Typically, many possible matches result when realistic experimental error is factored into a search. If the number of possibilities is large, the correct match may go unrecognized. Correcting the problem usually results in iterative searches requiring more or better observations and tighter match criteria. Done well, one ultimately obtains at most a few matches, including the correct one, to be confirmed ultimately by success in indexing one or more diagnostic zone axis patterns. Done badly, tightened criteria for the observations lead to failure by the elimination of even the correct match, a common pitfall when working with solid solutions or compounds which contain undetected low atomic number elements.

This chapter covers the acquisition, organization, and interpretation of compositional and crystallographic data obtained by analytical electron microscopy of thin specimens, and their successful application to phase identification by manual and computer-based search/match methods. The characteristics of several available reference collections, indexes, and computer databases are also discussed.

5.2 EXPERIMENTAL DATA

The utility of the analytical electron microscope (AEM) for phase identification arises from the fact that it can provide compositional, crystallographic, and morphological information about individual objects larger than a few tens of nanometers. Compositional and crystallographic information is used directly to identify unknowns. Morphological information is sometimes helpful in the later stages of confirming an identification.

5.2.1 Instrumentation

In the typical AEM, the operator may switch freely between the flooding or parallel beam TEM mode and the scanning transmission mode (STEM mode), where a finely-focused electron beam is scanned across the specimen to form the image. AEMs usually have facilities for compositional analysis by either energy dispersive x-ray spectrometry (EDS) or electron energy loss spectrometry (EELS) and crystallographic analysis by selected area electron diffraction (SAED) and convergent beam electron diffraction (CBED). All these analytical techniques (except SAED) can take advantage of the small electron probe focused on a moderately thin specimen (50-100 nm) to obtain a lateral spatial resolution of analysis on the order of 10-50 nm. The typical spatial resolution for SAED is about 0.5-1μm, although objects as small as 10nm can be identified when dark-field imaging is applied to overlapped patterns.

Most AEMs operate at 100-400 keV and employ a heated LaB_6 or tungsten electron source to produce a 10-50 nm (FWHM) diameter electron probe at the specimen with 0.1-1 nA of current. These instruments are usually TEM columns modified to produce a small electron probes and are called TEM/STEMs. Some specialized instruments equipped with field-emission electron sources can produce 1 nm probes with equivalent probe current. The latter instruments are termed dedicated STEMs if they don't have the ability to produce a flooding beam TEM image.

Obtaining compositional data requires ancillary equipment to be attached to the electron column. The typical energy-dispersive x-ray spectrometer (EDS) detects all elements heavier than Na (Lyman, Williams, and Goldstein, 1989). The EELS spectrometer also can identify elements and provide much additional information from very thin (typically, 10-50 nm) specimens (Egerton, 1986). However, the ease of elemental identification at all specimen thicknesses with the EDS usually makes it the technique of choice when analyzing unknown phases. Only EDS for compositional analysis will be discussed further.

Facilities for SAED are standard on all commercial TEMs, and CBED patterns may be obtained on most TEM/STEMs. While not necessary for diffraction analysis of fine crystalline powders, the freedom to tilt the thin specimen to particular zones is a requirement for single crystal analysis. The eucentric goniometer stage is considered a necessity for the diffraction work described below. Diffraction patterns are usually recorded on film in several exposures, to cover the large dynamic range of intensity, and analyzed at a later time after wet processing of the film. However, there is considerable advantage in digitally encoding data or patterns observed on the screen for analysis in a computer while the operator is still at the microscope.

5.2.2 Compositional Data

Qualitative knowledge of the elements present provides an enormous advantage in identifying an unknown by electron diffraction. Quantitative knowledge of elemental amounts is usually not necessary. While the EDS detector can detect elements in concentrations below 1 wt%, elements present at such low concentrations are probably solid solution impurities or surface contamination that would not cause detectable changes in the electron diffraction pattern.

Spatial Resolution and Elemental Detectability. The electron beam generates characteristic and continuum x-rays in a small volume along the path of the beam through the specimen as shown in Figure 5.1. Characteristic x-rays of the elements present provide the signal for compositional data while continuum x-rays contribute only to the background. The beam spreads as it traverses the specimen, but in the thin foils normally analyzed (50-200 nm thick) the size of the excited volume remains small. Typical values for the spatial resolution of analysis are 50 nm, 20 nm, and 5 nm for the tungsten hairpin, LaB_6, and field-emission sources, respectively. All AEMs, regardless of electron source, can detect elements present in amounts $\geq$0.2 wt% (for $Z>20$) provided that the specimen is thick enough to provide an adequate number of x-ray counts (Goldstein et al., 1990); however, the elements of interest in phase identification will be present in concentrations well above the detectability limit.

EDS Detectors. X-ray detection is accomplished by a Li-drifted Si detector that not only counts x-ray photons but also measures their

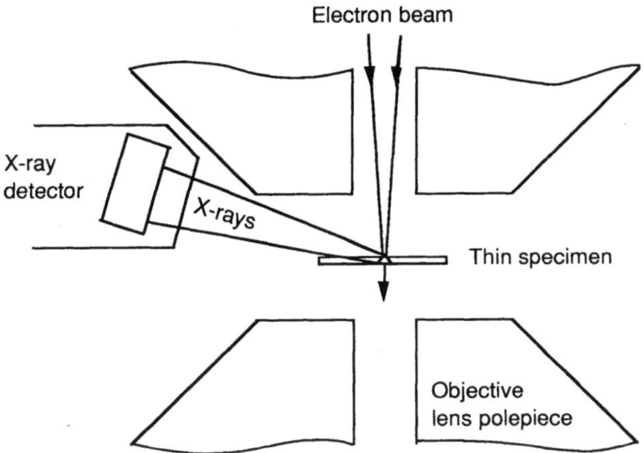

Figure 5.1. Schematic cross-sectional diagram through the objective lens, thin specimen, and energy-dispersive x-ray detector (EDS) for a typical analytical electron microscope (AEM).

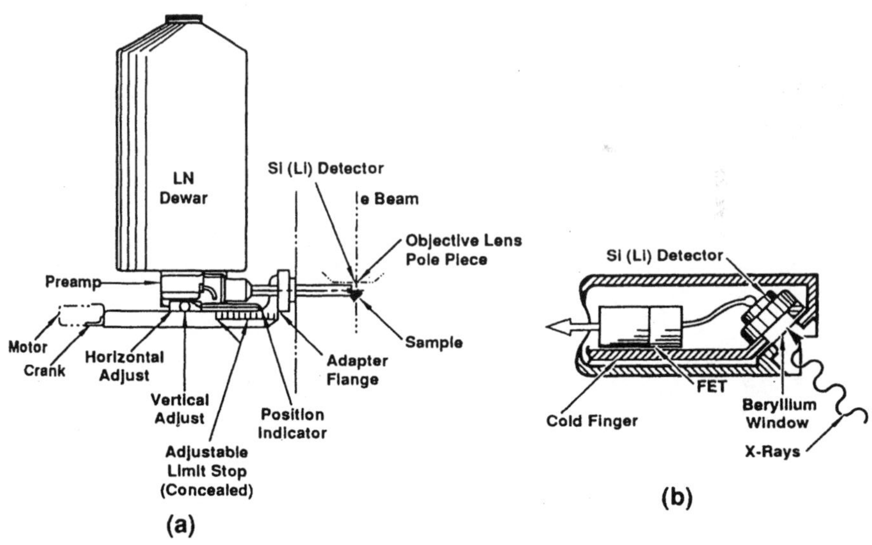

Figure 5.2. Typical EDS detector for an AEM. (a) Retractable detector and preamplifier electronics. (b) Detail of Si(Li) detector assembly showing Be window (from Goldstein et al., 1992; courtesy Plenum Press).

energy for elemental identification (Figure 5.2). Most detectors in use today have an 8-µm thick Be window between the specimen and the detector crystal, which, because of absorption, limits the detected x-rays to those with energies above ~1keV (i.e Na Kα, Zn Lα). This severely limits the information that one can obtain about the so-called 'light elements', namely, boron, carbon, nitrogen, oxygen, and fluorine. Some instruments are equipped with windowless or ultra-thin window (0.1-1µm-thick windows of low Z materials) detectors that permit detection of light elements. For higher energy x-rays, above 20 keV, detection efficiency drops because x-rays tend to pass through the Si detector without being counted. Thus, EDS systems are usually set to analyze x-rays only over the range 0-10 or 0-20 keV. The detected x-ray photon counts are routed to a multichannel analyzer, sorted by energy, and displayed as a histogram of the number of photons detected versus energy.

Obtaining Accurate EDS Data. The major barrier to qualitative EDS data suitable for matching routines is the recording of x-ray peaks from sources other than the phase under analysis. It is important to obtain spectra from clean, isolated areas of foil where the beam intercepts only a single phase. The area of the specimen under analysis must have a clear view of the EDS detector, and the objective aperture must be removed to minimize spurious x-rays generated on it by scattered electrons. While selected area electron diffraction is most conveniently done in TEM mode, EDS spectra often contain fewer spurious x-rays when the AEM is operated in STEM mode rather than TEM mode. There may be artifactual background peaks introduced from the specimen holder (usually Cu or Al), specimen grid (usually Cu), or the objective lens polepieces (usually Fe). Solvent residues, etching or polishing residues, and even silicone pump oils in evaporators can contribute interfering peaks. Obtaining a clean spectrum from an isolated small precipitate can be difficult unless the precipitate by chance protrudes into the hole without a coating of matrix material. Other methods of obtaining a clean precipitate spectrum include removing the precipitate from the matrix by extraction replication, or subtracting the matrix spectrum from the combined precipitate/matrix spectrum. It is also important to remember that with a Be window between the specimen and the detector, elements lighter than Na will not be detected even though they are present in great quantity. For example, the spectrum from aluminum oxide will look the same as the spectrum from pure aluminum metal. Fortunately, this lack of light element data generally does not hinder the search/match procedure.

Qualitative Analysis Procedure. Most commercial EDS systems offer some form of automatic peak identification, however, the results of such a procedure can lead to incorrect elemental identification if the operator is unaware of the basic principles of qualitative analysis with an EDS detector. Correct elemental identification in the AEM requires some knowledge of x-ray physics and a great deal of common sense. Measurement of the peak energies is sufficient for identification of most elements. The analyst should have access to a complete table of x-ray energies such as that given by Bearden (1967). However, because of the modest energy resolution (about 150 eV), severe peak overlaps may occur in EDS spectra. The general strategy is to identify a peak on the basis of all detectable x-ray lines for an element, rather than from the energy of a single peak. For example, the K-series of sulfur, the L-series of molybdenum, and the M-series of lead are each detected as a single peak at the nearly same energy. Each of these peaks contain several x-ray lines that are unresolved with the energy resolution of the EDS. Thus, when EDS data is collected over an energy range of only 0-10 keV, it is impossible to tell whether the single peak at 2.3 keV is due to S, Mo, Pb, or some combination of these (Figure 5.3a). This ambiguity can be removed by collecting x-rays over the energy range 0-20 keV and looking for the presence of the Mo Kα or the Pb Lα at 17.4 and10.6 keV, respectively (Figure 5.3b). Additional guidelines for qualitative analysis of EDS data may be found in Goldstein, et al. (1992).

Quantitative Analysis Procedure. Experience has shown that it is generally better not to use quantitative compositional information in search routines, since it may prejudice the search against valid matches for many solid solutions. This is particularly true in the early or screening stage of a search/match process where a preliminary match may be made to an isomorphic compound with a composition differing substantially from the unknown compound under investigation. Knowledge of the relative amounts of elements can be helpful in confirming matches, however, since the crystal structure of a proposed match must be able to accommodate the observed composition. Procedures for converting x-ray intensities from thin specimens to actual compositions may be found in Williams (1984) and in Joy, Romig, and Goldstein (1986).

5.2.3 Crystallographic Data

A key advantage of the AEM is that crystallographic information may be obtained from the very same region for which EDS compositional data is obtained. This combination gives the AEM unmatched capabilities for identification of particles and phases under 10 μm in diameter. Several diffraction techniques are available in a modern AEM, each providing different kinds information. However, for

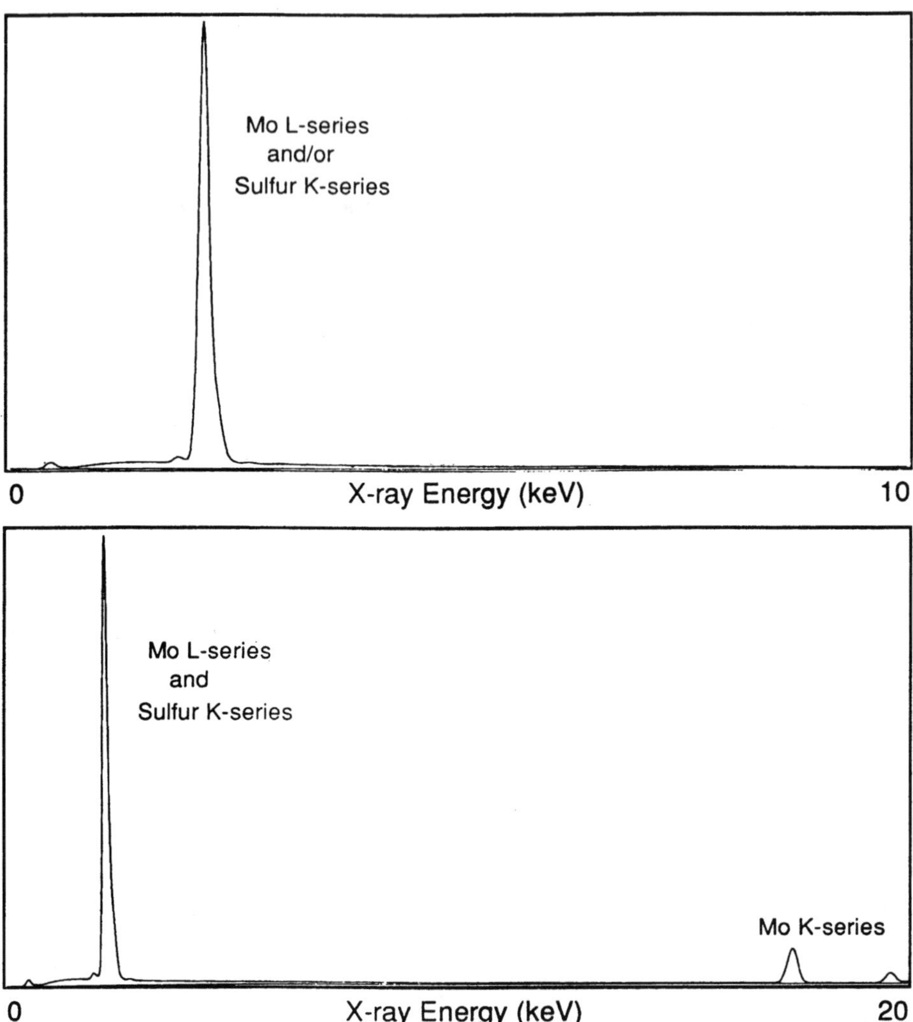

Figure 5.3. Typical EDS x-ray spectrum for molybdenum disulfide taken with different spectrometer energy range settings. (a) Using energy range 0-10 keV, the molybdenum L-series peak cannot be distinguished from the sulfur K-series peak at the same energy. (b) Using energy range 0-20 keV, the Mo K-series peaks are displayed leaving confirming the presence of Mo.

the identification of isolated single crystals larger than 0.5-1 μm, or oriented sets of finer crystals, such as solid state precipitates, the conventional method of selected area electron diffraction (SAED) is the most convenient. In fact, since all transmission electron microscopes have this facility, many analyses described in this chapter may be performed without an AEM by collecting the diffraction information on a particle in the conventional TEM and the compositional information on the same particle in an scanning electron microscope (SEM) equipped with an EDS detector.

Setup of the Tilting Stage. A modern AEM is usually equipped with an eucentric goniometer specimen stage which allows tilting ±30-60° about the principal tilt axis or x-tilt of the stage (usually the long axis of the specimen holder rod). An adjustment of the specimen height is provided to position the tilt axis so that the image of the specimen does not shift laterally more than 200 nm over the full range of x-axis tilt. This alignment of the eucentric axis is imperative for each area examined. To reach any region of reciprocal space, the specimen holder is provided with either a second axis (y-tilt) in the plane of the specimen perpendicular to the main tilt axis (a double-tilt holder) or a rotation axis normal to the plane of the specimen (a tilt-rotation holder). Usually, no alignment is possible for the second axis, so the eucentricity of cross tilting or rotation is not very good. Large y-tilts or rotations require considerable manual dexterity on the part of the operator since the specimen must be translated after each increment of tilt or rotation. Even so, with a good modern specimen stage it is possible to tilt through and interrogate a major fraction of reciprocal space.

Selected Area Electron Diffraction. In this technique the selected area diffraction (SAD) aperture is inserted into the column at the image plane of the objective lens, effectively limiting the specimen area contributing to the diffraction pattern to roughly the region visible in the image through the SAD aperture (see Figure 5.4). Spherical aberration of the objective lens and column misalignments may cause up to twice the visible area within the aperture to contribute to the diffraction pattern. Apertures capable of selecting 0.5 μm or less may be used. With the specimen at the eucentric height position, the objective lens is focused on the specimen while the diffraction lens is focused on the SAD aperture. This provides a repeatable condition for which an accurate camera length can be measured. With the condenser lens defocused to give nearly parallel illumination, the diffraction pattern formed consists of dim but sharp spots or rings. Fine focus of the diffraction spots (rings) with the diffraction lens control may be required. From the Ewald sphere construction (see Volume1, Chapter 3), these spots can be interpreted as the intersections

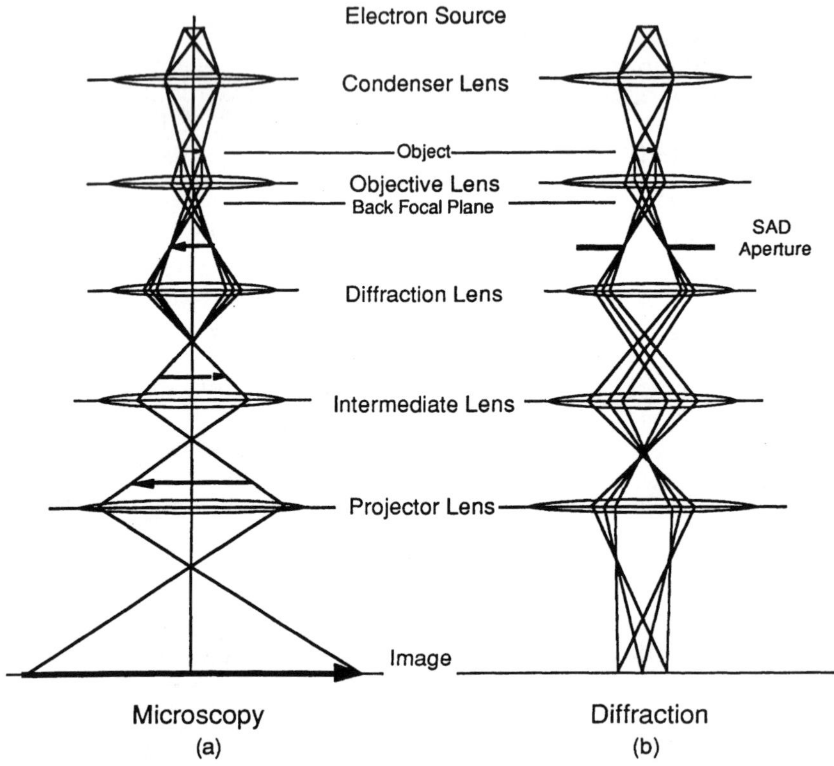

Figure 5.4. Schematic ray diagram of a typical TEM with four imaging lenses. (a) Imaging mode showing image reversals after each lens but not relative rotations. (b) Selected area electron diffraction (SAED) mode. Only one of two condenser lenses is shown.

between the reciprocal lattice and the sphere of reflection. The radius of curvature of the Ewald sphere, being inversely related to the wavelength of the diffracted electrons, is so large in electron diffraction that the diffraction pattern obtained is effectively a plane section through the reciprocal lattice.

Two kinds of information may be obtained from selected area diffraction patterns: interplanar spacings and interplanar angles. While d-spacing information can be obtained from both single crystal spot patterns and polycrystalline ring patterns, only spot patterns yield angular information.

Interplanar Spacings. Interplanar spacings can be determined from the distance between the center spot and each unique diffraction maxima in the diffraction pattern. The radial distance, r, measured from the center of the transmitted beam to the center of a diffracted spot is inversely related to the interplanar spacing, d, by

$$rd = \lambda L \tag{5.1}$$

where λ is the wavelength of electrons at the accelerating voltage (kV) employed, and L is the equivalent length of a simple diffraction camera as shown in Figure 5.5. Taken as a product, λL constitutes the camera constant of the instrument for a particular accelerating voltage and magnification. It is usually determined by placing a standard with known d-spacings at the eucentric stage position, obtaining a diffraction pattern under standard conditions, and measuring the r-distances in the pattern. For example, polycrystalline aluminum evaporated onto a carbon support film yields the ring pattern shown in Figure 5.6a. From the known lattice parameter of fcc Al (a = 4.0497Å) and the interplanar spacing equation for cubic crystals, the d-spacings may be determined for the first five Al reflections (Figure 5.6b). Typical camera constants range from 2.0 to 3.5 nm-mm (or Å-cm). The positions of rings and spots can be determined reliably to within ±0.2 mm over the usual range of r-spacings encountered: 1-30 mm at a nominal camera constant of λL = 2.5 nm-mm. A microscope will usually have several preset camera constants that each require calibration.

On a percentage basis, the precision of electron diffraction data is 1-2% in the middle of the usual range of r-spacings. Measurement precision is limited by several types of experimental errors (Andrews et al., 1967). More importantly, because the d-spacing is inversely related to the measured r-spacing, the precision of d-spacings determined by electron diffraction is nonlinear across the measurement range, being very poor at large d-spacings and

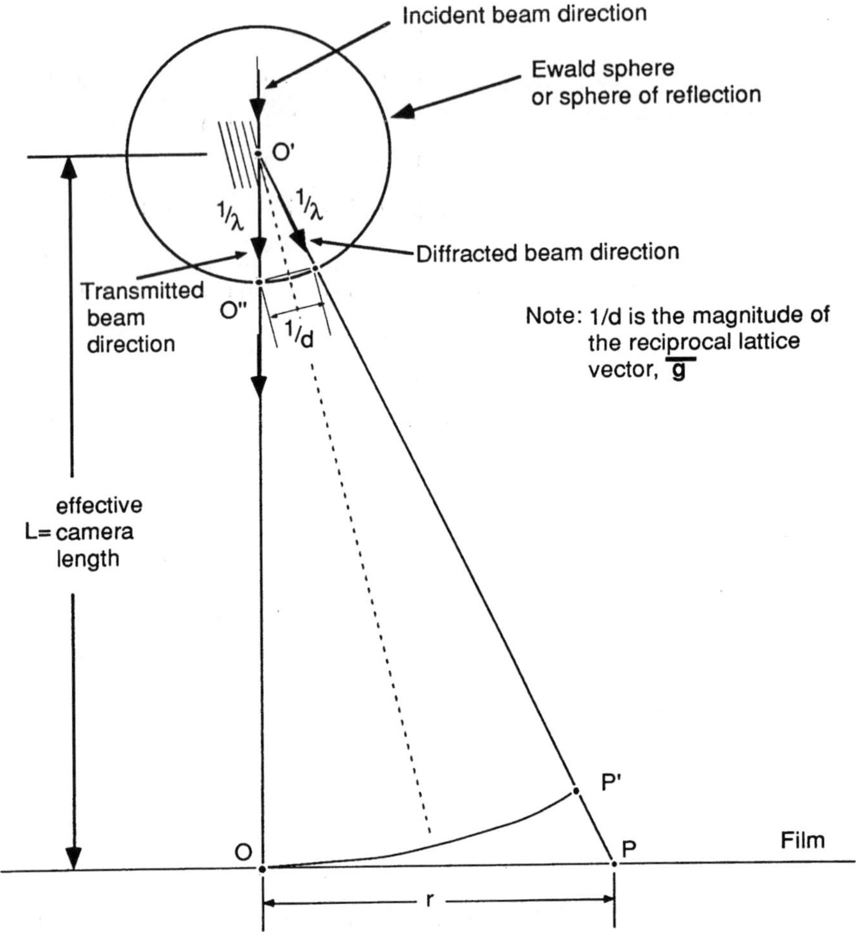

Figure 5.5. Schematic diagram showing the relationship between specimen crystal planes, the Ewald sphere, the reciprocal lattice vector (1/d-spacing), and the measured r-distance on the film. Effective camera length includes the magnification of the imaging lenses.

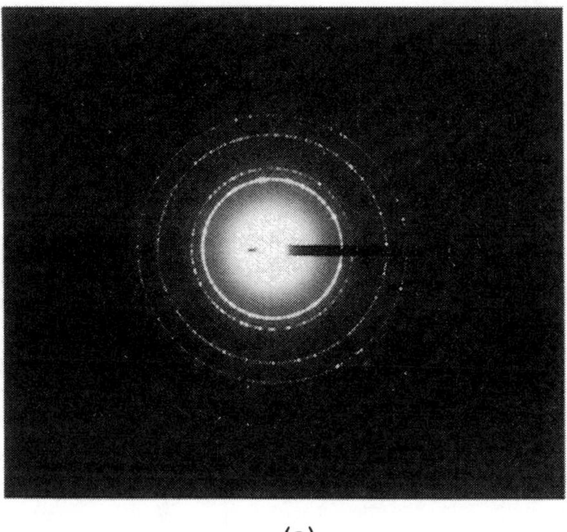

(a)

Use the cubic interplanar spacing formula

$$d_{hkl} = \frac{a}{\sqrt{h^2+k^2+l^2}}, \quad a = 0.40497 \text{ nm}$$

to calculate the d-spacings below for the first five reflections of aluminum:

r-distance	hkl	d-spacing	λL
10.3 mm	111	0.2338 nm	2.408 nm-mm
11.9	200	0.2025	2.409
16.8	220	0.1432	2.406
19.7	311	0.1221	2.405
20.6	222	0.1169	2.408
		Average =	2.407 nm-mm

(b)

Figure 5.6. Calibration of camera constant using a polycrystalline aluminum thin film diffraction standard. (a) Aluminum ring pattern. (b) Calculation of d-spacings for Al from lattice parameter and determination of the camera constant λL from r-distances measured on the ring pattern in (a).

improving somewhat at smaller d-spacings. For example, at large spacings 7Å and 8Å cannot be distinguished; whereas, at small spacings precision improves such that the ambiguity is between 1.25Å and 1.26Å. This has important effects on practical phase identification by polycrystal ring patterns, the electron diffraction equivalent of Debye-Scherrer patterns in x-ray diffraction. These patterns consist of concentric rings of greater or lesser perfection depending on the number of randomly oriented crystals contributing to the pattern. Ring patterns offer the advantage of displaying all possible diffraction vector lengths in a single pattern, but they harbor the dual disadvantages of lacking useful angular information for confirming matches and possibly disguising a fine grained mixture as a single phase. Ring patterns can be useful in practice because they are quick and because many 'unknowns' are actually simple structures. But when the unknown is a phase of modest to low symmetry, the many lines observed overlap those of so many candidate structures that a unique identification is not possible.

Interplanar Angles. Single crystal selected area diffraction patterns are useful because although they may not display all possible diffraction vectors in a single pattern, the angular relationships among planes are very apparent. In a typical single crystal pattern, all planes represented by the diffracted beams are parallel to the same direction in the crystal. This group of planes is called a zone, and the common direction is called the zone axis. A selected area diffraction pattern in which all the spots are in a periodic array is often referred to as a zone axis pattern (see Figure 5.7). Each diffraction vector $\boldsymbol{g}_{hkl}$ lies normal to a family of planes (hkl), and the angle between the normals to a pair of planes is equal to the angle between the planes themselves. Interplanar angles can be measured between vectors drawn from the transmitted beam to the diffracted beams. The same experimental errors that limit r-spacing precision also limit angular data to about $\pm 1\sim 2°$. Calculated interplanar angles can be tabulated for any given crystal structure (Andrews et al., 1967), but in any crystal the number of possible combinations of pairs of planes is very large, so phase identification on the basis of interplanar angles alone is not tractable for a large number of reference compounds. Angular information can be diagnostic, however, when combined with d-spacing information from specific diffraction vectors.

Zone Axis Patterns. Low-index zone axis patterns are the most diagnostic type of electron diffraction data since they are taken along a major crystallographic direction in the crystal, and because they contain information on both interplanar spacings and interplanar angles. But while some zone axis patterns can distinguish between two similar phases, others cannot as shown in Figure 5.8. When two

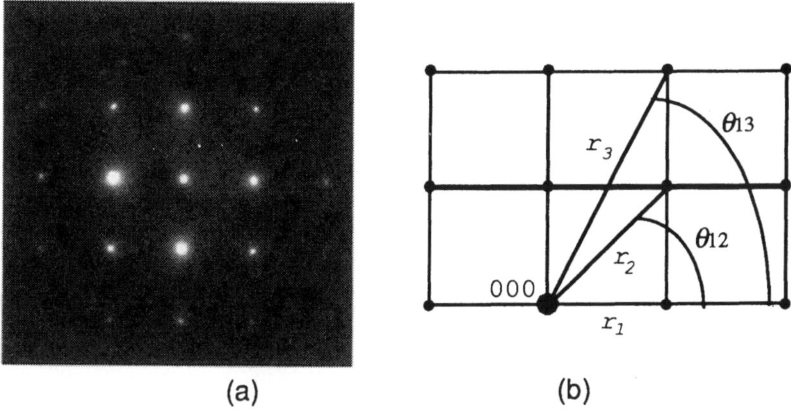

Figure 5.7. Zone axis SAED pattern of single crystal aluminum taken along [001]. (a) [001] pattern, (b) Selected r-distances along diffraction vectors and included angles.

or more individual zone axis patterns are obtained from the same crystal and the stage tilts or tilt/rotation settings are noted for each, spherical trigonometry can be used to obtain the interzonal angles. The interzonal angle between two zone axis directions can be calculated (Andrews et al., 1967) and tabulated, but, like interplanar angles, they are not used for wide-scale phase identification by themselves. Interzonal angles are very useful, however, for matching to small (i.e 'in-house') databases or to confirm a match resulting from a wide-scale search.

Two or more zone axis patterns may also be combined to obtain a basis triplet, three vectors that can be linearly combined to replicate the reciprocal lattice of the unknown crystal (Fraundorf, 1981). This basis triplet can be transformed to either the conventional unit cell or the reduced cell (Mighell and Himes, 1986), a useful diagnostic tool for phase identification (see Section 5.5.1).

Kikuchi Patterns. These sharp line diffraction patterns, formed from the diffraction of previously scattered electrons (Thomas,1970), can be used to obtain more precise angular information than from spot patterns (±0.1° versus ± 1°). Phases examined by this method must be strain-free and more than a few extinction distances thick at the analysis point. Interplanar spacings can be obtained from Kikuchi patterns by using the fact that the angular distance between an excess and defect pair is $2\theta_B$ which can be used to yield $d_{hkl} = \lambda/2\theta_B$ through the Bragg law. But this procedure is not recommended when spot or ring pattern data are available. Precise lattice parameter

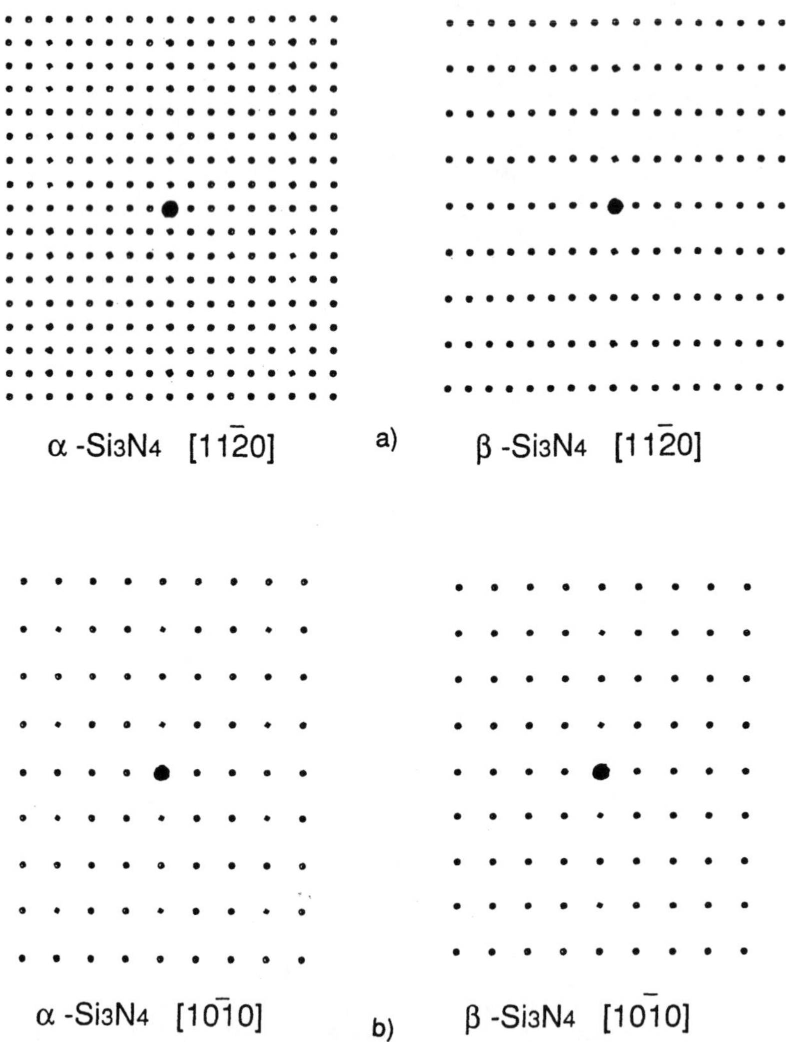

Figure 5.8. Zone axis patterns used to distinguish between α-silicon nitride and β-silicon nitride. The [1120] patterns (a) can be used to distinguish between the two phases, but the [10$\bar{1}$0] patterns (b) cannot.

determinations can also be made under certain favorable circumstances. Increased angular precision is often not critical to phase identification, but may be helpful when confirming an identification. Similar patterns may be obtained when stationary convergent beams illuminate thicker specimen regions.

Recently, wide-angle electron backscattering patterns, with a geometry similar to Kikuchi patterns, have been obtained from submicron regions of bulk SEM specimens (Dingley et al., 1989). These patterns provide interplanar spacing and angular information similar to that from selected area diffraction in the TEM but from typical SEM specimens. Most importantly, the low accelerating voltage used in the SEM results in a wide angular view (≤70°) of reciprocal space, which provides three-dimensional information about the size and shape of the unit cell in a single diffraction pattern. Electron backscattering patterns will undoubtedly be an important phase identification tool in the future.

Convergent Beam Electron Diffraction. In this technique a convergent electron probe is focused on the thin specimen rather than the parallel illumination used in SAED. The resulting diffraction pattern consists of disks rather than sharp points. The diameter of the disks is proportional to the convergence angle, which may be varied by the operator. The convergence angle usually is selected so that the disks just touch although there are reasons to use both larger and smaller convergence angles. The positions of the centers of the disks can be measured to obtain interplanar spacings and angular information similar to that obtainable from selected area diffraction patterns. Thus, single crystal CBED patterns yield information similar to SAED patterns but from regions about 100 times smaller (10-50 nm); however, the unique advantages of CBED lie elsewhere. The part of the pattern which yields information similar to SAED patterns is called the zero order Laue zone and is the region immediately around the direct beam. At small camera lengths, bright rings appear in CBED patterns at positions where the Ewald sphere intersects the other layers of the reciprocal lattice (first order and higher order Laue zones). The diameters of these bright rings give a measure of reciprocal lattice spacings normal to those in the plane of the CBED pattern, and hence can be used to obtain the lattice parameters and shape of the unit cell (Steeds, 1979; Raghavan et al., 1984). In addition, the fine structure of intensity variations within the disks can be used to determine crystal symmetry properties and to precisely measure lattice parameters. Since the convergent beam senses the three-dimensional aspects of the crystal, point group symmetry and often space group symmetry (Hahn, 1983) can be deduced from

CBED patterns taken on strain-free perfect specimens (see Volume 1, Chapter 7).

Obtaining Accurate Diffraction Data. There are four steps to accurate diffraction data:

1) Calibrate the microscope at one or more reproducible diffraction conditions. Set the specimen height to the eucentric position and obtain a focused image of polycrystalline Al or Au evaporated onto a carbon film. Focus the edge of the selected selected area aperture with the diffraction lens (Figure 5.4) and record a selected area diffraction pattern. Measure the first five ring diameters and divide by 2 to obtain a list of r-distances (Figure 5.6). Use tabulated or computed d-spacings for the first five reflections of Al to calculate five camera constants λL. Take the average the five camera constants even though there is a small systematic variation in the camera constant with r (Andrews et al., 1967).

2) Place the unknown specimen in the microscope under the exact same conditions. Before recording any diffraction patterns, tilt the specimen to find a low index zone axis pattern and make this pattern as symmetrical as possible before recording it.

3) Record additional different low index zone axis patterns from the same crystal and note the specimen tilt angles between each of the zones.

4) Measure the radial distances from the transmitted beam to 6-10 diffracted beams. If the available reference collection permits, work directly in r-spacings; otherwise, convert to d-spacings with Equation 5.1.

5.3 REFERENCE DATA

Efficient and successful phase identification requires accurate experimental data and an appropriate reference database. It also requires a clear understanding of both the value and the limitations of the experimental data in relation to the characteristics of the reference collection that is being searched. The type of information available to search against obviously affects the choice of analytical technique, but compositional information and d-spacing data are valuable in all cases. Conversely, it may be wise to select a database for searching whose contents match the output of the available analytical technique. Fortunately, several sources of reference material exist.

5.3.1 Data Record Information

Most databases contain similar information which is organized in different ways, while some databases contain special additional information that can be of great help in phase identification. The basic data required are the following:

- Chemical formula of substance
- Name of substance (written forms, mineral name)
- Unit cell (cell vectors ***a***, ***b***, ***c***; angles α, β, γ)
- d-spacings (or r-spacings at a given camera length)
- Space group, Pearson symbol, structure type
- Intensity data (x-ray powder diffraction intensities)
- Journal reference (including date, measurement temperature, material source and purity)
- Indexing number associated with a particular database.

These data may be found in three types of reference materials: "in-house" data from compounds routinely examined in a particular laboratory, traditional printed references on cards or in books, and computerized databases. The latter two types of resources are described in the following sections. In addition, there are several software packages that calculate electron diffraction patterns along any zone axis given the space group and the six unit cell parameters.

5.3.2 Reference Material in Printed Form

Tables of Unit Cell Data. Low index zone axis patterns often provide one or more lattice parameters for an unknown substance, especially for cubic, tetragonal, or orthorhombic crystals. In such cases, phase identification may be possible from the cubic lattice parameter or from the ratio of two lattice parameters listed in order. Tables for this type of determination are published as *Crystal Data: Determinative Tables* (Donnay and Ondik, 1972). As discussed above, the precision of lattice parameters determined by selected area electron diffraction is typically much worse than by x-ray diffraction. Thus, a lattice parameter for a substance measured by electron diffraction will lead to many possible matches from these tables; however, these matches can be narrowed to a manageable number when combined with compositional information. For example, if the lattice parameter for a cubic material were measured as 3.62±0.05Å, then the above tables would yield 38 possibilities. By knowing that only copper was detected in the EDS x-ray spectrum, the list is reduced to one candidate. Unit cell dimensions in three directions can be obtained from convergent beam electron diffraction by analysis of higher order Laue zone rings along the lowest-index zone directions. However, measuring just the lattice parameters may not be sufficient to confirm the identity of an unknown phase even when combined with compositional information. This is because double diffraction at forbidden reflections may cause an incorrect interpretation of the unit cell size.

Tables of Crystal Structures. Often there is a need for information about the crystal structure, the atomic positions within the unit cell, the space group, or other crystallographic and physical data for a compound that is known by name or crystal structure type. Three such compilations for inorganic crystals have been updated on a regular basis. *Crystal Structures* (R. W. G. Wyckoff, 1963) provides a listing of the space group and atomic positions within the unit cell as well as a visualization of many crystal structures through illustrations of atomic packing. *Structural Inorganic Chemistry* (A. F. Wells, 1984) also contains references to crystallographic data and illustrations of atomic packing for a wide variety of compounds. However, these two works do not necessarily cover the many of the thousands of intermetallic compounds. For these materials the standard reference is *Pearson's Handbook of Crystallographic Data for Intermetallic Phases* (Villars and Calvert, 1985). This compilation contains data on over 10,000 elemental, binary, ternary, and quarternary phases excepting halides and ternary and quarternary oxides. Compounds are tabulated alphabetically by element, and, in a separate section, the compound and its data may be found arranged according to Pearson symbol and space group number. A significant advantage of *Pearson's Handbook* is that following the entry for a particular structure type, such as the copper structure, most of the known phases with the same crystal symmetry (and Pearson symbol) are listed. For example, there are 375 phases listed with the copper entry as having the same symmetry and Pearson symbol (cF4). While less complete, the older version of this work (Pearson, 1967) is still very useful. These works may be used to identify unknowns when the lattice parameters are determined from zone axis patterns or as a check when using the methods of identification described below.

JCPDS-ICDD Powder Diffraction File. The most familiar reference collection for crystallographic phase identification is the Powder Diffraction File (PDF) compiled and distributed by the Joint Committee on Powder Diffraction Standards, a part of the International Centre for Diffraction Data (JCPDS-ICDD *Powder Diffraction File*, 1990). Originally, known to scientists and engineers as the ASTM Card File, these 3"x5" cards are a standard fixture in x-ray diffraction laboratories around the world. Each card contains d-spacings and peak intensities measured experimentally by x-ray diffraction (XRD) methods on a real specimen or calculated from crystallographic data.

New cards are issued annually in sets covering both inorganic and organic materials with each substance given a specific PDF card number. The information for the cards is derived largely from the open literature, and the cards are continually reviewed and edited to retain only the highest quality data. These patterns are very useful for phase identification by x-ray diffraction since they record the characteristics of the pattern that would be observed in any x-ray laboratory. However, the largest d-spacings may be absent from the PDF (see Figure 5.9) because published x-ray diffraction patterns may not have been recorded below 20° 2-theta or the x-ray background in this region may obscure a weak peak. But the largest d-spacings of a substance are the most diagnostic, and they will always be observable by electron diffraction. A convincing example of this is shown in Figure 5.10. Recently, a few powder patterns calculated from unit cell data have been entered into PDF which do include the largest d-spacings of the crystal, but they are by far the exception. The current PDF contains over 40,000 inorganic data records (through Set 42) that cover most of the materials of interest. However, many compounds have multiple entries because the patterns were taken on material from different sources or were measured to different degrees of precision. Most solid solutions are represented only by end members because of the large number of possible combinations.

There are several indexes or search manuals to the JCPDS Powder Diffraction File. The *Alphabetical Index* can be used to manually search on elements observed in the EDS spectrum or known to be in the unknown phase by some other means. The best known numerical index is the *Hanawalt Index* which orders the data in terms the d-spacings of the three strongest x-ray powder diffraction lines. This method makes use of the high precision available (better than 0.01%) in x-ray diffraction d-spacings to find a match to the experimental XRD pattern. Since electron diffraction intensities are unsuitable for analysis by this method, the Hanawalt search method is not used for phase identification by electron diffraction. The *Fink Index* orders the data in terms of descending d-spacing for the eight strongest x-ray diffraction lines. This index was originally developed to aid identification by electron diffraction but it strongly depends upon the accuracy of the measurement for the largest d-spacing. Since in electron diffraction the largest d-spacing will have the poorest measurement precision (and unreliable accuracy), this method by itself is difficult to use for identification. The recent *Elemental and Lattice Spacing Index* greatly improves the search procedure by combining compositional data from the EDS detector with d-spacing data obtained by electron diffraction (JCPDS-ICDD *Elemental and Lattice Spacing Index*, 1990). This index is the successor to the "Max-d" index (Anderson and Johnson, 1979) and now lists over 50,000

HKL	PDF 5-708 d-Spacing (Å)	2θ @λ=1.54 Å	Computed d-Spacing (Å)	r-Spacing @λL=2.5Å-cm (cm)
1 0 0	–	14.2°	6.22	0.40
2 0 0	–		4.40	0.56
1 0 1	–		4.03	0.62
2 1 0	–	22.3°	3.93	0.64
1 1 1	–		3.67	0.68
3 1 0	–		2.78	0.89
2 2 1	–	34.8°	2.57	0.97
3 0 1	–		2.46	1.01
3 1 1	–		2.37	1.05
0 0 2	2.26	38.6°	2.27	1.10
4 0 0	–		2.20	1.12
3 2 1	–		2.15	1.16
4 1 0	2.13	42.4°	2.13	1.17
1 1 2	2.13	42.4°	2.13	1.17
3 3 0	2.06	43.6°	2.07	1.20
2 0 2	2.02	44.8°	2.02	1.24

Figure 5.9. Example of missing low angle (large d-spacing) x-ray diffraction peaks on PDF card for Fe-Cr sigma phase. When the powder pattern is computed from the unit cell lattice parameters, it is clear that all of the missing peaks are easily detectable by electron diffraction.

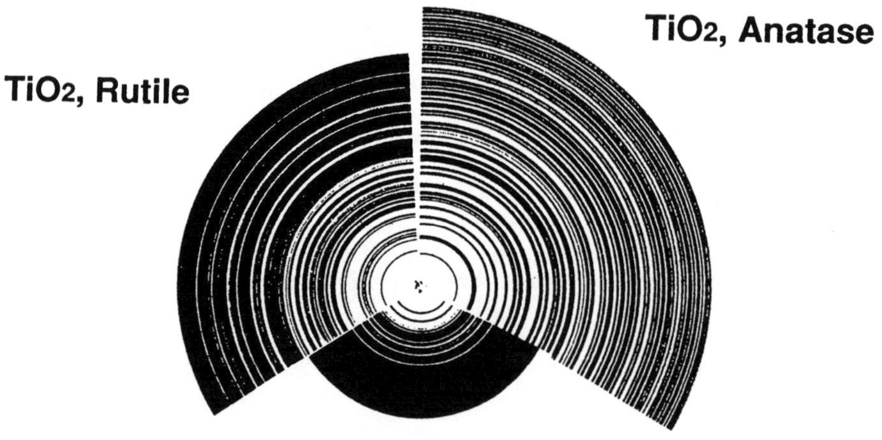

Figure 5.10. Sections of ring patterns of the three phases of TiO_2 (rutile, brookite, and anatase) showing that the largest d-spacings (smallest r-spacings) are the most diagnostic for identifying these compounds.

inorganic compounds. By first searching alphabetically for the elements present, the correct element or compound will be found on one or two pages of the book and can be identified from the largest five to eight d-spacings measured by electron diffraction. Not only is this search method efficient, but it also orders compounds such that oxides and hydroxides of compounds are found in the same short list of 20-50 possibilities. This index provides the JCPDS-ICDD PDF card number and/or the NIST Crystal Data File (CDF) number. The latter compilation of data has special advantages for phase identification by electron diffraction as described below.

Other Resources. While almost all compounds likely to be encountered are listed in the references above, it is still possible to come across new phases that have not yet been cataloged. Thus, it is always important to be aware of the recently published diffraction patterns of new phases listed in journals such as *Acta Crystallographica, Powder Diffraction, Journal of Solid State Chemistry,* etc.

5.3.3 Reference Material in Computer-Readable Form

Several of the computer databases below are now sold on magnetic tape and compact disk read-only memory (CD-ROM). The relatively low cost of a CD-ROM attached to an IBM PC or PC-compatible computer makes it possible to search a database for a match right in the AEM laboratory. With some AEMs, d-spacings may be read directly from a calibrated microscope making complete phase identification possible even during examination of the specimen in the microscope. These CD-ROM retrieval systems also can be used to access the literature on the thousands of compounds listed.

JCPDS-ICDD PDF-1 and PDF-2. These are computer readable versions of the Powder Diffraction File (JCPDS-ICDD Powder Diffraction File, 1990). PDF-1 contains d-spacings and x-ray diffraction intensities, the chemical formula and name, and the PDF number. PDF-2 was developed more recently when inexpensive computer storage allowed storage of all of the PDF-1 indexing information plus the supporting crystallographic and reference information that is listed on the PDF card. However, most of the programs for searching these databases use algorithms involving diffracted x-ray intensities and thus are not useful for identification by electron diffraction.

NIST Crystal Data File. This database (Mighell et al., 1987; Himes and Mighell, 1987; NBS CRYSTAL DATA, 1987) is an outgrowth of *Crystal Data* (Donnay and Ondik, 1972) and is under the auspices of the National Institute for Standards and Technology (NIST), formerly the National Bureau of Standards (NBS). This database contains crystallographic, physical, and reference information on 136,000 organic and inorganic materials. Each entry in the database includes both the conventional unit cell and a reduced cell for a given compound. The reduced cell is a mathematically unique, monomolecular (i.e., primitive) cell that nevertheless possesses all of the symmetry elements of the full conventional cell. For example, it is well known that a primitive rhombohedral cell exists within a common face-centered cubic cell (Cullity, 1978). The reduced cell has the very useful property that similar lattices have similar reduced cells, even though their conventional unit cells may appear very different. A slight distortion which may throw a cubic cell into the orthorhombic class or a hexagonal cell into the rhombohedral class cause the reduced cell to change only slightly in edge length or angle, with a negligible effect on reduced cell volume. The Crystal Data File is designed to be searched on the basis of reduced cell size, a method which is very tolerant of experimental errors, but which does require the analyst to determine at least a trial unit cell before starting the search/match procedure. Furthermore, the Crystal Data File contains many compounds that have been reported in the literature only as a unit cell (for example, from neutron diffraction or from TEM investigations of precipitation hardening), and without an x-ray powder pattern required for inclusion in the PDF. It therefore contains compounds that are not available in other reference sources. This database has some direct applications to phase identification by electron diffraction because it may be searched by conventional unit cell dimensions and space group, which makes it useful for locating isomorphs of a given compound.

NIST/Sandia/ICDD Electron Diffraction Database. Since this database lists compounds in terms of composition and reciprocal interplanar spacing, but not reflection intensity, it is the computer database of choice for phase identification by electron diffraction (Carr et al., 1989). By merging the unit cell data in both the PDF and the CDF, d-spacings have been calculated to insure that the diagnostic largest d-spacings are present. Even compressed for storage in a small computer, the data are as complete and as accurate as possible. The precision, however, has been reduced from better than 0.01% to a level more appropriate for electron diffraction work (about 1% at 1.5 Å), which also increases the speed of the search. This database currently contains a total of 71,142 entries (PDF-2 Sets 1-36). The space group, unit cell data, and calculated d-spacings are available for 59,612 of these entries (primarily from the CDF). This database is available for small microcomputers used with EDS spectrometers on AEMs and on CD-ROM for use with a personal computer.

A search algorithm (Carr et al., 1986; JCPDS-ICDD ED-PCSRCH, 1990) for this database relies on pattern matching between composition and r-spacing bit maps for the unknown and reference compounds. The tolerances for the screening are wide enough so that valid matches are not lost due to the imprecise nature of electron diffraction data. Double diffraction is also considered so that valid matches are not eliminated because of that effect. A typical search of 71,000 compounds takes about 2 minutes on an IBM PC/XT and extracts about 10-15 different matches or candidate compounds for a given set of input data. This software also has programs for calculating distances and angles in both real-space and reciprocal space as well as for indexing diffraction patterns.

Other Computer Databases. The German FIZ-4 Inorganic Crystal Structure Database (ICSD) contains crystal structure information for 24,000 inorganic compounds, except metals (Bergerhoff and Brown, 1987). In addition to searches on the compound name or chemistry, the ICSD provides a graphical display of the crystal structure. The National Research Council of Canada Metals Structure Database (MSD) contains information similar to the ICSD but primarily for the metals and their corrosion products (Rogers and Wood, 1987). Since electron beam damage can be severe for organic and organometallic materials, databases containing primarily organic compounds, such as the Cambridge Structural Database, have not been discussed here.

5.4 STRATEGY OF SEARCH/MATCH PROCEDURES

The goal of phase identification by search/match procedures is to list all the previously reported compounds that could possibly have produced the experimental observations from the unknown phase. Searches may be performed manually, but the need for objectivity and completeness leads immediately to computer-based search routines. Even when searching a computer database, successful phase identification requires the analyst to: (1) obtain accurate experimental data, (2) search the database for potential matches, (3) test the matches against the experimental data, and (4) confirm the identification with reliable tests. While these steps are important for all search procedures, they will be discussed here for the specific case of searching compositional and r-spacing (d-spacing) data.

5.4.1 Searches Using Composition and r-spacings

Step 1: Obtaining reliable data. A particle for phase identification is most easily selected using the EDS spectrum since particle morphology and electron diffraction may not quickly tell the analyst which particle is "different" from others in the same field of view. Compositional data is best obtained from isolated areas of the unknown such as fragments of the pure phase or from small particles collected on a clean extraction replica. When the unknown is a small particle embedded in a matrix, a reasonable qualitative analysis for the major elements may be obtained by subtracting a matrix-only spectrum from a matrix-plus-unknown spectrum. Without additional information to the contrary, all of the unobserved light elements ($Z<11$) should be tentatively considered possible components of the unknown phase. If the EDS spectrum shows no peaks at all from the unknown phase, the phase could still be BN, BeO, ice, etc. In such a case, unless some other method for detecting these light elements is used such as EELS, the analyst is totally dependent upon diffraction data for identification. Care should be taken to recognize, and eliminate from consideration, x-ray peaks from elements present in the instrument (grid, specimen holder, polepiece, apertures) and obvious contaminants (such as solvent residues). Most phases listed in the databases consist of six or fewer elements, and thus, the elements to be listed from the unknown phase should be present in concentrations large enough to be easily observed in the x-ray spectrum.

Diffraction data should not be recorded until the specimen height is aligned on the eucentric goniometer stage. It is best to record only low-index zone axis patterns, i.e., regular arrays of closely spaced spots. Such patterns should have spot intensities that are as symmetrical as possible about the transmitted beam. Spots closest to the central beam (small r-spacing and large d-spacing) are the most diagnostic for a given lattice. Symmetrical patterns suffer the least from the distortions and artifacts that can plague electron diffraction patterns. Six to ten spots of different *hkl* are usually sufficient for search purposes, and it is useful to obtain these from two or three different zone axis patterns on the same crystal. The amount of specimen tilt between each zone axis pattern should be recorded. While all the diffraction patterns should be photographed, the data of immediate use are the six to ten smallest r-spacings which may be read out directly on some microscopes.

Step 2: Searching the database for potential matches. Both manual and computer-based interrogations of the database begin with a search based on composition. Compositional information may be divided into three categories to aid the search:

1) *Observed elements*: This important category includes the elements that appear in the EDS spectrum of the unknown phase. Successful matches usually include one or more of these elements.

2) *Required elements*: This optional category is applied when certain elements are known to be present in the unknown. Such an element may not be detectable with the EDS system if Z<11. For example, carbon is likely to be present in certain precipitates in steels. However, setting carbon as a required element may cause elimination of the oxides, nitrides, and borides that are also common in steels. The required element category should be used with caution since often the unknown is an unexpected phase that may not contain a 'required' element.

3) *Excluded elements*: This optional category allows the computer to pass over large portions of the database that cannot possibly contain valid matches. A good rule of thumb is to initially exclude elements with Z≥11 (Na) that were not observed in the EDS spectrum.

The initial part of the search consists of retrieving from the data base only those compounds that contain one or more observed and required elements but no excluded elements. This search can be performed manually with the *Alphabetical Index* to the PDF (JCPDS-ICDD Powder Diffraction File, 1990) or the *Elemental and Lattice Spacing Index* (JCPDS-ICDD *Elemental and Lattice Spacing Index*, 1990). Clearly, computerized searching of the Powder Diffraction File, the Crystal Data File, or NIST/Sandia/ICDD Electron Diffraction Database (EDD) is the most efficient method. Since the EDD is designed specifically for searches on compositional data and electron diffraction data, it is currently the best database choice. Applying the above compositional criteria usually reduces the number of possible compounds from the 71,000 in the EDD to about 100.

The experimental diffraction data from the unknown (r-spacings or d-spacings) are then compared to the reference diffraction data for the compounds that passed the compositional screening. First, the number of experimental reflections not found in a given reference record are counted. If the number of missed reflections exceeds one-third to one-half of the total number of experimental reflections, that compound should be rejected. At this point no allowance for double diffraction should be made. Approximately one-third of the compounds that passed the compositional screening will be eliminated by this first diffraction criterion. Secondly, a computed figure-of-merit may be used to rank the remaining possible matches in terms of the goodness-of-fit to the experimental data. The method used should assign a higher relative weight to large d-spacing reflections and count a given reflection as a hit if it matches a reflection or matches a reflection at d/2 (or 2r). The latter criterion accounts for the most common aspect of multiple diffraction. Successful matches are those that exceed some empirically defined level of the figure-of-merit. At this point only a few compounds remain as possible matches, discounting multiple entries for the same compound. Those compounds that remain may be ranked in order of their goodness-of-fit to the experimental data, or simply ordered by increasing unit cell volume, since in the experience of the authors, a true match is more likely to be made with a smaller, simpler unit cell, in most cases.

Step 3: Testing matches against experimental data. A good test of the correctness of a match may be obtained by indexing the experimental zone axis patterns on the basis of the unit cell data for the surviving candidate compounds. Crystallographic databases that list unit cell data, and the indexed powder patterns calculated from them, are the most useful for this task. Tables of r-spacings or d-spacings for allowed reflections are prepared for each candidate compound, and from these data an attempt is made to index the experimental zone axis patterns in terms of each remaining candidate compound. Another way of indexing the experimental patterns is to compare them to compilations of standard zone axis patterns, but this will frequently fail unless the unknown happens to be cubic, tetragonal, or hexagonal. General methods for indexing electron diffraction patterns are given by Edington (1976) and Andrews et al. (1967). Several computer programs have been devised to either index diffraction patterns (an early example is by Rhodes, 1975) or compute patterns for crystals in particular orientations given space group and unit cell data.

Direct indexing of spot patterns can be accomplished manually and usually consists of the following tasks:

1) Assigning tentative Miller indices (*hkl*) to several of the reflections closest to the transmitted beam in a given pattern using d-spacing (r-spacing) tables computed for each candidate compound.
2) Comparing the d-spacings (or better, r-spacings) for the tentative (hkl) assignments with those computed for the candidate compounds.
3) Permuting the tentative indices and changing signs to arrive at a set of indices that are self-consistent when added vectorially.
4) Obtaining the vector cross product of any two indexed spots in a self-consistent set to find the zone axis [*uvw*] of the pattern.

5) Computing the diffraction pattern for the zone axis determined.
6) Checking the consistency between all of the spots in the experimental and computed patterns in terms of spacings and angles between spots.

After testing the indexing of the experimental pattern against all the candidate compounds, the list of candidate compounds usually will be reduced to a single match or to a few related or isomorphous compounds.

Step 4: Confirming the identification. Agreement between experimental zone axis patterns and computed patterns merely demonstrates consistency and should not be considered a positive identification. Confirmation of the identification requires systematic tilting the unknown phase in the microscope and recording two (or more) zone axis patterns and the tilt angles between them. Index each pattern successfully with the same unit cell and determine the zone axis for each by taking the cross product of two diffraction vectors. Quantitative agreement within 1-2° between the interzonal angles obtained in a systematic tilting experiment and those calculated for a candidate compound may be considered proof of identification. It is at this point that quantitative compositional analysis can be of significant benefit to determine if the unknown is a solid solution between two or more isomorphous compounds. This task can only be accomplished on the basis of composition because electron diffraction data are rarely precise enough to detect a change in lattice parameter with composition.

• *Example 1: Manual Search - Unknown phase in Pd-doped 13Cr-8Mo steel*

Approximately 0.5 weight% Pd was added to a heat of 13Cr-8Mo in an effort to produce a quench & tempered high strength steel with improved resistance to hydrogen embrittlement. When the experimental alloy did, in fact, exhibit improved toughness in a hydrogen environment, a TEM study was initiated to look for microstructural differences between the standard and modified 13-8 steels. It was immediately apparent that a new phase formed in the Pd-modified heat since globular particles were observed by TEM (Figure 5.11) containing only Pd & Al as detected by EDS (Figure 5.12a). Selected area diffraction patterns showed that the Pd-Al particles exhibited an orientation relationship with the ferrite matrix (Figure 5.12b). Measurements made on several such patterns wereconverted to accurate d-spacings by using the accompanying ferrite pattern (BCC, $a = 2.87$ Å) as an internal standard. These data were compared to data compiled for Al-Pd compounds in the

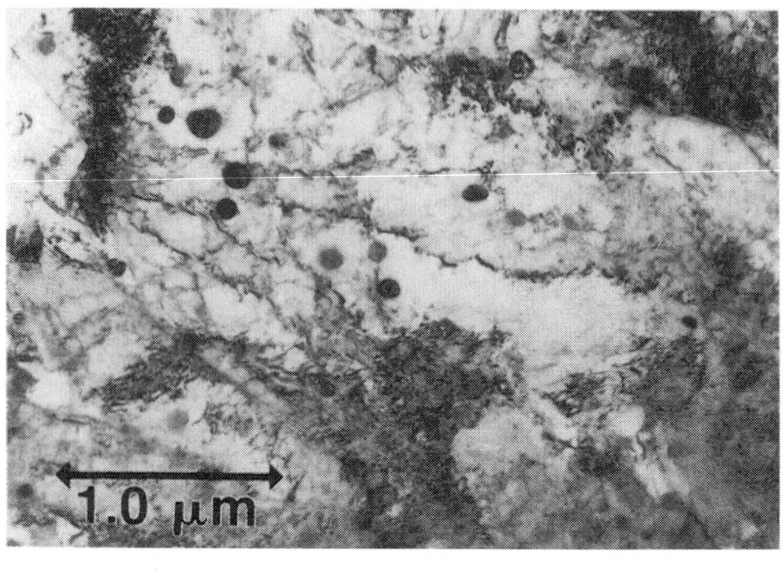

(a)

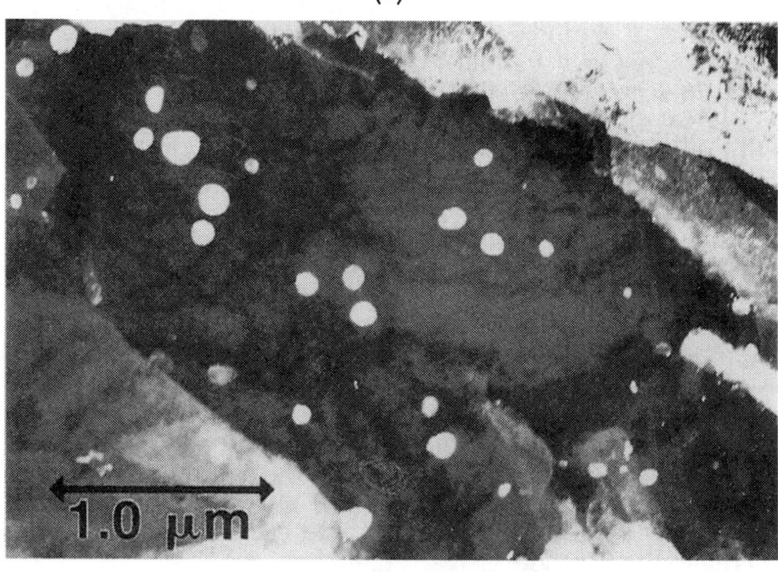

(b)

Figure 5.11. Bright-field (a) and dark-field (b) TEM images Pd-Al unknown phase in an experimental heat of 13Cr-8Mo steel.

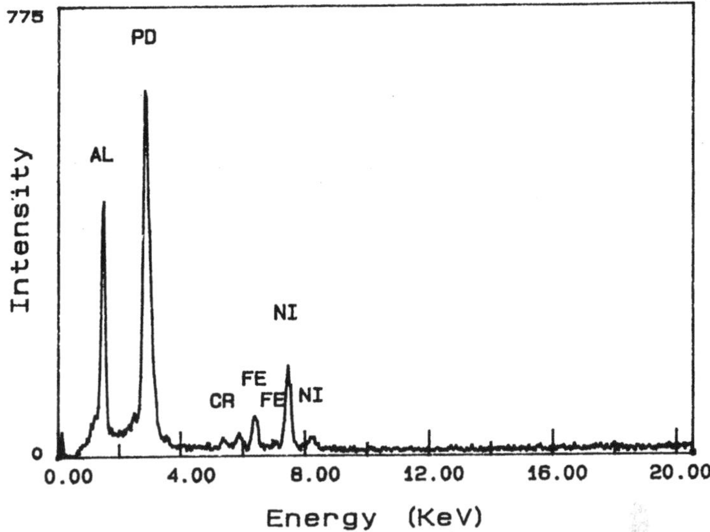

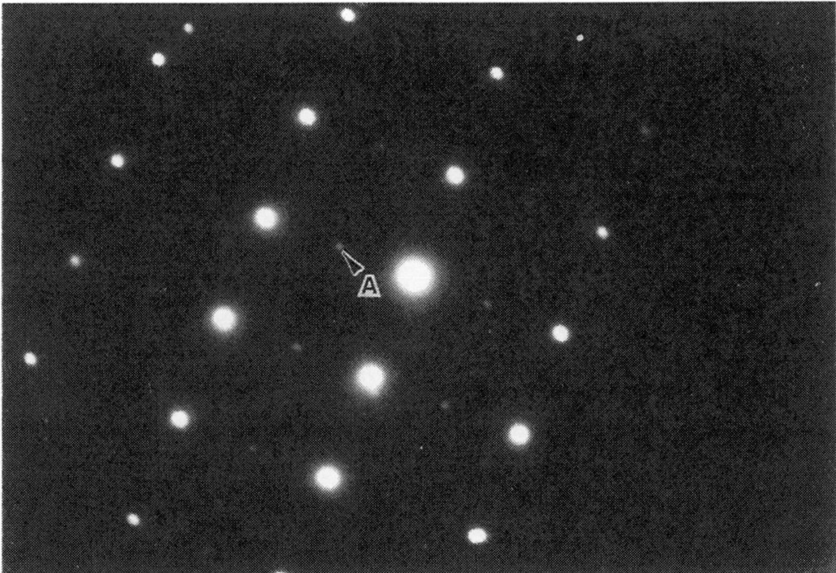

Figure 5.12. Data for analysis of unknown phase in Figure 5.11. (a) EDS spectrum shows Al and Pd as major elements. (b) Selected area diffraction pattern show all particles exhibit the same orientation relationship with surrounding grain.

a)

High Z			Low Z	Largest/Recorded "d"									
Al	Pb	Sr	O	7.66	6.63	5.93	5.41	4.69	4.42	4.00	3.83	3.68	3.54
Al	Pb	Sr		5.60	4.14	3.15	2.86	2.80	2.75	2.23	2.09	2.07	2.00
Al	Pb	Ta	O	6.07	3.72	3.17	3.03	2.63	2.41	2.15	2.02	1.86	1.78
Al	Pd			7.83	4.90	4.52	4.15	3.91	3.67	3.06	2.96	2.85	2.68
Al	Pd			5.21	4.76	4.03	3.73	3.19	3.08	2.91	2.74	2.68	2.60
Al	Pd			5.17	3.65	2.98	2.58	2.11	1.95	1.83	1.72	1.63	1.56
Al	Pd			5.16	4.21	4.13	3.77	3.62	3.54	3.33	3.21	2.78	2.71
Al	Pd			4.43	3.89	3.61	3.15	3.00	2.70	2.55	2.49	2.34	2.25
Al	Pd			3.44	2.81	2.43	2.18	1.99	1.72	1.62	1.54	1.47	1.41
Al	Pd			3.43	2.93	2.81	2.17	1.98	1.72	1.64	1.62	1.46	1.40
Al	Pd			3.41	2.78	2.41	2.16	1.97	1.70	1.61	1.52	1.45	1.39
Al	Pd			3.05	2.16	1.76	1.52	1.36	1.24	1.08	1.02	.964	.919
Al	Pd	Si		3.41	2.78	2.41	2.16	1.97	1.70	1.61	1.52	1.45	1.39
Al	Pd	Th		6.29	4.18	3.63	3.48	3.15	2.74	2.51	2.38	2.10	2.09
Al	Pr			8.03	5.68	4.64	4.01	3.59	3.28	2.84	2.68	2.54	2.42

Formula/Mineral Name	Code	Ratio	PDF#	CDF#
$Pb_{8.6}Sr_{0.4}Al_8O_{21}$	CP	13.2630		808240
Al_2Pb_2Sr	TI	2.5090	31–0023	709068
$AlPb_2TaO_6$	CF	10.5100		035276
AlPd	HR	0.3353	31–0027	123419
Al_3Pd_5	OP	0.5139	29–0064	706896
Al_3Pd_2	HP	1.2248	06–0654	026050
Al_3Pd	OP	0.9408	06–0705	007698
$AlPd_2$	OP	0.6950		107430
AlPd	CP	4.8680	34–0564	706897
AlPd	HP	1.4152	06–0625	026060
Al_3Pd_4	CP	4.8200	29–0066	706898
AlPd	CP	3.0490	06–0626	020475
Al_3Pd_4Si	CP	4.8200		032130
AlPdTh	HP	0.5754	21–0782	103206
Al_2Pr	C	8.0290		110541

b)

(partial) Unknown r-spacings cm	(partial) Unknown d-spacings Å	hex PdAl CDF 026060 PDF 06-0625	cubic PdAl CDF 706879 PDF34-0564	cubic PdAl CDF 020475 PDF06-0626
		3.43	3.44	...
0.84±.02*	3.05±.08	...	...	3.05
		2.93	...	...
		2.81	2.81	...
		...	2.43	...
1.20±.02	2.14±.04	2.17	2.18	2.16
		1.98	1.99	...
1.48±.02	1.73±.03	1.72	1.72	1.76
		1.64	...	...
		1.62	1.62	...
1.70±.02	1.50±.02	...	1.54	1.52
		1.46	1.47	...
		1.40	1.41	...
		...	...	1.36
2.09±.02	1.22±.02	...	...	1.24
		...	...	1.08
2.50±.02	1.03±.01	...	...	1.02
		...	...	.96

* very faint reflection

Figure 5.13. (a) Reproduction of part of a page from *Elemental and Lattice Spacing Index*. The compounds containing only Al and Pd are marked. Note that not all entries have both PDF and CDF numbers. (b) Measurements of r-spacings and calculation of d-spacings from the Al-Pd unknown. Three candidate matches are listed. The best match is with PDF06-0626.

HKL		d-Spacing (Å)	r-Spacing @ λL=2.56 Å-cm (cm)
1 0 0	w	3.0500	0.84
1 1 0	s	2.1567	1.19
1 1 1	w	1.7609	1.45
2 0 0	s	1.5250	1.68
2 1 0	w	1.3640	1.88
2 1 1	s	1.2452	2.06
2 2 0	s	1.0783	2.37
3 0 0	w	1.0167	2.52
2 2 1	w	1.0167	2.52
3 1 0	s	0.9645	2.66
3 1 1	w	0.9196	2.78
2 2 2	s	0.8805	2.91
3 2 0	w	0.8459	3.03
3 2 1	s	0.8151	3.14
4 0 0	s	0.7625	3.36
4 1 0	w	0.7397	3.46
3 2 2	w	0.7397	3.46
3 3 0	s	0.7189	3.56
4 1 1	s	0.7189	3.56
3 3 1	w	0.6997	3.66
4 2 0	s	0.6820	3.76

Figure 5.14. Allowed low index reflections, d-spacings, and r-spacings at a camera constant of 2.56 Å-cm for the compound PdAl (CsCl structure, a = 3.05 Å).

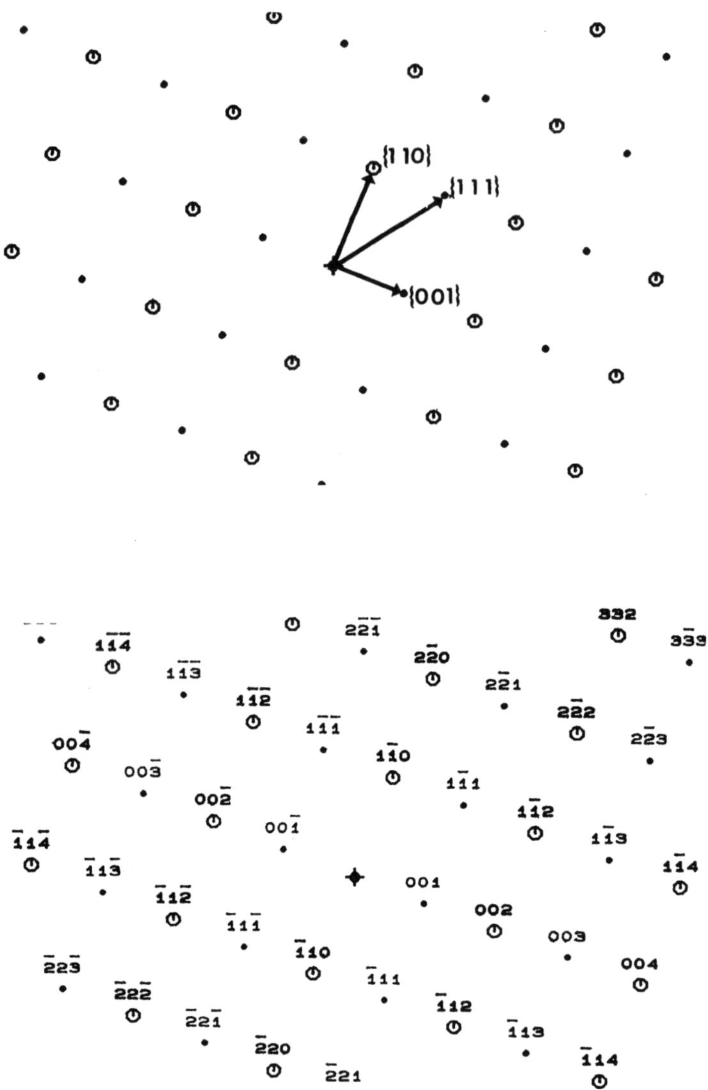

Figure 5.15. Indexing of the PdAl zone axis pattern in Figure 5-12b (a) Tentative indices assigned to each of several spots with a self consistent set of indices selected. (b) A computed pattern for the [110] zone.

Elemental and Lattice Spacing Index (Figure 5.13). Among several possibilities, the ordered cubic phase PdAl (CsCl structure, $a = 3.05$ Å) matched best, particularly when including a very faint (100) reflection detected in an overexposed [001] zone axis pattern taken to confirm the suspected cube-on-cube PdAl/ferrite orientation relationship. The lattice parameters for the best match were used to produce a trial indexing of Figure 5.12b. The first indexing step was to compute the allowed reflections, d-spacings, and r-spacings at the camera constant used in Figure 5.12b. These data are shown in Figure 5.14. Using these data, tentative indices can be assigned to diffraction spots on the basis of r-spacings (see Figure 5.15a). Note that several equivalent indices are possible for each diffraction spot. Next, one or more sets of indices are chosen from among the possibilities; the sets are chosen so that they are self-consistent when added and subtracted with one another. The vector cross product of the indices of any two diffraction vectors in a self-consistent set gives the zone axis direction of the full pattern. This zone axis direction was then used to compute a fully indexed pattern (Figure 5.15b) for comparison.

• *Example 2: Computerized Search - Contaminant on a Metallized Ceramic*

A thin layer of porous sintered molybdenum powder was bonded to a 94% alumina ceramic part to provide a substrate for a subsequent brazing operation (Figure 5-16). The metallized layer was produced by stenciling a slurry of molybdenum powder, titanium hydride, and an organic binder on to the ceramic part, then firing at 1495°C under a wet hydrogen atmosphere (dew point, 30°C). Normally, a clean molybdenum surface results, which is then nickel plated to finally produce a suitable surface for brazing. Occasionally, the molybdenum surface produced by the process was not clean, but rather was contaminated with blocky particles which did not accept the subsequent nickel plating. Weak braze joints resulted. It was decided that the first step in controlling the problem was to identify the blocky particles, hoping to understand their nature and prevent their occurrence. Specimens for analytical electron microscopy were prepared simply by scratching the contaminated surface with a pin and collecting some of the fragments on a carbon film over a copper support grid. Bright-field TEM images revealed the presence of two types of particles (Figure 5.17) which EDS analysis showed to be either molybdenum metal or an unknown phase X containing Al & Ti with apparently more Ti than Al. Diffraction patterns were obtained from several of the unknown particles, from which eight unique spacings were identified (Figure 5.18). These data were used to perform a computer search of the NIST/Sandia/ICDD Electron

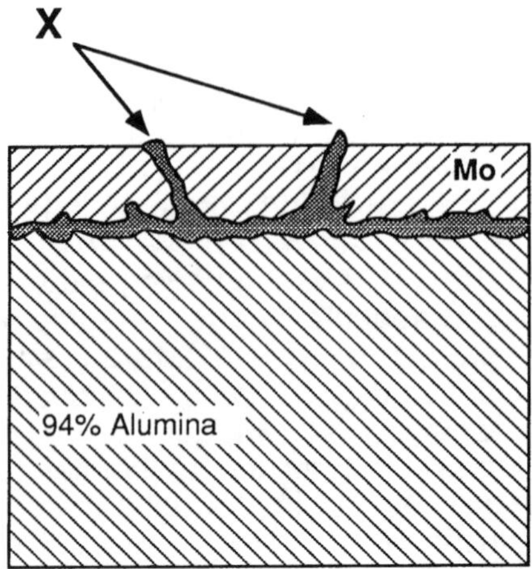

Figure 5.16. Schematic cross-section of metal/ceramic brazing assembly interface showing the metallized layer, reaction layer, and 94% Al_2O_3 substrate. Note that the unknown surface particles are connected to the reaction layer.

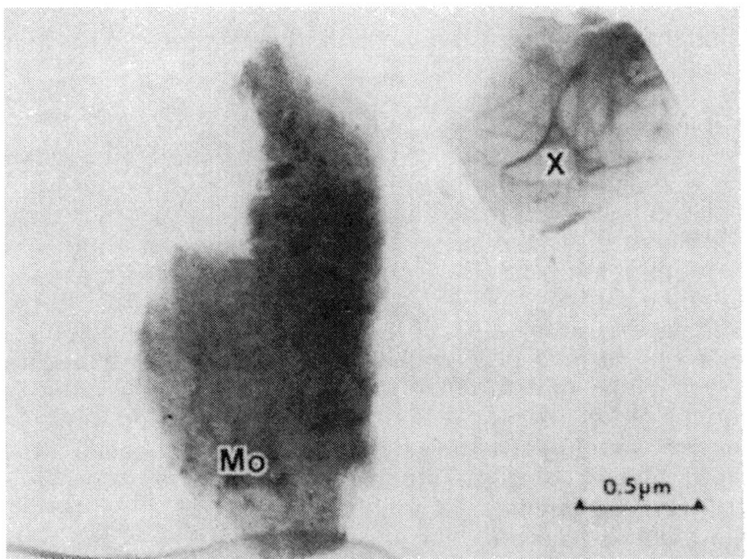

Figure 5.17. TEM bright-field image showing two types of particles on a carbon support film. EDS spectra are shown from a particle of the Mo metallization layer and for the unknown phase X. The unknown phase consists mostly of Ti and Al. Elements lighter than Na were not detected with the Be window detector.

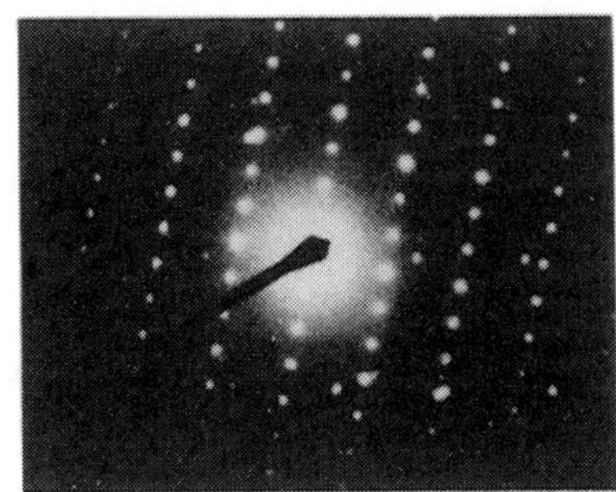

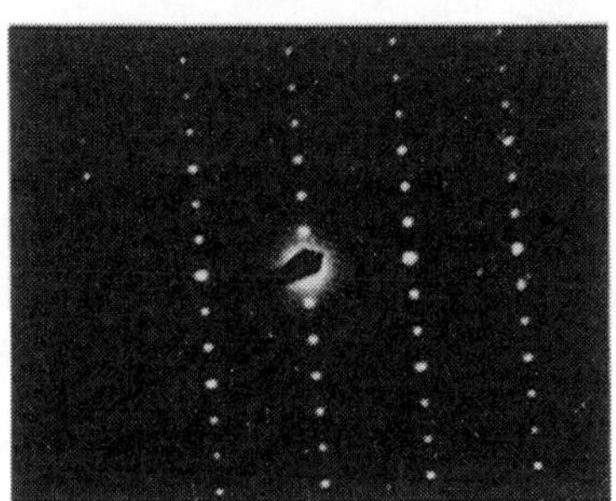

Selected r-Spacings @ λL = 2.49 Å-cm

0.25 (cm)	0.75
0.52	0.94
0.58	0.94
0.72	1.07

Figure 5.18. Symmetrical zone axis patterns obtained from different unknown phase X particles (above). Eight different r-spacings selected from the many reflections in the patterns. Original camera constant was 2.49 A-cm.

ID No.	Formula	Name
CDF 0027343	$Al(OH)_3$	Aluminum Hydroxide, Gibbsite
CDF 0102566	$Al(OH)_3$	Aluminum Hydroxide, Bayerite
CDF 0027845	(Ti, Al, C)	Titanium Aluminum Carbon, Carbide
CDF 0108326	Al_2O_3	Aluminum Oxide
CDF 0802241	Al_2TiO_5	Aluminum Titanate
CDF 0007735	Ti_3O_5	Titanium Oxide
CDF 0030844	TiO_2	Titanium Oxide, Anatase
CDF 0007739	AlB_{10}	Aluminum Boride
CDF 0703869	Al_3BO_6	Aluminum Borate
CDF 0023532	AlB_{12}	Aluminum Boride
CDF 0714227	$Al_8B_4C_7$	Aluminum Borocarbide
CDF 0713410	$Li_2Ti_3O_7$	Lithium Titanium Oxide
CDF 0031926	$Be_{12}Ti$	Beryllium Titanium

BEST FITS:

			a	b	c
Al_2TiO_5	orthorhombic	c2m	9.43	9.64	3.59
Ti_3O_5	orthorhombic	c2m	9.48	9.74	3.76

Figure 5.19. Sampling of candidate compounds matching the compositional and diffraction data from the unknown phase X. Unit cell dimensions of these two isomorphous orthorhombic (space group = c2/m) compounds are shown.

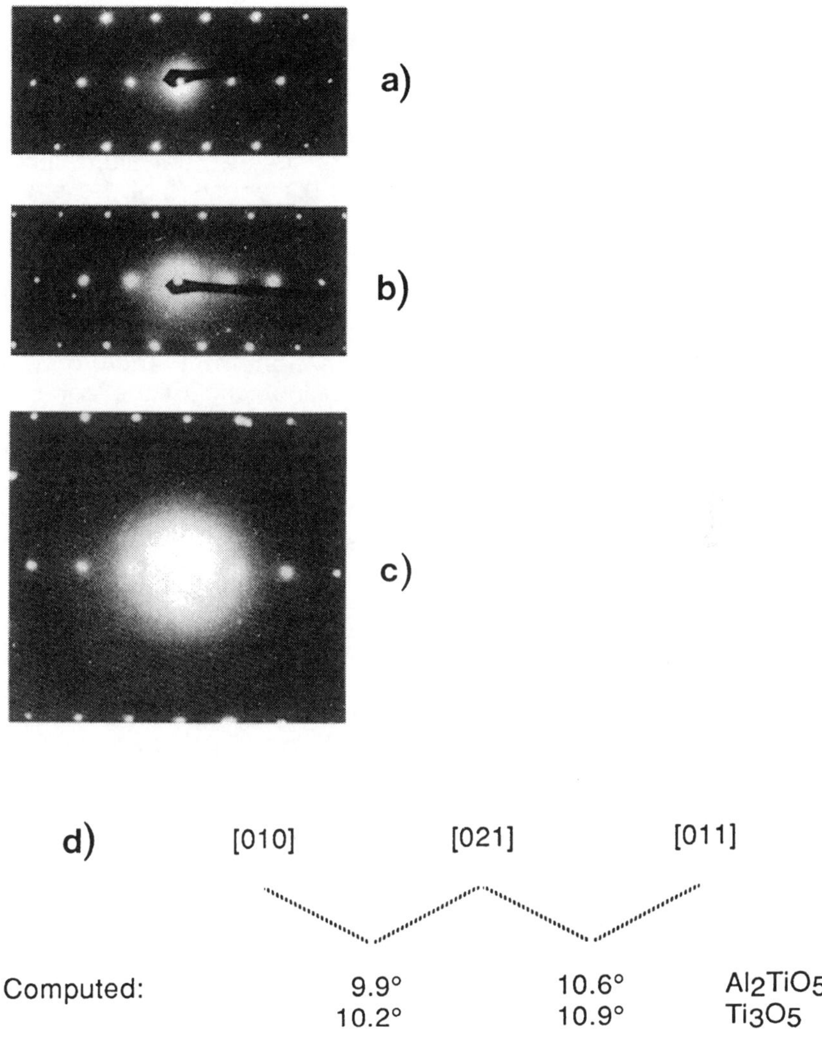

Measured: 9.9° 11.3°

Figure 5:20. Zone axis patterns obtained by tilting a single crystal about the normal to the (200) plane. Zone axes: a) [010], b) [021], c) [011]. A comparison of measured and calculated interzonal angles (d) confirms the identification.

Diffraction Database. Matches were constrained to compounds which contain Al or Ti, but no other elements with Z>11, and d-spacings corresponding to the eight observed spacings either directly or by double diffraction. The search yielded 69 candidate compounds as valid matches. Upon inspection, it was noted that many of the matches contained unlikely light elements such as beryllium, lithium, and boron which were not detected by bulk analysis of the starting materials or process gases. Restricting matches to compounds containing only hydrogen, oxygen, or no light elements at all, reduced the number of candidates to 24. The matches included some of the starting materials: Al_2O_3, TiH, several intermetallic compounds of Al and Ti, and a number of pure and mixed oxides or suboxides of Al and Ti (Figure 5.19). The unit cell information from each of these compounds was used to produce a trial indexing of the zone axis patterns in Figure 5.18. Compounds that did not produce a plausible indexing of all patterns were rejected. Only two of the 24 candidates could be used to successfully index all of the experimental zone axis patterns of Figure 5.18. These two, Ti_3O_5 and Al_2TiO_5, were isomorphous orthorhombic compounds of the pseudobrookite type (Fe_2TiO_5). To confirm the match, a series of zone axis patterns were photographed by tilting about a systematic row of spots common to all the patterns in the series, and the angles of tilt between the patterns were recorded. The individual patterns in the series were indexed and the interzonal angles were computed using the unit cell data from both Ti_3O_5 and Al_2TiO_5. The observed interzonal angles matched both cells very well (Figure 5.20), perhaps matching slightly better to Ti_3O_5. Although neither match contained the same proportions of Al and Ti measured in the unknown, the structures are similar enough to conclude that the unknown was a solid solution between the two.

5.5 OTHER SEARCH/MATCH METHODS

5.5.1 Searches Using Reduced Cell Edges

Originally search/match procedures for x-ray diffraction were developed for powder patterns only (Hanawalt et al., 1938), and ever since powder diffractionists have ground up crystals to obtain a fine powder or employed a Gandolfi camera to obtain a powder pattern. Unfortunately, the powder pattern may be consistent with more than one unit cell definition for a compound. However, the reduced primitive unit cell (Cullity, 1978) is unique and its dimensions and volume can be used in a search against a standard file of lattice parameter data derived from single-crystal experiments (NBS CRYSTAL DATA, 1987). The unit cell combined with the space group may be used to calculate a set of all possible d-spacings for a compound and this was the method used for generating most of the reference data in the NIST/Sandia/ICDD Electron Diffraction Database (Carr et al., 1989). But the Crystal Data File also may be searched directly using unit cell parameters measured from [100], [010], or [001] SAED or CBED zone axis patterns. If compositional data is available, this lattice-matching procedure will work acceptably even with the relatively poor precision of lattice parameters available from selected area electron diffraction (Mighell and Himes, 1986). The lattice matching program (Himes and Mighell, 1987) first transforms the unit cell data for the unknown phase into a reduced unit cell of minimum volume, then compares the cell lengths of this reduced cell to those in the database which are listed in increasing unit cell lengths. An example of such a search/match procedure is shown in Figure 5.21. The experimental data was in the form of a C-centered cell which was transformed to a primitive cell and reduced. The reduced cell was then compared with all the reduced cells in the Crystal Data File (NBS CRYSTAL DATA, 1987). The sample was found to be sodium sesquicarbonate dihydrate by a direct match of the reduced cell parameters. A missed reflection in an experimental diffraction pattern can easily lead to the assignment of a cell that is too large (supercell) or too small (subcell), but NBS*SEARCH (1987) can check this assignment against all possible multiple or subcells and arrive at the correct unit cell.

Initial Cell	Reduced Cell	CDF1	CDF2	CDF3
20.44 Å	***a*** = 3.49 Å	3.49 Å	3.49 Å	3.49 Å
3.49	***b*** = 10.33	10.13	10.31	10.43
10.33	***c*** = 10.37	12.08	10.35	12.17
90.00°	α=106.24°	90.0°	106.1°	90.0°
106.48	β = 99.69	90.0	99.7	90.0
90.0	γ = 90.0	90.0	90.0	90.0

Figure 5.21. Identification of an unknown compound using the NIST CRYSTAL DATA FILE. Initial C-centered cell transformed to reduced cell. Candidate compounds were found by searching on dimensions of the reduced cell using NBS*SEARCH. The three compounds listed are consecutive entries in the datafile not separate candidates. Best match is to CDF2 representing sodium sesquicarbonate dihydrate (from Mighell and Himes, 1986).

5.5.2 Searches Using Data from Convergent Beam Electron Diffraction Patterns

There are several ways that crystallographic data obtained from CBED patterns may be used in phase identification, especially when the search is combined with compositional data. First, as described above, lattice parameters determined from zone axis CBED patterns (Ayer, 1989) of cubic, tetragonal, or orthorhombic crystals may be searched manually in *Crystal Data: Determinative Tables* (Donnay and Ondik, 1972) or by computer in NIST CRYSTAL DATA (1987). Secondly, since the point group, and often the space group, may be determined from the symmetry of intensity variations within the CBED disks (see Volume 1, Chapter 7), searches on the point group or space group may provide an adequate identification when the composition is included. Third, a simple phase identification scheme employs fingerprinting analysis of the intensity variations in zone axis patterns taken from crystals of adequate thickness (Mansfield, 1989). These patterns generally show characteristic features that may be matched to standard patterns taken along the same zone axis in an atlas of CBED patterns (The Bristol Group, 1984). An important advantage of this method is that compounds differing slightly in all other aspects may be distinguished by differences in CBED pattern symmetry.

Microdiffraction patterns taken with a smaller beam convergence, somewhat intermediate between SAED and CBED, have also been used (Morniroli, 1990). These patterns also provide a route to the point group and the Bravais lattice even for small precipitates and particles. Here the search should be on composition (from the EDS spectrum) and r-spacings (d-spacings) plus point group/Bravais lattice (from the microdiffraction pattern).

In a computerized search of a crystallographic database, the listed data (unit cell edges, Bravais lattice, point group, space group, structure type, Pearson symbol) may be combined with the compositional information through the Boolean AND. Any database indexing on unit cell dimensions or space group may be used. Although one might expect that a search for cubic 'AND' aluminum 'AND' nickel would turn up many, perhaps hundreds, of compounds, surprisingly, of the ~52,000 compounds listed in the *Elemental and Lattice Spacing Index*, there are only two that meet these conditions. By varying the elements used in such a search, this procedure would locate isomorphs of the unknown phase.

5.5.3 Searches Using both Vectors and Included Angles

A method has been developed for searching on two diffraction vectors and the included angle (Lee, 1984). This method works well for rapid phase identification when the database to be searched contains only a small set of possibilities such as in asbestos identification.

Acknowledgements

This work was supported by the U.S. Dept. of Energy at Sandia National Laboratories under contract #DE-AC04-76DP00789 and at Lehigh University under contract #DE-FG02-86ER45269.

References

Ayer, R. (1989). *Journal of Electron Microscopy Technique* **13** 16-26.

Anderson, R. M. and Johnson, G. G. (1979). *37th Annual Proceedings of the Electron Microscopy Society of America*, G.W. Bailey (ed.), Claitor's Publishing, Baton Rouge, 444.

Andrews, K. W., Dyson, D. J. , and Keown, S. R. (1967). *Interpretation of Electron Diffraction Patterns*, Plenum Press, New York.

Bearden, J. A. (1967). X-ray wavelengths and x-ray atomic energy levels, NSRDS-NBS 14, National Bureau of Standards, U.S. Department of Commerce, Washington, DC. Also published in recent editions of the *CRC Handbook of Chemistry and Physics*, The Chemical Rubber Company, Cleveland, OH.

Bergerhoff, G. and Brown, I. D. (1987). Inorganic crystal structure database. In *Crystallographic Databases*, Allen, Bergerhoff, and Siever, (eds.). International Union of Crystallography, Chester, England, p. 77.

The Bristol Group (1984). *Convergent Beam Electron Diffraction of Alloy Phases*, by the Bristol Group under the direction of J. Steeds and compiled by J. Mansfield, Hilger, Bristol.

Carr, M., Chambers, W. F., and Melgaard, D. (1986). *Powder Diffraction* **1** 226-234.

Carr, M., Chambers, W. F., Melgaard, D., Himes, V. L., Stalick, J., and Mighell, A. D. (1989). *Journal of Research NIST* **94** 15-20.

Cullity, B. D. (1978). *Elements of X-ray Diffraction*, Addison-Wesley Publishing Company, Inc., Reading, MA.

Dingley, D. J., Mackenzie, R., and Baba-Kishi, K. (1989). *Microbeam Analysis-1989*, P. E. Russell, ed., San Francisco Press, San Francisco, 435-440.

Donnay, J. D. H., and Ondik, H. M., (general eds.), (1972). *Crystal Data: Determinative Tables*, 3rd. ed., U. S. Dept. of Commerce, National Bureau of Standards, Washington, DC.

Edington, J. W. (1976). *Practical Electron Microscopy in Materials Science*, Van Nostrand Reinhold, New York.

Egerton, R. F. (1986). *Electron Energy Loss Spectroscopy*. Plenum, New York.

Fraundorf, P. (1981). *Ultramicroscopy* **6** 227-236.

Goldstein, J. I., Newbury, D. E., Echlin, P., Joy, D. C., Romig, A. D., Lyman, C. E., Fiori, C., and Lifshin, E. (1992). *Scanning Electron Microscopy and X-ray Microanalysis*, Plenum, New York.

Goldstein, J. I., Lyman, C. E., and Zhang, J. (1990). *Microbeam Analysis-1990*, J. R. Michael and P. Ingram, eds., San Francisco Press, San Francisco, 265-271.

Hahn, T., (ed.) (1983). *International Tables for Crystallography*, Vol. A, Reidel Publishing, Dordrecht, Holland.

Hanawalt, J. D., Rinn, H. W., and Frevel, L. K. (1938). *Ind. Eng. Chem. Anal. Ed.* **10** 457-512.

JCPDS-ICDD ED-PCSRCH (1990). Software to search the Cd-ROM containing the NIST/Sandia/ICDD Electron Diffraction Database available from JCPDS-International Centre for Diffraction Data, 1601 Park Lane, Swarthmore, PA 19081.

JCPDS-ICDD *Elemental and Lattice Spacing Index* (1990). This index is available in printed form from JCPDS-International Centre for Diffraction Data, 1601 Park Lane, Swarthmore, PA 19081.

JCPDS-ICDD *Powder Diffraction File* (1990). A database of x-ray diffraction data issued annually with several search manuals that index the data in various ways. Database is available as a paper product (cards and books), on microfiche, and as a computer database on magnetic tape and CD-ROM. JCPDS-International Centre for Diffraction Data, 1601 Park Lane, Swarthmore, PA 19081.

Joy, D. C., Romig, A. D., and Goldstein, J. I. (eds.) (1986). *Principles of Analytical Electron Microscopy*, Plenum Press, New York.

Lee, R. J. (1984). Private communication.

Lyman, C. E., Williams, D. B., and Goldstein, J. I. (1989). *Ultramicroscopy* **28** 137-149.

Mansfield, J. (1989). *Journal of Electron Microscopy Technique* **13** 3-15.

Mighell, A. D., Stalick, J. K., and Himes, V. L. (1987). NBS Crystal Data: Database description and applications. In *Crystallographic Databases*, Allen, Bergerhoff, and Siever (eds.), International Union of Crystallography, Chester, England, p.134.

Mighell, A. D. and Himes, V. L. (1986). *Acta Crystallographica* **A42** 101-105.

Morniroli, J. P. (1990). EMAG-MICRO 89, Inst. Phys. Conf. Ser. No. 98, Institute of Physics, London, p. 87.

NBS CRYSTAL DATA (1987). The database of chemical and crystallographic data compiled and evaluated by the NIST Crystal Data Center, National Institute of Standards and Technology, Gaithersburg, MD 20899 [formerly NBS, National Bureau of Standards]. The full database is available on CD-ROM and magnetic tape by license only. Portions of the data are available in book form, in several volumes, as *Crystal Data: Determinative Tables* (Donnay and Ondik, 1972).

Himes, V. L. and Mighell, A. D. (1987). NBS*SEARCH: A program to search NBS CRYSTAL DATA , NIST Crystal Data Center, Reactor Radiation Division, National Institute for Standards and Technology, Gaithersburg, MD 20899.

Pearson, W. B. (1967). *A Handbook of Lattice Spacings and Structures of Metals and Alloys*, Pergamon Press, Oxford.

Raghavan, M., Scanlon, J., and Steeds, J. W. (1984). *Metallurgical Transactions* **15A** 1299.

Rhoades, B. L. (1975). *Micron* **6** 123-127.

Rogers, J. D. and Wood, G. H. (1987). NRCC Metals Crystallographic Data File. In *Crystallographic Databases*, Allen, Bergerhoff, and Siever (eds.), International Union of Crystallography, Chester, England, p. 96.

Steeds, J. W. (1979). Convergent beam electron diffraction. In *Introduction to Analytical Electron Microscopy*, J. J. Hren, J. I. Goldstein, and D. C. Joy, eds., Plenum Press, New York, 387-422.

Thomas, G. (1970). Kikuchi electron diffraction and applications. In *Modern Diffraction and Imaging Techniques in Materials Science*, S. Amelinckx et al. (eds.), North Holland, Amsterdam.

Villars, P. and Calvert, L. D. (1985). *Pearson's Handbook of Crystallographic Data for Intermetallic Phases*, American Society for Metals, Metals Park, OH.

Wells, A. F. (1984). *Structural Inorganic Chemistry*, 5th edition, Oxford University Press, New York.

Williams, D. B. (1984). *Practical Analytical Electron Microscopy in Materials Science*, Philips Electron Optics, Mahwah, New Jersey.

Wyckoff, R. W. G. (1963). *Crystal Structures*, Wiley Interscience, New York.

Index